NEW CONCEPTS
IN THE PATHOGENESIS
OF NIDDM

ADVANCES IN EXPERIMENTAL MEDICINE AND BIOLOGY

Recent Volumes in this Series

NEW CONCEPTS IN THE PATHOGENESIS OF NIDDM

Edited by

Claes Göran Östenson
Suad Efendić

Karolinska Hospital
Stockholm, Sweden

and

Mladen Vranic

University of Toronto
Toronto, Ontario, Canada

SPRINGER SCIENCE+BUSINESS MEDIA, LLC

Library of Congress Cataloging-in-Publication Data

Toronto-Stockholm Symposium on Perspectives in Diabetes Research (2nd
  : 1992 : Stockholm, Sweden)
    New concepts in the pathogenesis of NIDDM / edited by Claes Göran
Östenson, Saud Efendić, and Mladen Vranic.
       p.   cm. -- (Advances in experimental medicine and biology ; v.
334)
    "Proceedings of the Second Toronto-Stockholm Symposium on
Perspectives in Diabetes Research, held September 13-16, 1992, in
Stockholm, Sweden"--T.p. verso.
    Includes bibliographical references and index.
    ISBN 978-1-4613-6262-3   ISBN 978-1-4615-2910-1 (eBook)
    DOI 10.1007/978-1-4615-2910-1
    1. Non-insulin-dependent diabetes--Pathogenesis--Congresses.
I. Östenson, Claes Göran. II. Efendić, Suad. III. Vranic, Mladen.
IV. Title. V. Series.
RC660.A15T67 1992
616.4'62--dc20   93-29031
                                                        CIP

Proceedings of the Second Toronto–Stockholm Symposium on Perspectives in Diabetes Research, held September 13–16, 1992, in Stockholm, Sweden

ISBN 978-1-4613-6262-3

© 1993 by Springer Science+Business Media New York
Originally published by Plenum Press New York in 1993
Softcover reprint of the hardcover 1st edition 1993

PREFACE

The pathogenesis of non-insulin-dependent diabetes mellitus (NIDDM) has attracted the interest of our group during the last three decades. As early as 1969, a Nobel Symposium dealing with this topic was organized in Stockholm. This was followed in 1987 by a Nobel Conference devoted to the same subject. The main purpose of these meetings was to bring together the most distinguished scientists from all over the world and present theories on molecular and genetic mechanisms responsible for the development of glucose intolerance in NIDDM. This idea was followed also in the present symposium, "New Concepts in the Pathogenesis of NIDDM," organized with diabetologists from Toronto in Canada. Our purpose is to biannually organize international meetings covering important aspects of diabetes research, hoping that this type of interaction may result in new concepts and treatment alternatives. For us, participating in this symposium in September 1992, the meeting in Stockholm was very stimulating and innovative. It is a special pleasure that almost all invited lecturers submitted manuscripts. Thus, the publication of the proceedings of the symposium makes it possible for all interested in diabetes research to share new ideas and findings presented at the meeting.

Claes-Göran Östenson
Suad Efendic
Mladen Vranic

CONTENTS

REGULATION OF INSULIN SECRETION

INSULIN, INSULIN RECEPTORS AND GLUCOSE TRANSPORTERS

POTENTIAL IMPACT OF NEW CONCEPTS IN NIDDM ON DELIVERY OF CARE TO DIABETIC POPULATIONS

IMPAIRED GLUCOSE-INDUCED INSULIN SECRETION:
STUDIES IN ANIMAL MODELS WITH SPONTANEOUS NIDDM

Claes-Göran Östenson, Akhtar Khan, and Suad Efendic

Department of Endocrinology and
Rolf Lufts Center for Diabetes Research
Karolinska Hospital and Institute
S-104 01 Stockholm, Sweden

INTRODUCTION

Non-insulin-dependent diabetes mellitus (NIDDM) is a hereditary, chronic disease which is characterized by hyperglycemia. In most patients with manifest NIDDM, signs of hormonal dysbalance as well as insulin resistance are evident (1,2). Of the hormonal disturbances, impaired glucose-stimulated insulin response from the B-cells stands out as the most important defect (1,3). However, altered A- and D-cell secretion of glucagon and somatostatin, respectively, may also contribute to the diabetic state. Thus, the glucagon response to amino acids given i.v. and mixed meal is exaggerated and the suppressibility of glucagon release by glucose administered orally or i.v. is reduced (4,5). As in case of insulin release, glucose-induced somatostatin release is impaired or abolished (6). Since normalization of blood glucose levels by insulin infusion restored somatostatin response in NIDDM patients (7), the defect in D-cell function appears secondary to hyperglycemia and/or hypoinsulinemia. The mechanisms behind the altered A-cell response in NIDDM are more complex. Thus, after normalization of glycemia in artificial pancreas normal glucagon responses were found when patients were given intravenous arginine (8) and glucose (9). Conversely, even after normalization of glucose control in artificial pancreas, a mixed meal induced an exaggerated glucagon response, while oral glucose failed to suppress glucagon levels (7). Interestingly, in patients with insulin-dependent diabetes mellitus treatment with artificial pancreas normalized glucagon responses irrespective whether glucose or amino acids were given i.v. or orally (10,11).

There is controversy concerning which defect that is primary in NIDDM, that is whether an impaired insulin release, or decreased glucose usage in muscle, or enhanced hepatic glucose production constitutes an initial defect in NIDDM. It is possible that the impact of these three defects varies in different subgroups of NIDDM patients. Thus, the pathogenetic background of the disease may be different in the population of normal weight patients, included in our studies, as compared to grossly obese patients among Pima indians.

In our subgroup of NIDDM patients, an impaired stimulus-secretion coupling in the B-cells is proposed to constitute the initial defect, which then leads to hyperglycemia and insulin resistance (1,3). In obese NIDDM patients, insulin resistance in muscle and/or liver may represent the primary inherited defect. When the B-cells are incapable to secrete enough insulin to compensate for insulin resistance, hyperglycemia develops (2,12,13). Subjects with intact B-cell secretory capacity can cope with a major insulin resistance. Among obese individuals with marked insulin resistance, only a small group exhibited impaired glucose tolerance (14). After weight reduction they demonstrated an impaired insulin response ("low insulin response") to a standardized glucose infusion test. Conversely, the subjects who had a normal glucose tolerance in the obese state, maintained a normal, high insulin response after weight reduction. Similarly, among patients with acromegaly only low insulin responders developed a decreased glucose tolerance (15). Thus, defective B-cell responsiveness appears a prerequisite for development of NIDDM.

ANIMAL MODELS OF NIDDM

We have recently investigated islet function in a nonobese rat model of hereditary NIDDM, the GK rat, with the purpose to reveal defects of importance for the pathogenesis of NIDDM. From these studies, aspects on insulin secretion and islet glucose metabolism will be presented. In addition, we will discuss islet function in another rodent model of non-insulin-dependent diabetes, the obese-hyperglycemic (*ob/ob*) mouse.

Glucose Intolerance in GK Rats

The GK (Goto-Kakizaki) rat model originated in 1973 by repeated, selective breeding of nondiabetic Wistar rats, using high-normal blood glucose concentrations during an oral glucose tolerance test as selection index (16,17). The sixth generation (F_5) of rats had impaired glucose tolerance and a suppressed insulin response to glucose in the isolated perfused pancreas (16). After nearly 40 generations, the model exhibited further impairment of glucose tolerance and insulin response (18). The rats are however still mildly diabetic with fasting blood glucose levels of 7-8 mmol/l, and without ketonemia and significant aberration of body weights (18-20). Glucose intolerance develops early in GK rats, since significant hyperglycemia can be demonstrated in one week old animals with similar body weights as control rats (21). Preliminary studies have indicated moderate insulin resistance in liver as well as muscle in adult GK rats (22), but the precise role of this factor in the development of hyperglycemia has not yet been studied. In a series of studies by the original Japanese group, GK rats have been shown to develop morphological and functional changes as seen in diabetic late complications, e.g. neuropathy and nephropathy (20,23,24).

Insulin Release in GK Rats

The Stockholm colony of GK rats started with 5 breeding pairs of F40 rats, obtained from Japan. We characterized the kinetics of insulin response to glucose in the isolated, perfused pancreas of 2-month-old, male F_{42}-F_{43} GK rats (19). Basal insulin secretion, at 3.3 mmol/l glucose, was 10-fold higher (p < 0.01) in GK than in control (Wistar) pancreata (Fig. 1). During perfusion with 16.7 mmol/l glucose, the normal biphasic insulin response, as seen in control perfusions, was entirely absent in GK perfusions, or rather was replaced by a tendency to inhibition by glucose. When the period with high glucose perfusion passed into a final period with 3.3 mmol/l, a transient increase ("off-response") in insulin secretion was observed (Fig. 1). These results demonstrate a profound impairment of B-cell function *in vitro* in male GK rats. Interestingly, in the perfused pancreas of female GK rats neither glucose-induced inhibition of insulin secretion, nor the off-response were seen (unpublished). The reason for the paradoxical responses to glucose in the males is not clear, although some hypothetical explanations have been suggested, based on studies in other diabetic models with similar B-cell response to glucose. Thus, abnormal regulation by glucose of the cytosolic Ca^{2+} (25) and entry of water into the diabetic B-cell after a switch from high to low medium osmolarity (26), may cause glucose-induced inhibition of insulin release and the off-response, respectively. Alternatively, the inhibitory effect of glucose on insulin release in the GK pancreata could be due to a transient drop in the cytoplasmic concentration of ATP, reflecting increased consumption of this nucleotide during the glucose phosphorylation process (27). The drop in ATP would in turn induce increased activity of the ATP-regulated K^+ channels, hyperpolarization of the B-cell membrane, a decrease in cytosolic free calcium and inhibition of insulin release (28).

Although glucose-induced insulin secretion was abolished in the perfused pancreas, it was significant but minute in isolated pancreatic islets of male GK rats (Fig. 2). In these experiments, the release at 16.7 mmol/l was twice that at 3.3 mmol/l glucose albeit much lower than from control islets. Similar results have been obtained previously when comparing insulin responses in the perfused pancreas and isolated islets of nSTZ rats (29,30).

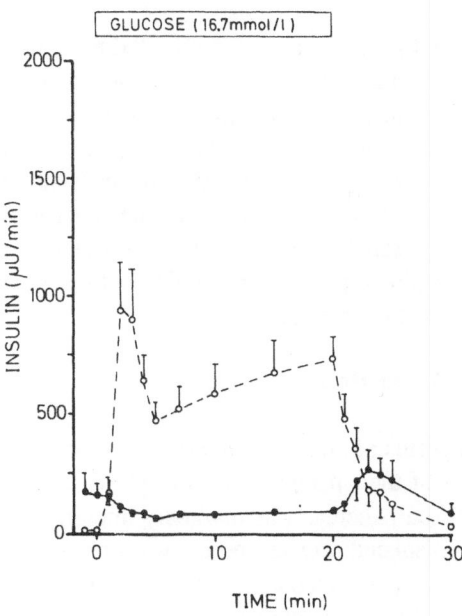

Fig. 1. Insulin responses to 16.7 mmol/l glucose in the perfused pancreas of male non-diabetic control (o---o; n=6) and diabetic GK (●—●; n=6) rats. Results (mean ± SEM) given as μU of insulin released per min. (Reproduced with permission from ref. 19.)

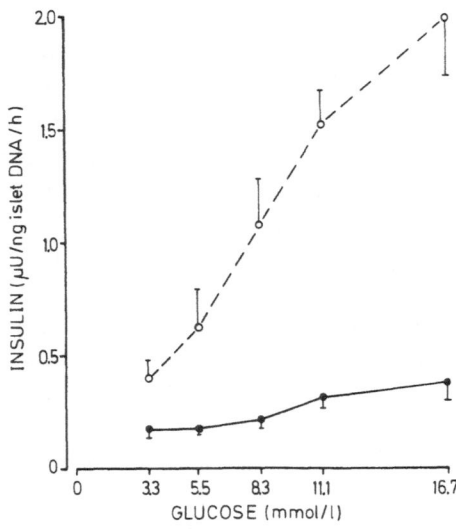

Fig. 2. Glucose dependency of insulin release from pancreatic islets isolated from non-diabetic control (o---o; n=6) and diabetic GK (●—●; n=6) rats. Results (mean ± SEM) are expressed as μU of insulin released per ng islet DNA content per h. (Reproduced with permission from ref. 19.)

The mechanism behind this discrepancy is not known. The suggestion that glucagon accumulating during static incubations could enhance insulin release (29), was not supported by experiments in which nSTZ islets were incubated in the presence of anti-glucagon antiserum (30).

Pancreatic Insulin Content in GK Rats

Insulin content in GK rat pancreas has been reported to be decreased by 15-40% in 4- to 6-month old animals (17,20,22). However, in 2-month old GK rats we found an insulin content identical to that in control rats (31). Also in isolated islets from the latter rats, insulin content per DNA content was normal (19). This is in concert with the observation of a normal B-cell frequency in the islets of the 2-months old GK rats (unpublished). With increasing age, some but far from all islets appear affected by fibrosis with disturbed architecture, i.e. "starfish-shaped" islets. Altogether, these findings indicate that the impairment of insulin responsiveness is due neither to decreased stores of the hormone nor to deranged islet morphology.

F_1-Hybrids of GK and Wistar Rats

Two-month old F_1-hybrids of GK and nondiabetic Wistar rats exhibit glucose intolerance intermediate to that of their parents (21). In hybrid rats, glucose-stimulated insulin release from the perfused pancreas was markedly impaired (21), whereas insulin-stimulated glucose uptake in isolated skeletal muscle was only slightly decreased as compared with control Wistar rats (32). In one-month old F_1-hybrids, glucose tolerance was also impaired but the insulin-sensitive glucose uptake in muscle was not affected. **Thus, it seems that defective B-cell secretion is the early and probably most important factor in development of glucose intolerance in the hybrids of GK rat.**

The Obese-Hyperglycemic (*ob/ob*) Mouse

The obese-hyperglycemic syndrome in the mouse is recessively inherited in animals homozygous for the *ob*-gene (33). In addition to increased body weight, due to excessive deposition of fat, the manifest syndrome is characterized by hyperglycemia and hyperinsulinemia. At 4 weeks of age, body weights of *ob/ob* mice are significantly higher than of their lean litter mates. The majority of *ob/ob* mice develop increased fasting blood glucose levels between 4 and 7 months, and demonstrate normalization of glycemia after 12 months of age (33). Their pancreatic islets are functionally hyperactive and greatly enlarged, of which about 90% are B-cells. Insulin release from isolated islets of *ob/ob* mice is high at basal glucose levels (5.5 mmol/l) and only stimulated 2-3-fold by 16.7 mmol/l glucose (34), indicating impaired insulin response to glucose. B-cell responsiveness to glucose was similar in 6- and 12-month-old *ob/ob* mice (unpublished), suggesting that the degree of insulin resistance, or other factors, are of decisive importance for regulation of glucose tolerance.

Stimulus-Secretion Coupling in Normal and Diabetic B-cells

Stimulus-secretion coupling in the B-cell is complexely regulated by substrates and a series of endocrine, neurocrine, paracrine and autocrine factors (35,36). However, in case of glucose, the metabolism of this substrate seems to play a key role. Under normal conditions, glucose is rapidly transported across the B-cell membrane, and this step is governed by specific glucose transporter molecules, GLUT-2, localized to the cell membrane (37,38). Glucose is then phosphorylated by glucokinase to glucose 6-phosphate (39). A rise in the extracellular glucose concentration thus immediately increases the metabolism of glucose in the B-cell, leading to enhanced oxidation of the hexose and increased intracellular ATP/ADP ratio. This closes the ATP-regulated K^+ channels, and when most of these channels (>99%) are closed, the B-cell depolarizes and voltage-dependent Ca^{2+} channels open, giving rise to action potentials. The resulting increase in cytoplasmic free Ca^{2+} leads to exocytosis of insulin from the B-cell (40) (Fig. 3).

The increase in cytoplasmic free calcium also leads to activation of adenyl cyclase (AC) and the formation of cAMP. The activity of AC is furthermore under the influence of various receptor ligands, e.g. hormones, neuropeptides and other neurotransmitters, via receptors coupled to the enzyme through stimulatory (GS) and inhibitory (GI) GTP-binding proteins. Other receptors are linked to the phospholipase C (PLC) system via a G-protein. Activation of the latter system leads to the formation of diacylglycerol (DAG) and inositol 1,4,5-triphosphate ($InsP_3$), thereby activating protein kinase C (PKC) and mobilizing intracellularly bound calcium, respectively (see 40 for review).

We have studied glucose metabolism in islets of GK rats and *ob/ob* mice. In the GK rat, three major alterations in islet metabolism of glucose have been demonstrated: 1) increased glucose utilization but unchanged glucose oxidation, 2) increased glucose cycling, and 3) decreased activity of glycerol phosphate shuttle.

Utilization of glucose was determined from the production of ^3H_2O from islets incubated in [5-^{3}H]glucose, and glucose oxidation rates were obtained by measuring the yield of $^{14}CO_2$ from [U-^{14}C]glucose (19). In GK islets incubated at 16.7 mmol/l, glucose utilization was 2.4-fold higher than in control islets (3.12 ± 0.66 vs 1.28 ± 0.11 pmol/ng islet DNA/h; p<0.01), while glucose oxidation was not significantly en hanced (0.54 ± 0.07 vs 0.40 ± 0.05 pmol/ng islet DNA/h). These data contrast to those obtained in islets from nSTZ rats, in which glucose utilization was close-to-normal and glucose oxidation decreased by at least 40% (41,42).

Glucose cycle is a futile cycle in which glucose is phosphorylated by glucokinase to glucose 6-phosphate, which then directly is dephosphorylated to glucose by glucose-6-phosphatase (43). For every molecule of glucose turning in the glucose cycle, one molecule of ATP is consumed. Glucose-6-phosphatase is present in islet B-cells (44), and glucose cycle is indeed operating not only in the liver but also in pancreatic B-cells

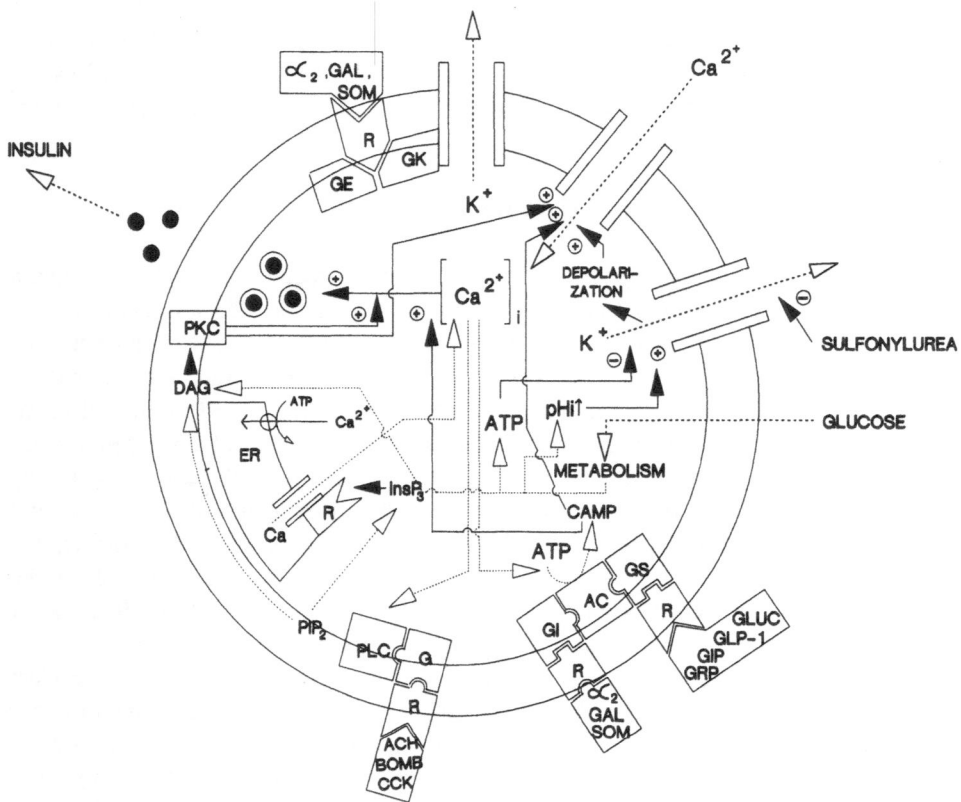

Fig. 3. Mechanisms involved in the regulation of insulin secretion. See text for explanation

(45). Hence, an increased glucose cycling in B-cells could contribute to a decreased cytoplasmic ATP/ADP ratio and decreased insulin release.

Glucose cycling was determined from the incorporation of ^{3}H into carbon 2 of medium glucose after incubation of islets with ^{3}H$_2$O, and 5.5 and 16.7 mmol/l unlabelled glucose (45). Estimates depend on the assumptions that before each molecule of glucose 6-phosphate is hydrolyzed to glucose, a hydrogen from the medium equilibrates with the hydrogen bound to carbon 2 of glucose 6-phosphate, because of the rapid equilibration between glucose 6-phosphate and fructose 6-phosphate (45). The latter equlibration is likely to occur, since phosphohexose isomerase activity is high in pancreatic islets (46). To the extent the isotopic equilibration is incomplete, glucose cycling is underestimated.

In GK islets, the amount of ^{3}H incorporated into medium glucose was 6-fold higher than in control islets (0.56 ± 0.10 vs 0.09 ± 0.01 picoatom/ng islet DNA/h; $p < 0.001$). Glucose cycling, estimated as the percent of total phosphorylated glucose that was dephosphorylated, was 2.5-fold enhanced in GK islets as compared to control islets (16.4 ± 3.4 vs $6.4 \pm 1.0\%$; $p < 0.01$). In nSTZ islets, incubated under identical conditions, we have found a similar rate of glucose cycling (15.7%) (41).

A third abnormality in glucose metabolism of GK islets is assigned to a reduced activity of the glycerol phosphate shuttle. Through this shuttle, controlled by the FAD-linked glycerophosphate dehydrogenase, NADH is oxidized to NAD. Coupled to this reaction the protons/electrons released will be transferred bound to FAD into the mitochondria and enter the respiratory chain (aerobic glycolysis). The activity of FAD-linked glycerophosphate dehydrogenase was decreased by 40-50%, while some other mitochondrial enzymes had similar activities in GK as compared with control islet homogenates (unpublished; cf the chapter by Malaisse). It is likely that, in islets of GK rat, the oxidation of NADH occurs at an increased rate of lactate production, yielding the NAD required for further augmentation of the anaerobic glycolysis. Hypothetically, an impaired glycerol phosphate shuttle, resulting in decreased oxidation of NADH, may be a primary event triggering increased anaerobic glycolysis. A decreased glycerol phosphate shuttle as well as increased glucose cycling result in a diminished ATP/ADP ratio. It is not clear whether this occurs to such an extent that it leads to impaired insulin release. It is possible that a significant decrease in ATP/ADP ratio occurs in an adenine nucleotide pool in the vicinity of the cell membrane and with an exquisite role for regulation of ATP-sensitive K^+ channels. At present, techniques are not available for measurements of the postulated cytoplasmic pools of ATP and ADP.

We have also studied glucose metabolism in islets of *ob/ob* mice. Glucose utilization at 5.5 mmol/l glucose was 24.1 ± 2.6 pmol/islet/h by islets from lean mice and 117.6 ± 11.2 pmol/islet/h by islets from *ob/ob* mice (47). At 16.7 mmol/l glucose, glucose utilization increased about 3-fold in islets from lean and *ob/ob* mice. Total glucose phosphorylation, comprising glucose utilization and glucose cycling, was 6 times higher in the islets of *ob/ob* mice than in the islets of lean animals at both glucose concentrations. Glucose cycling was about 3% in lean mouse islets at both 5.5 and 16.7 mmol/l glucose. In *ob/ob* mouse islets, glucose cycling was markedly increased to 17% and 30% at 5.5 and 16.7 mmol/l glucose, respectively.

Glucocorticoids are known to augment activity of glucose-6-phosphatase and thereby increase the rate of glucose cycle. We have therefore investigated the effect of dexamethasone on glucose cycling and insulin release in islets of *ob/ob* animals. It appeared that treatment for 48 h of *ob/ob* mice *in vivo* with dexamethasone (25 μg/d) further increased the rates of glucose cycling both at 5.5 mmol/l and 16.7 mmol/l glucose (24% and 56%, respectively) (34). Islet utilization and oxidation of glucose were not affected by the glucocorticoid. In parallel, dexamethasone treatment inhibited glucose-stimulated insulin release by approximately 60% at both glucose concentrations. Thus, it is likely that the inhibitory effect of dexamethasone on insulin secretion from *ob/ob* islets is due to enhanced glucose cycling. The above findings suggest that an increased rate of glucose cycling, leading to a decreased ATP/ADP ratio, may result in an inappropriate insulin response and glucose intolerance in *ob/ob* mice.

CONCLUSIONS

It is generally accepted that an impaired B-cell secretory response is an important feature of NIDDM. This defect is most likely of genetic background. At least in some subgroups of NIDDM patients, the B-cell defect ("low insulin response") seems to precede development of insulin resistance. This appears to be the case also in the GK rat, where we demonstrated a severely impaired insulin response paralleled by normal insulin sensitivity in skeletal muscle. Studies of the stimulus-secretion coupling in GK islets revealed three defects, such as increased glucose utilization and glucose cycling as well as decreased α-glycerol phosphate shuttle. In *ob/ob* mouse islets, a very high rate of glucose cycle, as compared to normal mice, was a predominant defect. The derangements of glucose metabolism in GK and *ob/ob* islets may hypothetically lead to decreased ATP/ADP ratio, resulting in incomplete closure of ATP-regulated K^+ channels and impaired exocytosis of insulin. An indirect evidence supporting this hypothesis is provided by experiments demonstrating that a short-term treatment of *ob/ob* mice with dexamethasone, known to augment glucose-6-phosphatase activity, increased glucose cycling to such a high level to comprise 56% of phosphorylated glucose, and in parallel decreased glucose-stimulated insulin release by 60%. In young GK rats a normal insulin content as well as a normal frequency of B-cells in the pancreas were found. This further strengthens the above hypothesis that the impaired stimulus-secretion coupling is a main defect responsible for the defective insulin release in GK rats.

REFERENCES

1. S. Efendic, R. Luft, and A. Wajngot, Aspects of the pathogenesis of type 2 diabetes, *Endocr.Rev.* 5:395-409 (1984).
2. R.A. DeFronzo, R.C. Bonadonna, and E. Ferrannini, Pathogenesis of NIDDM. A balanced overview, *Diabetes Care* 15:317-68 (1992).
3. S. Efendic, V. Grill, R. Luft, and A. Wajngot, Low insulin response: a marker of prediabetes. *Adv. Exp. Med. Biol.* 246:167-74 (1988).
4. R.H. Unger, E. Aquilar-Parada, W.A. Müller, and A.M. Eisentraut, Studies of pancreatic alpha cell function in normal and diabetic subjects, *J. Clin. Invest.* 49:837-48 (1970).
5. S. Efendic, M. Gutniak, and V. Grill, Hormonal release in NIDDM, *in:* "Pathogenesis of Non-Insulin Dependent Diabetes Mellitus", V. Grill, and S. Efendic, eds, Raven Press, New York, pp. 271-84 (1988).
6. V. Grill, M. Gutniak, A. Roovete, and S. Efendic, A stimulating effect of glucose on somatostatin release is impaired in non-insulin-dependent diabetes mellitus, *J. Clin. Endocrinol. Metab.* 59:293-7 (1984).
7. M. Gutniak, V. Grill, and S. Efendic, Effects of insulin on fasting and meal-stimulated somatostatin-like immunoreactivity in noninsulin-dependent diabetes mellitus: Evidence for more than one mechanism of action, *J. Clin. Endocrinol. Metab.* 62:77-83 (1986).
8. R. Kawamori, M. Shichiri, Y. Yamasaki, and H. Abe, Perfect normalization of excessive glucagon response to intravenous arginine in human diabetes mellitus with the artificial beta cell, *Diabetes* 29:762-5 (1980).
9. P. Raskin, Y. Fujita, and R.H. Unger, Effect of insulin-glucose infusions on plasma glucagon levels in fasting diabetics and non-diabetics, *J. Clin. Invest.* 56:1132-8 (1975).

10. M. Gutniak, V. Grill, K.-L. Wiechel, and S. Efendic, Basal and meal-induced somatostatin-like immunoreactivity in healthy subjects and in IDDM and totally pancreatectomized patients. Effects of acute blood glucose normalization, *Diabetes* 36:802-7 (1987).

11. M. Gutniak, V. Grill, A. Roovete, and S. Efendic, Impaired somatostatin response to orally administered glucose in NIDDM patients entails both somatostatin-28 and -14 and is associated with deranged metabolic control, *Acta Endocrinol* 121:322-6 (1989).

12. S.J. Lillioja, B.L. Nyomba, M.F. Saad, R. Ferraro, C. Castillo, P.H. Bennett, and C. Bogardus, Exaggerated early insulin release and insulin resistance in a diabetes prone population: a metabolic comparison of Pima Indians and Caucasians, *J. Clin. Endocrinol. Metab.* 73:866-76 (1991).

13. G.M. Reaven, Banting Lecture. Role of insulin resistance in human disease, *Diabetes* 37: 1595-607 (1988).

14. R. Luft, E. Cerasi, and B. Andersson, Obesity as an additional factor in the pathogenesis of diabetes, *Acta Endocrinol. (Copenh.)* 59:344-52 (1968).

15. R. Luft, E. Cerasi, and C.A. Hamberger, Studies on the pathogenesis of diabetes in acromegaly, *Acta Endocrinol (Copenh.)* 56:593-607 (1967).

16. K. Kimura, T. Toyota, M. Kakisaki, M. Kudo, K. Takebe, and Y. Goto, Impaired insulin secretion in the spontaneous diabetic rats, *Tohoku J. Exp. Med.* 137:453-9 (1982).

17. Y. Goto, K. Suzuki, M. Sasaki, T. Ono, and S. Abe, GK rat as a model of nonbese, noninsulin-dependent diabetes, *in:* "Frontiers in Diabetes Research. Lessons from Animal Diabetes II", E. Shafrir, and A.E. Renold, eds, John Libbey, London, pp. 301-3 (1988).

18. B. Portha, P. Serradas, D. Bailbe, K. Suzuki, Y. Goto, and M.-H. Giroix, ß-Cell insensitivity to glucose in the GK rat, a spontaneous nonobese model for type II diabetes, *Diabetes* 40:486-91 (1991).

19. C.-G. Östenson, A. Khan, S.M. Abdel-Halim, A. Guenifi, K. Suzuki, Y. Goto, and S. Efendic, Abnormal insulin secretion and glucose metabolism in pancreatic islets from the spontaneously diabetic GK rat, *Diabetologia* 36:3-8 (1993).

20. K. Suzuki, Y. Goto, and T. Toyota, Spontaneously diabetic GK (Goto-Kakizaki) rats, *in:* "Frontiers in Diabetes Research. Lessons from Animal Diabetes IV", E. Shafrir, ed, Smith-Gordon, London, pp. 107-16 (1992).

21. C.-G. Östenson, S.M. Abdel-Halim, V. Grill, A. Guenifi, P. Jalkanen, M. Sundén, and S. Efendic, Impact of diabetic inheritance on glucose tolerance and insulin secretion in spontaneously diabetic GK-Wistar rats, *Diabetologia* 34 (Suppl. 2):A76 (1991).

22. B. Portha, D. Bailbe, P. Serradas, O. Blondel, and M.-H. Giroix, Insulin resistance in the GK rat, a spontaneous non obese model for non-insulin-dependent diabetes, *Diabetologia* 33:A229 (1990).

23. Y. Goto, M. Kakizaki, and S. Yagihashi, Neurological findings in spontaneously diabetic rats, *Excerpta Medica ICS,* No. 581:26-38 (1982).

24. K. Suzuki, Y.-C. Hsiao, T. Toyota, el al., The significance of nerve sugar levels for the peripheral nerve impairment of spontaneously diabetic (Goto-Kakizaki) rats. *Diabetes Res.* 14:21-5 (1990).

25. B. Hellman, C. Berne, E. Grapengiesser, V. Grill, E. Gylfe, and P.-E. Lund, The cytoplasmic Ca^{2+} response to glucose as an indicator of impairment of the pancreatic ß-cell function, *Europ. J. Clin. Invest.* 20:S10-7 (1990).

26. A. Marcström, P.-E. Lund, and B. Hellman, Regulatory volume decrease of pancreatic ß-cells involving activation of tetraethylammonium-sensitive K^+ conductance, *Mol. Cell Biochem.* 96:35-41 (1990).

27. W.J. Malaisse, J.C. Hutton, S. Kawazu, A. Herchuelz, I. Valverde, and A. Sener, The stimulus-secretion coupling of glucose-induced insulin release. XXXV. The links between metabolic and cationic events, *Diabetologia* 16:331-41 (1979).

28. P. Arkhammar, T. Nilsson, P. Rorsman, and P.-O. Berggren, Inhibition of ATP-regulated K^+ channels precedes depolarization-induced increase in cytoplasmic free Ca^{2+} concentration in pancreatic ß-cells, *J. Biol. Chem.* 262:5448-54 (1987).

29. P.A. Halban, S. Bonner-Weir, and G.C. Weir, Elevated proinsulin biosynthesis *in vitro* from a rat model of non-insulin-dependent diabetes mellitus, *Diabetes* 32:277-81 (1983).

30. V. Grill, and C.-G. Östenson, The influence of a diabetic state on insulin secretion: studies in animal models of non-insulin dependent diabetes, *in:* "Pathogenesis of Non-Insulin Dependent Diabetes Mellitus", V. Grill, and S. Efendic, eds, Raven Press, New York, pp. 93-106 (1988).

31. S.M. Abdel-Halim, A. Guenifi, S. Efendic, and C.-G. Östenson, Both somatostatin and insulin responses to glucose are impaired in the perfused pancreas of the spontaneously non-insulin-dependent diabetic GK (Goto-Kakizaki) rats, *Acta Physiol. Scand.*, in press.

32. L.A. Nolte, I.K. Martin, S.M. Abdel-Halim, A. Guenifi, C.-G. Östenson, and H. Wallberg-Henriksson, Effect of elevated blood glucose on insulin-stimulated glucose transport in rat muscle, *Diabetologia* 35 (Suppl. 1):A80 (1992).

33. S. Westman, Development of the obese-hyperglycemic syndrome in mice, *Diabetologia* 4:141-9 (1968).

34. A. Khan, C.-G. Östenson, P.-O. Berggren, and S. Efendic, Glucocorticoid increases glucose cycling and inhibits insulin release in pancreatic islets of *ob/ob* mice, *Am. J. Physiol.* 263:E663-6 (1992).

35. B. Ahrén, C.-G. Östenson, and S. Efendic, Other islet peptides, *in:* "The Endocrine Pancreas", E. Samols, ed, Raven Press, New York, pp. 153-73 (1991).

36. P.-O. Berggren, P. Rorsman, S. Efendic, C.-G. Östenson, P. Flatt, T. Nilsson, P. Arkhammar, and L. Juntti-Berggren, Mechanisms of action of entero-insular hormones, islet peptides and neural input on the insulin secretory process, *in:* "Nutrient Regulation of Insulin Secretion", P. Flatt, ed, Portland Press, London, pp. 289-318 (1992).

37. J.H. Johnson, C.B. Newgard, J.L. Milburn, H.F. Lodish, and B. Thorens, The high K_m glucose transporter of islets of Langerhans is functionally similar to the low affinity transporter of liver and has an identical primary sequence, *J. Biol. Chem* 265:6548-51 (1990).

38. L. Orci, B. Thorens, M. Ravazzola, and H.F. Lodish, Localization of the pancreatic beta cell glucose transporter to specific plasma membrane domains, *Science* 245:295-7 (1990).

39. M.D. Meglasson, and F.M. Matschinsky, New perspectives on pancreatic islet glucokinase, *Am. J. Physiol.* 246:E1-13 (1984).

40. S. Efendic, H. Kindmark, and P.-O. Berggren, Mechanisms involved in the regulation of the insulin secretory process, *J. Internal Med.* 229 (Suppl. 2):9-22 (1991).

41. A. Khan, V. Chandramouli, C.-G. Östenson, H. Löw, B.R. Landau, and S. Efendic, Glucose cycling in islets from healthy and diabetic rats, *Diabetes* 39:456-9 (1990).

42. B. Portha, M.-H. Giroix, P. Serradas, N. Welsh, C. Hellerström, A. Sener, and W.J. Malaisse, Insulin production and glucose metabolism in isolated pancreatic islets of rats with NIDDM, *Diabetes* 37:1226-33 (1988).

43. J. Katz, and R. Rognstad, Futile cycles in the metabolism of glucose, *Curr. Top. Cell. Regul.* 10:237-89 (1976).

44. I.-B. Täljedal, Presence, induction and possible role of glucose-6-phosphatase in mammalian pancreatic islets, *Biochem. J.* 114:387-94 (1969).

45. A. Khan, V. Chandramouli, C.-G. Östenson, B. Ahrén, W.C. Schumann, H. Löw, B.R. Landau, and S. Efendic, Evidence for the presence of glucose cycling in pancreatic islets of *ob/ob* mouse, *J. Biol. Chem.* 264:9732-3 (1989).

46. R. Anjaneyulu, K. Anjaneyulu, A.R. Carpinelli, A. Sener, and W.J. Malaisse, The stimulus secretion coupling of glucose-induced insulin release: enzymes of mannose metabolism in pancreatic islets, *Arch. Biochem. Biophys.* 212:54-62 (1981).

47. A. Khan, V. Chandramouli, C.-G. Östenson, P.-O. Berggren, H. Löw, B.R. Landau, and S. Efendic, Glucose cycling is markedly enhanced in pancreatic islets of obese hyperglycemic mice, *Endocrinology* 126:2413-16 (1990).

PERTURBATION OF ISLET METABOLISM AND INSULIN RELEASE IN NIDDM

Willy J. Malaisse

Laboratory of Experimental Medicine
Erasmus Medical School
Brussels Free University, Brussels, Belgium

INTRODUCTION

The perturbation of islet metabolism and insulin release in NIDDM can be viewed in a dual perspective.

Some of these anomalies may represent the consequence of sustained hyperglycemia in a phenomenon of so-called B-cell glucotoxicity. It was recently proposed that two rather specific features of B-cell glucotoxicity consist in a transient paradoxical fall of insulinemia and an altered anomeric preference of insulin release in response to intravenous D-glucose administration.[1,2] Both anomalies might be attributable, in part at least, to the accumulation of glycogen in the pancreatic B-cell.[3-5]

Other manifestations of B-cell dysfunction in NIDDM may represent primary anomalies and, hence, represent the cause rather than consequence of chronic hyperglycemia.

The major aim of the present report is to propose that a site-specific defect of FAD-linked glycerophosphate dehydrogenase (m-GDH) in the B-cell mitochondria represents a far-from-uncommon determinant, or at least, contributive factor, of type-2 diabetes. It is further argued that such an enzymic deficiency may account for a preferential alteration of the B-cell secretory response to D-glucose, as distinct from other nutrient or non-nutrient secretagogues.

These proposals are based on both the role currently ascribed to m-GDH in the normal process of glucose-stimulated insulin secretion and the perturbation of such a process in NIDDM.

New Concepts in the Pathogenesis of NIDDM, Edited by
C. G. Östenson *et al.*, Plenum Press, New York, 1993

PARTICIPATION OF m-GDH TO THE GLUCOSE-SENSING DEVICE OF THE PANCREATIC B-CELL

It is currently believed that the capacity of D-glucose to stimulate insulin release is tightly linked to the catabolism of the hexose in the pancreatic B-cell *via* an increase in ATP generation rate.

Several specific features of D-glucose catabolism in islet cells are well suited to optimalize the yield of ATP in islets exposed to increasing concentrations of the hexose.[6] In the perspective of the present report, the most relevant of these specific features consists in a preferential stimulation of oxidative relative to total glycolysis in response to a rise in extracellular D-glucose concentration.[7,8]

Oxidative glycolysis is defined as the modality of D-glucose conversion to pyruvate, which is not coupled with the subsequent conversion of the 2-keto acid to L-lactate. It implies that NADH generated in the reaction catalyzed by glyceraldehyde-3-phosphate dehydrogenase is reoxidized to NAD^+ at the intervention of a suitable shuttle for the transfer of reducing equivalents from the cytosol into the mitochondria. We have recently documented that, in isolated islets exposed to $[2-^3H]$glycerol, the flux through the glycerol phosphate shuttle is close to the rate of oxidative glycolysis.[9] The accelerated circulation in the glycerol phosphate shuttle observed at high concentrations of D-glucose appears attributable to activation of m-GDH by Ca^{2+}.[10]

The activity of m-GDH relative to cell protein content is much higher in islet than in liver or spleen homogenates. In the islets, however, it is at least one order of magnitude lower than that of the cytosolic NAD-dependent glycerophosphate dehydrogenase, provided that the activity of m-GDH is measured in the presence of FAD rather than an artificial electron acceptor such as iodonitrotetrazolium.[10,11] The mitochondrial FAD-linked glycerophosphate dehydrogenase may thus be looked upon as the key regulatory enzyme of the glycerol phosphate shuttle.

A decrease in the activity m-GDH could thus conceivably be responsible for a preferential alteration of the B-cell secretory response to D-glucose, as distinct from other nutrient or non-nutrient secretagogues. At this point, it should be emphasized that an increase in glucose-derived pyruvate oxidation relative to total glycolytic flux, as observed in normal islets exposed to increasing concentrations of D-glucose, can also be viewed as a direct consequence of m-GDH activation by cytosolic Ca^{2+}. In other words, the mitochondrial oxidation of glucose-derived pyruvate is only possible if suitably coupled with oxidative glycolysis.

DEFICIENCY OF ISLET m-GDH IN STZ RATS

Animals injected with streptozotocin during the neonatal period or later in life, here referred to as STZ rats, display a severe decrease in m-GDH activity in islet, but not liver, homogenates.[12] This coincides with (i) an altered flux through the glycerol phosphate shuttle as judged from either the absolute values or the glucose-induced increment relative to basal reading for ^{3}HOH production by intact islets exposed to [2-^{3}H]glycerol,[13,14] (ii) a decreased contribution of oxidative to total glycolysis and (iii) a preferential impairment of the B-cell secretory response to D-glucose, as distinct from other secretagogues.[15] No obvious alteration of anaerobic glycolysis is observed in the islets of STZ rats. However, the ratio in total glycolysis at high/low hexose concentration is also lower in islets from STZ than control rats.[16]

Several complementary experiments were conducted to assess the significance of these findings.

First, it was verified by morphometric measurements that the enzymic defect found in islet homogenates could not be attributable solely to a decrease in the relative contribution of B-cells to total islet mass.[14] This is an important consideration since the activity of m-GDH, when expressed relative to DNA content, is about one order of magnitude higher in B than non-B islet cells.[17] Moreover, a decrease in m-GDH activity, relative to DNA content, was also documented in purified B-cells cultured for 2 days after exposure to streptozotocin in vitro.[17]

Second, it was shown that the enzymic defect in the islet of STZ rats does not represent a secondary manifestation of B-cell glucotoxicity. The islet activity of m-GDH is not decreased in islets exposed for several days to high concentrations of D-glucose, whether *in vitro* or *in vivo*.[11,18]

Third, it was documented that the decrease in m-GDH activity does not merely reflect an overall alteration of the activity of all mitochondrial dehydrogenases. For instances, the activity of glutamate dehydrogenase is not decreased in islet homogenates prepared from STZ rats.[12,19]

Last, none of the alternative possible explanations for the preferential alteration of glucose-induced insulin release in this model of type-2 diabetes could be substantiated. The anomaly of hexose transport in the islets of STZ rats was interpreted as reflecting a decrease in the distribution space readily accessible to 3-0-methyl-D-glucose rather than a delayed equilibration of hexose concentrations across the B-cell plasma membrane.[14] No decrease of hexokinase or glucokinase activity was found in islet homogenates.[15] Likewise, no primary anomaly was found in either pyruvate oxidative decarboxylation or in the ratio between the oxidation in the Krebs cycle of glucose-derived acetyl residues and their generation in the reaction catalyzed by pyruvate dehydrogenase when intact islets of STZ rats were exposed to suitably ^{14}C-labelled D-glucose.[13]

As a matter of fact, in our most recent study,[20] the exposure of intact islets from STZ rats to non-glucidic nutrient secretagogues, such as L-leucine, its deamination product 2-ketoisocaproate, its non-metabolized analogue 2-bicylo[2,2,1]heptane-2-carboxylic acid (BCH) or 3-phenylpyruvate, was found to unveil, at a low concentration of D-glucose (2.8 mM), the same oxidative defect as that otherwise seen at a high concentration of the hexose (16.7 mM). This is illustrated in the upper panel of Fig. 1, which reveals that, in the presence of 2.8 mM D-glucose, the integrated generation of ^{14}CO$_2$ from D-[2-^{14}C]glucose and D-[6-^{14}C]glucose is, as a rule, lower in islets from STZ than control rats provided that the incubation medium also contains a non-glucidic nutrient secretagogue. The rationale of this approach resides in the fact that these non-glucidic nutrient secretagogues preferentially stimulate circulation in both the glycerol phosphate shuttle and Krebs cycle in islets from normal rats.[9,21]

DEFICIENCY OF ISLET m-GDH IN INHERITED ANIMAL MODELS OF DIABETES

In the light of the results collected in STZ rats, we decided to investigate whether a comparable situation may prevail in animals with inherited, rather than acquired, diabetes.

At the time of completing this report, this was investigated in three of such models, namely in GK rats, db/db mice and BB rats.

In the GK rats, a deficiency in D-[3,4-^{14}C]glucose oxidation relative to D-[5-^{3}H]glucose utilization was first observed in intact islets.[16] Further work revealed a decreased activity of m-GDH in islet homogenates (unpublished observation). Once again, the enzymic defect did not appear to be attributable to any sizeable decrease in the contribution of B-cells to total islet mass.[22]

In db/db mice, the experiments were so far restricted to the measurement of enzymic activities in islet homogenates and again revealed a decreased activity of m-GDH.[23]

Incidentally, in both STZ and GK rats, the m-GDH defect curiously coincided with an abnormally high ratio between glutamate-pyruvate and glutamate-oxalacetate transaminase activities in islet homogenates (Table 1), as if a genomic link may exist between the two enzymatic anomalies.

It should also be stressed that the m-GDH activity failed to display abnormally low values in the liver of either GK rats or db/db mice.

A vastly different picture was obtained when comparing diabetes prone BB to control BW rats.[24] In this case, the activity of m-GDH was unexpectedly higher in islets from BB than BW rats. The relevance of the latter finding to the autoimmune process of B-cell aggression in this animal model of type-1 diabetes should not be ignored. It was indeed previously speculated that, in STZ rats, the cytotoxic agent may preferentially affect a subpopulation of B-cells or progenitor cells characterized by a high activity of m-GDH.[25]

Taken as a whole, these findings indeed suggest that a site-specific deficiency of m-GDH in the pancreatic B-cell could well represent a far-from-uncommon feature in animal models of type-2 diabetes.

POSSIBLE ROLE OF m-GDH IN THE PATHOGENESIS OF TYPE-2 DIABETES IN HUMAN SUBJECTS

To which extent can we extrapolate data collected in animal models of type-2 diabetes to the situation found in human patients with NIDDM ?

It is presently not possible, for methodological and deontological reasons, to have access to fresh pancreatic material collected from a large number of patients with type-2 diabetes.

It is conceivable, however, that in some of these patients, the postulated deficiency in m-GDH activity would not be restricted to the B-cell, at variance with the situation apparently prevailing in GK rats or db/db mice. Therefore, we have initiated a study on the m-GDH activity in extrapancreatic cells, namely T-lymphocytes, obtained from type-2 diabetics.[26] Incidentally a deficiency of m-GDH generalized to several cell types is not necessarily incompatible with an isolated functional anomaly of the pancreatic B-cell. The activity of m-GDH relative to protein content is indeed much higher in the B-cell than in other cell types so far examined for such a purpose. In these other cells, a decreased activity of m-GDH may have little if any unfavourable metabolic and functional consequences.

Table 1. Ratio between glutamate-pyruvate and glutamate-oxalacetate transaminase activitivies in islets from STZ and GK rats

Expt. 1	Control rats : 13.0 ± 1.0 %	STZ rats : 24.5 ± 3.9 %
Expt. 2	Control rats : 6.4 ± 1.2 %	GK rats : 23.2 ± 3.6 %

The results so far collected in human lymphocytes, which were isolated from 10 ml blood samples and then cultured for 2 or 3 weeks in order to obtain a homogenous population of T-lymphocytes in sufficient amount for enzymic determinations, are illustrated in Fig. 2. They reveal that, out of 32 type-2 diabetics, 12 subjects were found to display an abnormally low activity of m-GDH as assessed by both a colorimetric procedure based on the use of iodonitrotetrazolium as electron acceptor and a radioisotopic assay based the generation of ^{3}HOH from L-[2-^{3}H]glycerol-3-phosphate. A comparable situation characterized by a deficient activity of m-GDH in the two assay procedures was only observed once in 26 other subjects, who included 11 healthy volunteers, 9 non-diabetic patients, 5 type-1 diabetics and one pancreatectomized subject. It should be noted, however, that a dissociated behaviour of m-GDH characterized by a normal activity in the colorimetric assay but a low value in the radioisotopic procedure was observed in 2 out of 5 type-1 diabetics and a further subject with type-2 diabetes.

The type-2 diabetics with either normal or low m-GDH activity could not be distinguished from one another by such criteria as the duration and severity of diabetes.

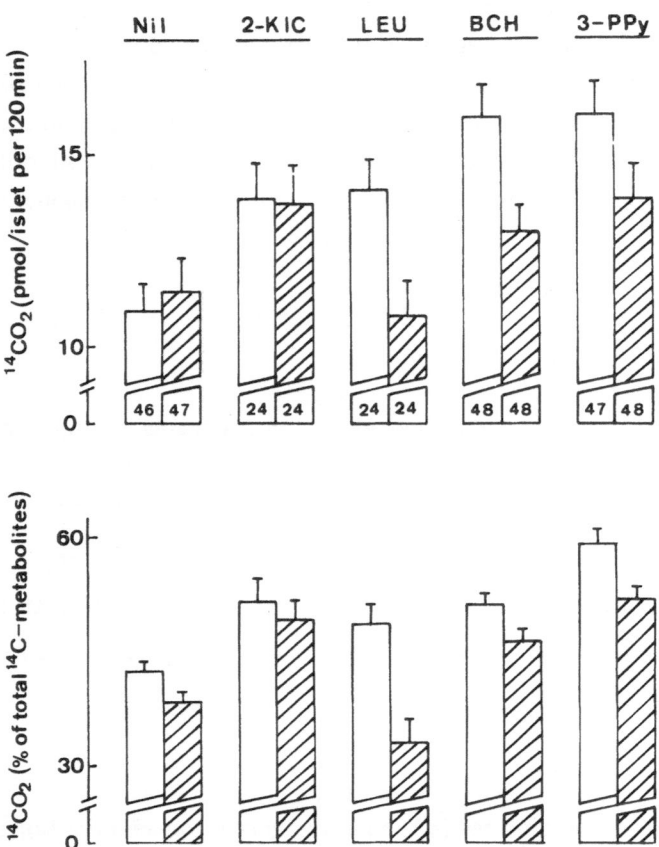

Figure 1. Absolute values for $^{14}CO_2$ output (upper panel) and paired ratios between $^{14}CO_2$ output and the total generation of ^{14}C-labelled metabolites (lower panel) in islets from control (open columns) and STZ rats (hatched columns) exposed to 2.8 mM D-glucose in the absence (Nil) or presence of non-glucidic nutrient secretagogues, including 2-ketoisocaproate (2-KIC), L-leucine (LEU), BCH and 3-phenylpyruvate (3-PPy). Mean values ($\pm$ SEM) refer to the integrated values for the oxidation of D-[2-^{14}C]glucose and D-[6-^{14}C]glucose, as derived from the number of primary measurements mentioned at the bottom of each column in the upper panel.

Interestingly an abnormally low ratio between glutamate-oxalacetate and glutamate-pyruvate transaminase activities in T-lymphocytes, as already observed in islets of STZ and GK rats, was also found in 8 type-2 diabetics with either normal or abnormal m-GDH activity (Fig. 3). This suggests, once again, a possible genetic coupling between these two distinct enzymic perturbations.

It remains obviously to assess whether the enzymic defect found in T-lymphocytes also affects the pancreatic B-cell of patients with type-2 diabetes.

The approach of this issue by molecular biology is presently hampered by the unavailability of the relevant DNA probe. Hence, a first step in this perspective is to purify the enzyme. This work was indeed undertaken using a mitochondria-rich subcellular fraction of rat liver as starting material. The question cannot be ignored, however, whether islet m-GDH and liver m-GDH are identical or may represent distinct isoenzymes. Preliminary experiments conducted in crude homogenates suggest that the m-GDH activity found in distinct organs (such as islet, liver and spleen), may display significant differences in terms of affinity for L-glycerol-3-phosphate, relative extent of activation by Ca^{2+} and change in reaction velocity when iodonitrotetrazolium is substituted to FAD as the electron acceptor.[27]

A further difficulty could be encountered in the study of the genetic aspects of m-GDH deficiency. The data so far available indeed suggest that, in some cases at least, it is the specific expression of the m-GDH gene in the pancreatic B-cell which is impaired.

CONCLUDING REMARKS

The present report summarizes the present status of our study on the possible role of a m-GDH deficiency in the pancreatic B-cell as a causal or contributing factor in the pathogenesis of type-2 diabetes. It could be objected, quite wrightly, that the significance of this issue remains to be convincingly documented. Much experimental work indeed remains to be carried out to assess fully the validity of our working hypothesis. Even if biased by personal enthusiasm, I feel that such a work deserves, nevertheless, to be completed.

Acknowledgements

The work here under review was supported by grants from the Belgian Fondation for Scientific Medical Research and a Concerted Research Action of the French Community of Belgium. I wish to underline the contribution of Drs. M.-H. Giroix (Paris), C.-G. Östenson (Stockholm), L. Herberg (Düsseldorf), S. Kukel and U. Reinhold (Bonn) and their colleagues to this work. I am most indebted to F. Malaisse-Lagae, J. Rasschaert and A. Sener for their essential collaboration. I am also grateful to C. Demesmaeker for secretarial help.

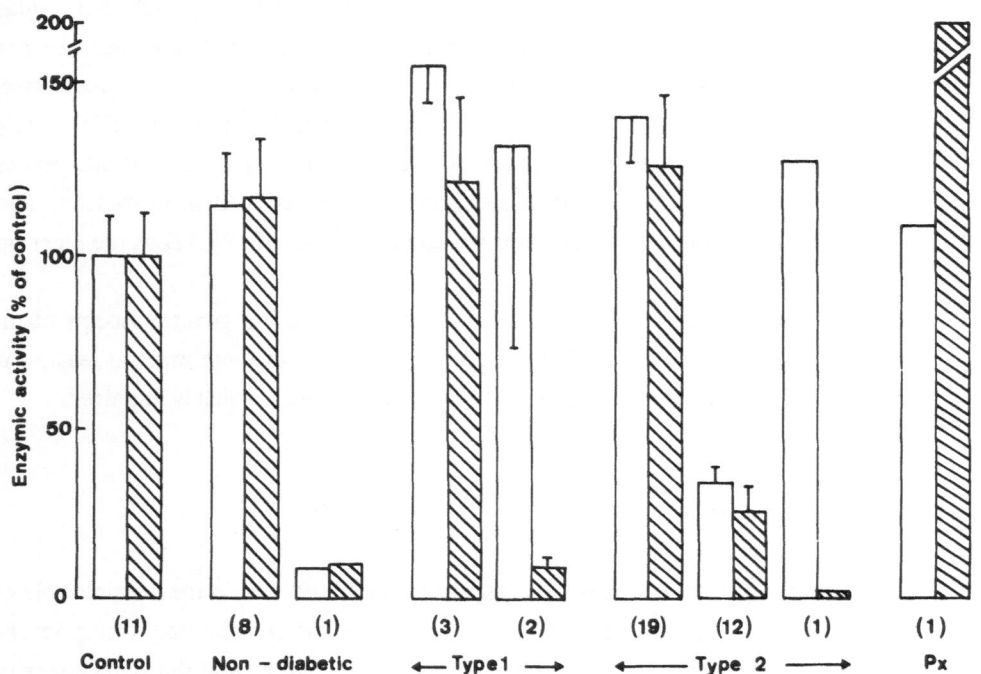

Figure 2. Mean values ($\pm$ SEM) for the activity of m-GDH in T-lymphocytes, as measured by either a colorimetric procedure (open columns) or radioisotopic assay (shaded columns), in 11 healthy volunteers, 9 non-diabetic patients, 5 subjects with type-1 diabetes, 32 type-2 diabetics and one pancreatectomized (Px) subject are expressed relative to the mean corresponding value found in the 11 control subjects. A low activity in both assay systems was observed in 12 out of 32 type-2 diabetics, but only once in the other 26 subjects. Dissociated results in the colorimetric and radioisotopic assay were observed in two type-1 and one type-2 diabetic patients.

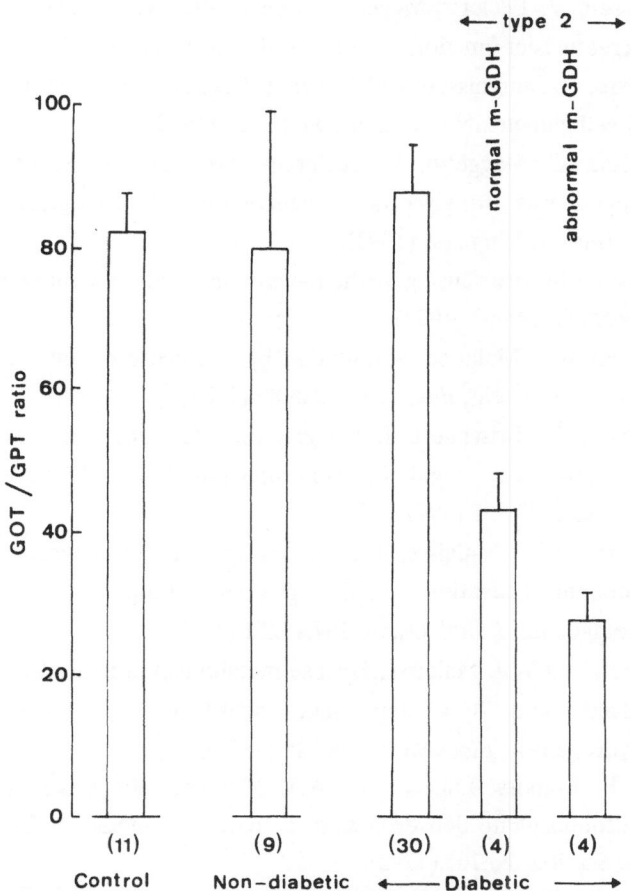

Figure 3. Mean absolute values (± SEM) for the ratio between glutamate-oxalacetate and glutamate-pyruvate transaminases activities in T-lymphocytes of 11 healthy volunteers, 9 non-diabetic patients and 38 diabetic subjects. A low ratio was found in 8 type-2 diabetics with either normal or abnormal m-GDH activity.

REFERENCES

1. W.J. Malaisse, Physiology of insulin secretion and its alteration in diabetes : the concept of glucotoxicity, *in*: "Diabetic Complications : Epidemiology and Pathogenic Mechanisms", D. Andreani, J.L. Gueriguian and G.E. Striker, eds., Raven Press, New York, pp 3-23 (1991).

2. W.J. Malaisse, The anomeric malaise : a manifestation of B-cell glucotoxicity, *Horm. Metab. Res.* 23:307 (1991).

3. G. Marynissen, V. Leclercq-Meyer, A. Sener, and W.J. Malaisse, Perturbation of pancreatic islet function in glucose-infused rats, *Metabolism* 39:87 (1990).

4. W.J. Malaisse, G. Marynissen, and A. Sener, Possible role of glycogen accumulation in B-cell glucotoxicity, *Metabolism* 41:814 (1992).

5. W.J. Malaisse, C. Maggetto, V. Leclercq-Meyer, and A. Sener, Interference of glycogenolysis with glycolysis in pancreatic islets from glucose-infused rats, *J. Clin. Invest.* 91:in press (1993).

6. W.J. Malaisse, Glucose-sensing by the pancreatic B-cell : the mitochondrial part, *Int. J. Biochem.* 24:693 (1992).

7. A. Sener, and W.J. Malaisse, Stimulation by D-glucose of mitochondrial oxidative events in islet cells, *Biochem. J.* 246:89 (1987).

8. W.J. Malaisse, J. Rasschaert, I. Conget, and A. Sener, Hexose metabolism in pancreatic islets. Regulation of aerobic glycolysis and pyruvate oxidation, *Int. J. Biochem.* 23:955 (1991).

9. A. Sener, and W.J. Malaisse, Hexose metabolism in pancreatic islets. Ca^{2+}-dependent activation of the glycerol phosphate shuttle by nutrient secretagogues, *J. Biol. Chem.* 267:13251 (1992).

10. J. Rasschaert, and W.J. Malaisse, Hexose metabolism in pancreatic islets. Glucose-induced and Ca^{2+}-dependent activation of FAD-glycerophosphate dehydrogenase, *Biochem. J.* 278:335 (1991).

11. A. Sener, F. Malaisse-Lagae, and W.J. Malaisse, Pancreatic islet FAD-linked glycerophosphate dehydrogenase activity in a model of B-cell glucotoxicity, *Med. Sci. Res.* 20:701 (1992).

12. M.-H. Giroix, J. Rasschaert, D. Bailbe, V. Leclercq-Meyer, A. Sener, B. Portha, and W.J. Malaisse, Impairment of the glycerol phosphate shuttle in islets from rats with diabetes induced by neonatal streptozotocin, *Diabetes* 40:227 (1991).

13. M.-H. Giroix, J. Rasschaert, A. Sener, V. Leclercq-Meyer, D. Bailbe, B. Portha, and W.J. Malaisse, Study of hexose transport, glycerol phosphate shuttle and Krebs cycle in islets of adult rats injected with streptozotocin during the neonatal period, *Mol. Cell. Endocrinol.* 83:95 (1992).

14. M.-H. Giroix, D. Baetens, J. Rasschaert, V. Leclercq-Meyer, A. Sener, B. Portha, and W.J. Malaisse, Enzymic and metabolic anomalies in islets of diabetic rats : relationship to B-cell mass, *Endocrinology* 130:2634 (1992).

15. M.-H. Giroix, A. Sener, D. Bailbe, B. Portha, and W.J. Malaisse, Impairment of

mitochondrial oxidative response to D-glucose in pancreatic islets from adult rats injected with streptozotocin during the neonatal period, *Diabetologia* 33:654 (1990).

16. A. Sener, M.-H. Giroix, B. Portha, and W.J. Malaisse, Preferential alteration of oxidative relative to total glycolysis in islets of rats with inherited or acquired non-insulin-dependent diabetes, *Diabetologia* 35 (suppl. 1):A78 (1992).

17. J. Rasschaert, Z. Ling, and W.J. Malaisse, Effect of streptozotocin and nicotinamide upon FAD-glycerophosphate dehydrogenase activity and insulin release in purified pancreatic B-cells, *Mol. Cell. Biochem.*, in press (1992).

18. J. Rasschaert, D.L. Eizirik, and W.J. Malaisse, Long term *in vitro* effects of streptozotocin, interleukin-1 and high glucose concentration upon the activity of mitochondrial dehydrogenases and the secretion of insulin in pancreatic islets, *Endocrinology* 130:3522 (1992).

19. J. Rasschaert, and W.J. Malaisse, Streptozotocin-induced FAD-glycerophosphate dehydrogenase suppression in pancreatic islets. Relationship to severity and duration of hyperglycemia and resistance to insulin and riboflavin treatment, *Acta Diabetol.*, in press (1992).

20. A. Sener, M.-H. Giroix, F. Malaisse-Lagae, D. Bailbe, V. Leclercq-Meyer, B. Portha, and W.J. Malaisse. Metabolic response to non-glucidic nutrient secretagogues and enzymic activities in pancreatic islets of adult rats after neonatal streptozotocin administration, *Biochem. Med. Metab. Biol.*, in press (1993).

21. W.J. Malaisse, and A. Sener, Hexose metabolism in pancreatic islets. Activation of the Krebs cycle by nutrient secretagogues, *Mol. Cell. Biochem.* 107:95 (1991).

22. C.-G. Östenson, S. Abdel-Halim, J. Rasschaert, F. Malaisse-Lagae, S. Meuris, A. Sener, S. Efendic, and W.J. Malaisse, Deficient activity of FAD-linked glycerophosphate dehydrogenase in islets of GK rats, submitted for publication.

23. A. Sener, L. Herberg, and W.J. Malaisse, FAD-linked glycerophosphate dehydrogenase deficiency in pancreatic islets of mice with hereditary diabetes, *FEBS Letters*, in press (1993).

24. J. Rasschaert, and W.J. Malaisse, Increased activity of FAD-linked glycerophosphate dehydrogenase in pancreatic islets of BB rats, submitted for publication.

25. J. Rasschaert, and W.J. Malaisse, Streptozotocin-induced suppression of FAD-linked glycerophosphate dehydrogenase in pancreatic islets of adult rats, *Biochem. Int.* 23:707 (1991).

26. W.J. Malaisse, F. Malaisse-Lagae, S. Kukel, U. Reinhold, and A. Sener, Could non-insulin-dependent diabetes mellitus be attributable to a deficiency of FAD-linked glycerophosphate dehydrogenase ?, submitted for publication.

27. J. Rasschaert, and W.J. Malaisse, Intrinsic properties of FAD-linked glycerophosphate dehydrogenase in islets from normal and streptozotocin-induced diabetic rats, *Diab. Res.*, in press (1993).

REGULATION OF CYTOPLASMIC FREE Ca^{2+} IN INSULIN-SECRETING CELLS

Per-Olof Berggren, Per Arkhammar, Md Shahidul Islam, Lisa Juntti-Berggren, Akhtar Khan, Henrik Kindmark, Martin Köhler, Kerstin Larsson, Olof Larsson, Thomas Nilsson, Åke Sjöholm, Jaroslaw Szecowka and Qimin Zhang

The Rolf Luft Center for Diabetes Research, Department of Endocrinology, Karolinska Institute, Stockholm, Sweden

Introduction

The cytoplasmic free Ca^{2+} concentration($[Ca^{2+}]_i$) has a fundamental role in the β-cell stimulus-secretion coupling and is regulated by a sophisticated interplay between nutrients, hormones and neurotransmitters. Metabolism of glucose and other nutrients leads to ATP generation, closure of ATP-regulated K$^+$-channels, depolarization, opening of voltage-activated L-type Ca^{2+}-channels, increase in $[Ca^{2+}]_i$ and insulin release [1,2]. Hormones and neurotransmitters affect the β-cell through the activation of receptors coupled to various effector systems, such as the adenylate cyclase (AC) or phospholipase C (PLC) system [2]. Upon activation of these systems, cAMP is formed or phosphatidyl inositol 4,5-bisphosphate is hydrolysed, resulting in the formation of inositol 1,4,5-trisphosphate (InsP$_3$) and diacylglycerol (DAG). Whereas InsP$_3$ mobilizes intracellularly bound Ca^{2+}, most probably from the endoplasmic reticulum, DAG activates protein kinase C (PKC) [1-3]. Although InsP$_3$ increases $[Ca^{2+}]_i$, there is little effect on insulin release, suggesting that the trisphosphate is not primarily involved as a signal for exocytosis in the β-cell [3]. With regard to PKC, the physiological role is more clear and this enzyme is involved as a modulator of multiple steps in the β-cell signal-transduction pathway [1-3].

The approach to measure $[Ca^{2+}]_i$ at the single cell level revealed that the β-cell, in accordance to what has been demonstrated for a number of other cell types, exhibits $[Ca^{2+}]_i$-oscillations upon stimulation. It is likely, that oscillations in $[Ca^{2+}]_i$ are not only beneficial in terms of a more efficient signal-transduction and the ability to avoid cytotoxic effects of

Ca^{2+}, but also constitute the molecular basis for oscillations in insulin release. To date, little is known about the molecular mechanisms regulating [Ca^{2+}]$_i$-oscillations in the pancreatic β-cell. We have recently demonstrated that activation of a novel low conductance Ca^{2+}-dependent K$^+$-current produce a transient membrane repolarization in glucose-stimulated β-cells (4). Membrane repolarization, induced by activation of this K$^+$-current, results in closure of the voltage-dependent Ca^{2+}-channels. K$^+$-currents of the same type are evoked by hormones and neurotransmitters present in the β-cell micromilieu and as well by intracellular application of compounds mobilizing Ca^{2+} from InsP$_3$-sensitive Ca^{2+}-stores. This suggests that also the excitable pancreatic β-cell exhibits InsP$_3$- and maybe Ca^{2+}-mediated (5) periodic increases in [Ca^{2+}]$_i$, associated with changes in membrane conductance. Hence, regulation of [Ca^{2+}]$_i$-oscillations in this cell type most likely represents a complex interplay between plasma membrane and intracellular Ca^{2+}-transport. In the present chapter we discuss the molecular mechanisms involved in the regulation of both Ca^{2+}-transport over the plasma membrane and intracellular Ca^{2+}-handling, in relation to [Ca^{2+}]$_i$-oscillations in the pancreatic β-cell.

Plasma membrane Ca^{2+}-transport through voltage-dependent Ca^{2+}-channels

The resting [Ca^{2+}]$_i$ in the β-cell is maintained at approximately 100 nM, similar to what has been found in other cell types, despite the 10 000-fold higher concentration of Ca^{2+} in the extracellular space. Thus, small changes in membrane permeability to Ca^{2+} produce a large increase in influx of the ion and, consequently, Ca^{2+}-channels in the plasma membrane play a central role in regulating [Ca^{2+}]$_i$. Upon depolarization of the cell, specific voltage-dependent Ca^{2+}-channels are activated. Functional studies have identified several types of such channels, distinguished by pharmacology, electrophysiology, and tissue localization. More recently, molecular cloning has revealed an even greater diversity among Ca^{2+}-channels, arising from multiple genes and alternative splicing (6,7). Here we concentrate on those voltage-dependent Ca^{2+}-channels that have been found in the pancreatic β-cell, using the nomenclature proposed by Nowycky et al. (8).

The importance of Ca^{2+}-channels in the β-cell signal-transduction pathway was first established applying conventional electrophysiology (9-11). These studies suggested that Ca^{2+}-influx was voltage-dependent and associated with the characteristic pattern of electrical activity in the β-cell. With the introduction of the patch-clamp technique, it has been possible to characterize the voltage-dependent Ca^{2+}-channels in detail. Using the whole-cell configuration of the patch-clamp technique, Ca^{2+}-currents have been recorded in β-cells from a variety of species. Two subtypes of voltage-dependent Ca^{2+}-channels, conferring different characteristics, have been found (12). The predominant Ca^{2+}-current in β-cells is going through L-type Ca^{2+}-channels, so named because of the long-lasting nature of the current ($t_{1/2}$ of several 100 ms) (8). This channel has been observed in all types of β-cell preparations. The characteristics of the β-cell L-type Ca^{2+}-channel are similar to those of

L-type Ca^{2+}-channels described in other cells, e.g. skeletal and cardiac muscle and neurons (8). Whole-cell recordings of L-type Ca^{2+}-currents from β-cells are observed following depolarizations to potentials more positive than -50 mV and reverse at about +50 mV, with peak currents generally observed at -20 to -10 mV (12). Changing the holding potential between -90 and -50 mV does not alter the I-V relationship. The L-type Ca^{2+}-channel is sensitive to dihydropyridines (DHP). Whereas the DHP antagonists nifedipine, nimodipine and nitrendipine cause an inhibition of the current, the agonist BAY-K 8644 augments the Ca^{2+}-current.

A second type of Ca^{2+}-channel, the T-type Ca^{2+}-channel (transient), has also been found in β-cells (13,14). The T-type channel can be distinguished from the L-type channel in that it inactivates much more rapidly, is completely inactivated at holding potentials more positive than -40 mV and is insensitive to DHP (8). Hence, the T-type Ca^{2+}-channel is unlikely to contribute to the generation of action potential depolarization in the β-cell and its function in the insulin secretory process is not yet understood.

Regulation of voltage-dependent Ca^{2+}-channels

Voltage-dependent Ca^{2+}-channels are regulated by protein phosphorylation (15). Such regulation is mediated by phosphorylation either directly of the channel protein or of associated regulatory proteins and may alter ion-channel properties. Such alterations affect the electrical response pattern of the cell. In this context, it is of interest to note that compounds, like Mg-ATP, cAMP and the catalytic subunit of protein kinase A, that promote protein phosphorylation, slow down time-dependent run-down of Ca^{2+}-currents (16,17). Whereas cAMP, in the pancreatic β-cell, does not seem to affect peak Ca^{2+}-currents, it may affect the rate of inactivation and thereby increasing the net influx of Ca^{2+} (12,18). With regard to effects of PKC on transmembrane Ca^{2+}-transport, most studies have been performed by measuring [Ca^{2+}]$_i$ in clonal insulin-producing RINm5F cells. The interpretation of these studies varies. Whereas Di Virgilio et al proposed that PKC may have an inhibitory effect on the voltage dependent Ca^{2+}-channel (19), Yada et al reported membrane depolarization and Ca^{2+} influx upon stimulation of the enzyme with phorbolester (20). Using the patch-clamp technique, Rorsman et al reported that acute stimulation of RINm5F cells with the phorbol ester TPA, leads to an increase in Ca^{2+}-currents (21). In normal mouse β-cells, we have demonstrated that down-regulation of PKC results in a marked reduction of the voltage-dependent Ca^{2+}-current (Figure 1), associated with a delayed increase in [Ca^{2+}]$_i$.

Glucose may influence Ca^{2+}-influx through voltage-dependent Ca^{2+}-channels in at least two ways. Namely, either by regulating the β-cell membrane potential or by biochemically modulating the channel protein itself. Whereas the former effect is well-established, the latter has been suggested only recently and is likely to depend on glucose metabolism (22,23). A direct effect of glucose on the voltage-dependent Ca^{2+}-channel may be mediated by diacylglycerol and accounted for by PKC-induced phosphorylation (24).

There is good evidence that Ca^{2+} in itself acts as an inactivator of the L-type voltage-dependent Ca^{2+}-channel. To prevent this inactivation, EGTA is often included in the pipette solution during patch-clamp recordings. Alternatively, Ca^{2+} is exchanged for Ba^{2+}, which leads to a significant reduction of inactivation. Inactivation of the voltage-dependent Ca^{2+}-channel by $[Ca^{2+}]_i$, may be mediated by a Ca^{2+}-dependent protein phosphatase (16). Whereas

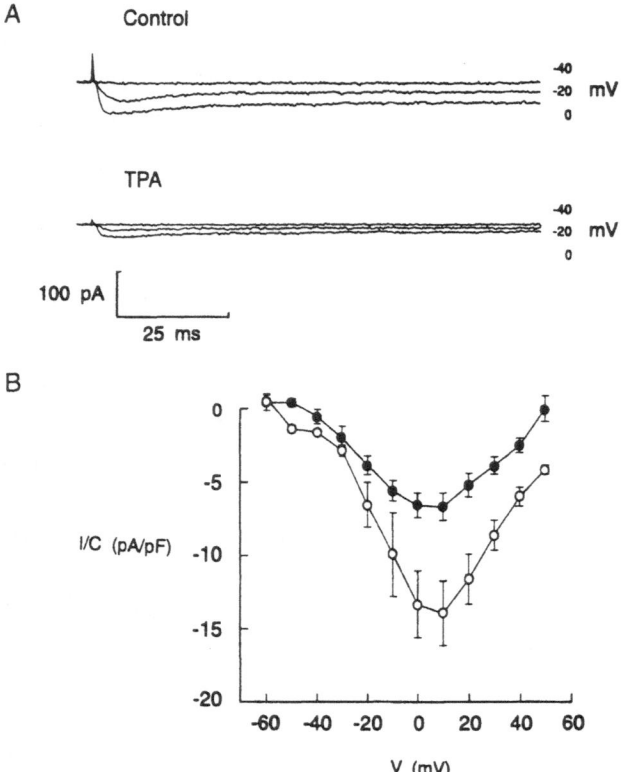

Figure 1. Effect of PKC down-regulation, treatment with 200 nM TPA for 12 h, on whole-cell Ca^{2+}-currents in normal mouse β-cells. The results demonstrate a 50 % reduction in the voltage-dependent Ca^{2+}-currents, compared to control cells. Membrane currents were recorded using the whole-cell configuration of the patch-clamp technique. Currents were recorded during depolarizing voltage-steps to membrane potentials between -60 and +50 mV, from a holding potential of -70 mV. A and B, show current traces from control (DMSO) and TPA-treated cells, respectively. C, I-V relationship in control (open circles) and TPA-treated (filled circles) cells. Data are presented as means ± S.E.M., n=6.

relatively much is known about the mechanisms underlying phosphorylation-induced effects on ion-channel activity, considerably less is known with regard to possible effects of protein phosphatases (15). Among the protein phosphatases, a lot of interest has recently been focused on the serine/threonine protein phosphatases. These phosphatases can be divided

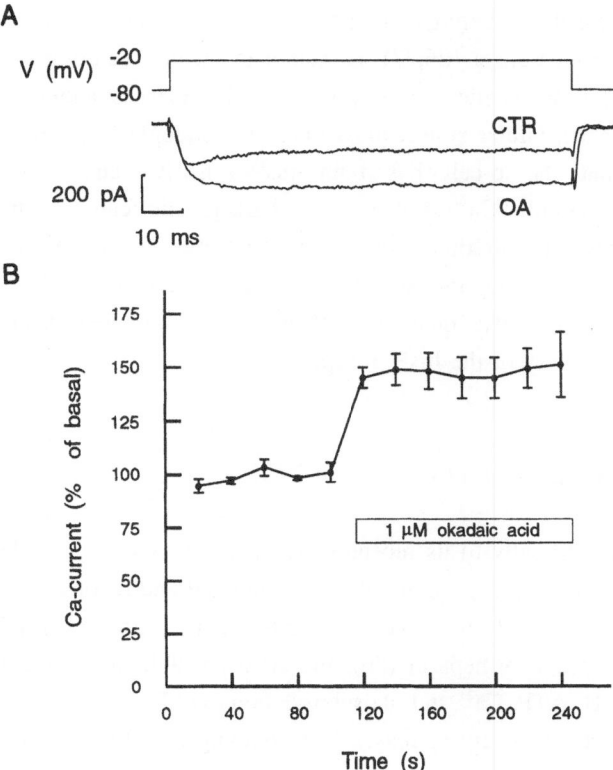

Figure 2. Effects of extracellular application of 1 μM OA on voltage-dependent Ca^{2+}-currents. Membrane currents were recorded in RINm5F cells, using the whole-cell configuration of the patch-clamp technique. Repetitive depolarizing voltage-steps (100 ms) were applied to -20 mV, from a holding potential of -80 mV. Test pulses were given every 20 s. A, typical effect of OA on membrane currents. The traces are from the same cell and the control current is recorded 40 s before application of the phosphatase inhibitor. B, compiled data showing that OA increases the magnitude of the Ca^{2+} currents with approximately 50 %. The 100 % level is calculated by averaging the first five test pulses for each cell before addition of OA. Data are presented as means ± S.E.M. for 3 cells.

into type 1, type 2A, 2B and 2C and type 3, based on the use of specific inhibitors, substrate specificity as well as biochemical characteristics (15). The toxin okadaic acid (OA), produced by the marine plankton dinoflagellates, is a potent and specific inhibitor of protein phosphatase type 1, type 2A and type 3, whereas type 2B (calcineurin) is only slightly affected and type 2C not at all (25). We have recently found that OA affects the activation of voltage-dependent Ca^{2+}-channels in RINm5F cells (Figure 2), suggesting that protein phosphorylation indeed plays a central role in modulating $[Ca^{2+}]_i$ and thereby the β-cell stimulus-secretion coupling.

Intracellular Ca^{2+}-pools and regulation of $[Ca^{2+}]_i$

During recent years, much effort has been devoted to understand the molecular mechanisms regulating intracellular Ca^{2+}-transport and the intracellular pools involved in this regulation in insulin-secreting cells. Shortly after the identification of InsP$_3$ as the Ca^{2+}-mobilizing second messenger (26,27), its action in insulin-secreting cells was confirmed (28,29). In normal pancreatic mouse β-cells, InsP$_3$ releases about 30 % of the Ca^{2+} sequestered into intracellular stores, most likely the endoplasmic reticulum (ER), clearly demonstrating that the β-cell has both InsP$_3$-sensitive and InsP$_3$-insensitive non-mitochondrial intracellular Ca^{2+}-pools (30). A characteristic feature of InsP$_3$-induced Ca^{2+}-release is that a low concentration of the trisphosphate rapidly releases only part of the Ca^{2+}, instead of slowly depleting the actual Ca^{2+}-pool completely (31,32). This interesting phenomenon has been called "quantal" Ca^{2+}-release or "increment detection" and appears to be an unique property of the InsP$_3$-receptor.

Regulation of the InsP$_3$-receptor

InsP$_3$ binds specifically to its receptor, which includes the Ca^{2+}-release channel (33). The InsP$_3$-receptor has been purified, cloned, expressed and reconstituted in lipid bilayers and its regulation has been studied in considerable detail (33-35). InsP$_3$-induced Ca^{2+}-release is completely inhibited by heparin (36). Probable physiological regulators of the InsP$_3$-channel include pH, ATP, Ca^{2+} and phosphorylation (37,38). An increase in pH, within the physiological range, markedly increases InsP$_3$-binding (37). In the ß-cell this may be of physiological significance, since glucose-stimulated insulin release is accompanied by a rise in cytosolic pH (39). ATP allosterically regulates the InsP$_3$-receptor and there is a high affinity binding site for the nucleotide in the receptor protein (40). Low micromolar concentrations of ATP (1-10 µM) markedly augments the ability of InsP$_3$ to release Ca^{2+} (40). At higher concentrations, like in the resting cell (0.1-1 mM), the enhancing effect of ATP is lost. These effects of ATP are not mediated by phosphorylation (40). The extent to which changes in intracellular ATP concentration, subsequent to glucose stimulation, affect InsP$_3$-induced Ca^{2+}-release in the β-cell merits further investigations. Also Ca^{2+} itself affects InsP$_3$-induced Ca^{2+}-release. In this case high concentrations of the ion reduce binding of the trisphosphate to its receptor (41). Indeed, Ca^{2+}, within the physiological range of [Ca^{2+}]$_i$, has been shown to both inhibit and potentiate InsP$_3$-induced Ca^{2+}-release (41-43), effects that may be of importance in the regulation of [Ca^{2+}]$_i$-oscillations. Although the Ca^{2+}-concentration in the ER lumen has been suggested to regulate the sensitivity of the InsP$_3$-receptor to InsP$_3$, thereby being responsible for "quantal" Ca^{2+}-release (44), several studies, including those with reconstituted InsP$_3$-receptors, indicate that this may be a fundamental property of the receptor itself. The InsP$_3$-receptor can be specifically phosphorylated by cAMP-dependent protein kinase, protein kinase C and by Ca^{2+}-calmodulin-dependent protein kinase II (37,38). All three enzymes phosphorylate serine residues but at three different sites (37,38). Hence, phosphorylation provides a means for fine regulation of the InsP$_3$-receptor channel by different signalling systems. Interestingly however, in RINm5F cells, when PKC is activated with the phorbolester TPA, there is neither an interference with InsP$_3$-induced release of Ca^{2+} nor subsequent buffering of the ion (45).

InsP₃ and Ca²⁺-sequestration in permeabilized cells

While Ca^{2+}-release by $InsP_3$ is well established, a possible role for the trisphosphate in promoting Ca^{2+}-sequestration by ER is not as obvious. In this context it should be noted, that inositol 1,3,4,5-tetrakisphosphate ($InsP_4$) has been reported to promote Ca^{2+}-sequestration in one cell type (46). In permeabilized RINm5F cells, we have demonstrated that the ambient free Ca^{2+}-concentration obtained, subsequent to $InsP_3$-induced Ca^{2+}-release, is lower than that before addition of the trisphosphate (47 and Figure 3). A similar effect was obtained in the presence of heparin, suggesting the existence of high affinity $InsP_3$-receptors continuously operating in the insulin-secreting cell. Whereas a high concentration of $InsP_3$ will desensitize such receptors, they are inhibited by heparin, thus explaining the increased sequestration observed in the presence of these compounds (47). However, we could not observe any effect of $InsP_4$ on Ca^{2+}-sequestration in RINm5F cells.

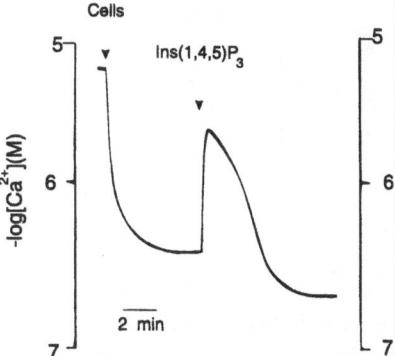

Figure 3. Effects of $InsP_3$ on Ca^{2+}-release and re-uptake in permeabilized insulin-secreting RINm5F cells. Clonal insulin-secreting RINm5F cells were permeabilized by exposure to high voltage-discharges. Experiments were carried out at room temperature and the cell suspension was stirred continuously, using a small magnetic bar. Changes in the ambient free Ca^{2+}-concentration were recorded using a Ca^{2+}-selective minielectrode. $InsP_3$ (5 μM, final concentration) was added as indicated.

Ca²⁺ mobilization by ER Ca²⁺-ATPase inhibitors

One way of characterizing intracellular Ca^{2+}-pools, is by using specific inhibitors of the pump that loads Ca^{2+} into the ER (ER Ca^{2+}-ATPase). Several highly specific and potent inhibitors of ER Ca^{2+}-ATPase are currently available. These include thapsigargin and 2,5-di-(tert-butyl)-1,4-benzohydroquinone (tBuBHQ). An interesting and useful feature of these agents, is that they can mobilize Ca^{2+} from the $InsP_3$-sensitive intracellular Ca^{2+}-pool, without affecting InsP$_3$-metabolism. Ca^{2+}-mobilization by these agents has been characterized in many cell types but until recently, little was known about their actions in insulin-secreting cells. In electropermeabilized RINm5F cells, we have demonstrated that both thapsigargin and tBuBHQ mobilize Ca^{2+} from intracellular pools in a dose-dependent manner (Islam and Berggren, unpublished data). Although thapsigargin is effective in nm concentrations and tBuBHQ is effective in μm concentrations, a maximally effective concentration of the latter is marginally more effective than the former in mobilizing Ca^{2+}. Both Ca^{2+}-ATPase inhibitors release Ca^{2+} predominantly from the InsP$_3$-sensitive pool and our data suggest that thapsigargin has greater specificity than tBuBHQ in this respect.

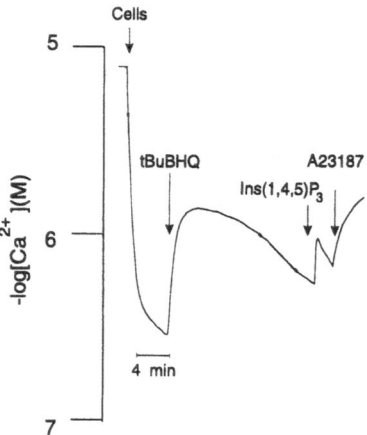

Figure 4. In permeabilized RINm5F cells, tBuBHQ releases Ca^{2+} predominantly from the InsP$_3$-sensitive Ca^{2+}-pool. This Ca^{2+}-ATPase inhibitor can, however, not empty the trisphosphate-sensitive pool readily. For experimental details, see figure 3. At points indicated, tBuBHQ (25μM), InsP$_3$ (5 μM) and A23187 (4 μM) were added.

Nevertheless, the magnitude of Ca^{2+}-increase induced by either of the two Ca^{2+}-ATPase inhibitors is always smaller than that evoked by InsP$_3$ and neither of them can empty the trisphosphate-sensitive pool completely (Figure 4). A possible implication of these findings is, that the additional Ca^{2+}-release evoked by InsP$_3$ under these conditions originates from a different trisphosphate-sensitive pool, which is insensitive to the Ca^{2+}-ATPase inhibitors. The mechanism of Ca^{2+}-uptake into this putative second InsP$_3$-sensitive pool is not known. Neither the protonophore carbonylcyanide m-chlorophenylhydrazone (CCCP) nor Bafilomycin A$_1$, a specific inhibitor of V-type ATPase, abolished this additional InsP$_3$-response, indicating that it does not require a proton gradient for maintaining its Ca^{2+}. Moreover, this finding may be an indication of compartmentalization within the InsP$_3$-sensitive pool (see below).

The InsP$_3$-sensitive organelle

It is evident that InsP$_3$ mobilizes Ca^{2+} from some intracellular non-mitochondrial membrane-bound structures, closely associated with ER. Early studies, utilizing subcellular fractionation, have shown a close correlation of InsP$_3$-induced Ca^{2+}-release with ER markers in insulinoma cells (28). Since ER is an extensive structure having many metabolic and synthetic functions to perform, separation of its role in Ca^{2+}-signalling from its other functions can be anticipated. Localized structural and functional specialization are known to occur in membranes in general and in ER in particular. Whether the whole or only some compartments of ER act as a Ca^{2+}-storing organelle is not clear. Measurements of the Ca^{2+}-concentration in the ER lumen is at present difficult, due to the relatively low saturation points of available fluorescent Ca^{2+}-indicators. Nevertheless, InsP$_3$-receptors do not appear to be present throughout the ER.

Our experiments with thapsigargin and tBuBHQ suggest the existence of separate uptake and release compartments in the InsP$_3$-sensitive pool (Islam and Berggren, unpublished observations and Figure 5). Thus these two Ca^{2+}-ATPase-inhibitors apparently release Ca^{2+} predominantly from the uptake compartment, whereas Ca^{2+} located in the release compartment can not easily be mobilized by this means. This is clearly shown in experiments where prolonged treatment with thapsigargin or tBuBHQ fails to deplete this pool completely. Under such conditions, InsP$_3$ readily releases the remaining Ca^{2+}. The mechanism of translocating Ca^{2+} from the uptake to the release compartment in vivo is not known, but GTP and membrane trafficking G-proteins may play a role. At least two different studies have demonstrated that GTP releases Ca^{2+} from islet cells (48,49). Interestingly, in one of these studies (48) Ca^{2+} release was evoked in the absence of polyethylene glycol, an agent which is often used in combination with GTP and is supposed to promote membrane fusion (48). So far, we have not been able to demonstrate GTP-induced Ca^{2+}-release in either RINm5F cells or normal pancreatic mouse β-cells.

Ca^{2+}-induced Ca^{2+}-release

Ca^{2+}-induced Ca^{2+}-release (CICR), first discovered in skeletal muscle, is a phenomenon whereby an increase in [Ca^{2+}]$_i$ causes further release of Ca^{2+}, by acting on specific receptor-operated channels located in the sarcoplasmatic reticulum (50). Conventionally the receptor for CICR is called the "ryanodine" receptor, because it binds the plant alkaloid ryanodine specifically (50). Caffeine is another alkaloid which potentiates CICR (50). Since direct demonstration of CICR is difficult, most studies have resorted to the use of pharmacological agents like ryanodine or caffeine and Ca^{2+} release induced by these agents is generally believed to indicate the existence of CICR in any particular cell. Indeed, CICR is not limited to the skeletal muscle, but rather a more general phenomenon documented in smooth muscle cells, neuronal cells, in nearly all excitable endocrine cells and even in non-excitable cells (51-53). Ca^{2+} is believed to be the physiological agonist for the ryanodine receptor, although this may be an oversimplification. In this context, cyclic ADP ribose is an interesting compound that has been shown to activate CICR in sea urchin eggs and in pituitary cells (54,55). However, cyclic ADP ribose did not mobilize Ca^{2+} in permeabilized RINm5F-cells (Islam and Berggren, unpublished observations). Hence, to demonstrate CICR in insulin-producing cells, we used a different approach (56). The sarcoplasmic reticulum ryanodine receptor contain "critical" sulphydryl-groups (SH-groups). Oxidation of these groups, by a variety of sulphydryl-reagents, causes conformational change of the ryanodine receptor protein, resulting in Ca^{2+}-release (57,58). Thimerosal is a sulphydryl-reagent that, in low micromolar concentrations, has been shown to sensitize CICR in non-muscle cells (59). In electropermeabilized RINm5F cells, thimerosal released Ca^{2+} from a non-

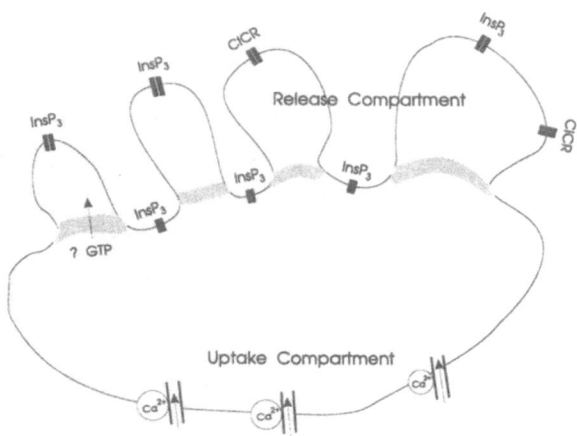

Figure 5. A tentative model of the compartmentalized $InsP_3$-sensitive Ca^{2+}-pool in insulin-secreting cells. In addition to depicting the different uptake and release compartments, the model also shows CICR (see text).

mitochondrial intracellular Ca^{2+}-pool in a dose dependent manner (56). This release was reversed after addition of the reducing agent dithiothreitol. Ca^{2+} was released from the $InsP_3$-insensitive pool, since release was observed even after depletion of the $InsP_3$-sensitive pool by a supramaximal dose of inositol 2,4,5-trisphosphate (Figure 6), as well as in the presence of heparin. The effect of $InsP_3$ and thimerosal were additive and the $InsP_3$-sensitive pool remained essentially unaltered by thimerosal. Furthermore, thimerosal-induced Ca^{2+}-release was potentiated by caffeine, in keeping with an effect of the sulphydryl-reagent on the CICR-channel. Although the physiological significance remains to be settled, these findings indeed suggest the existence of CICR also in insulin-secreting cells.

Oscillations in $[Ca^{2+}]_i$

So far, most studies trying to identify the molecular mechanisms involved in regulating oscillations in $[Ca^{2+}]_i$ have been performed in non-excitable cells, where Ca^{2+}-mobilization induced by $InsP_3$ seems to have a profound effect (60). In an excitable cell, like the pancreatic β-cell, these mechanisms are bound to be much more complicated, most likely involving a sophisticated interplay between Ca^{2+}-influx through voltage-gated Ca^{2+}-channels and Ca^{2+}-mobilization from intracellular stores (4,61). Subsequent to stimulation with glucose, single β-cells or small β-cell aggregates demonstrate slow 2-5 min $[Ca^{2+}]_i$-

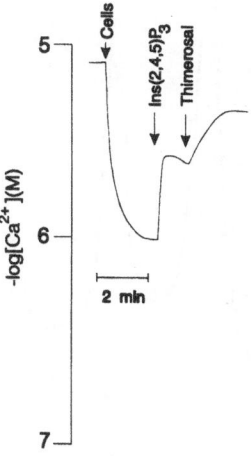

Figure 6. Thimerosal induces Ca^{2+}-release from an $InsP_3$-insensitive Ca^{2+}-pool in permeabilized RINm5F cells, since release was obtained even after exposure of the cells to a supramaximal dose of inositol 2,4,5-trisphosphate. For experimental details, see figure 3. At points indicated, $Ins(2,4,5)P_3$ (25 μM) and thimerosal (25 μM) were added.

oscillations (Figure 7), that can be prevented by activation of PKC (62). The latter effect probably reflects both stimulated efflux of Ca^{2+} from the β-cell, due to activation of the plasma membrane Ca^{2+}-pump, and direct inhibition of the PLC pathway (61). Noteworthy is, that the slow $[Ca^{2+}]_i$-oscillations are inhibited also in the absence of extracellular Ca^{2+} and in the presence of blockers of L-type voltage-activated Ca^{2+} channels (63). Interestingly, the slow $[Ca^{2+}]_i$-oscillations are much more irregular in cells subjected to PKC down-regulation (62). This probably reflects the fact that PKC is needed also in the phosphorylation of the voltage-dependent Ca^{2+}-channels, enabling them to conduct Ca^{2+}-influx more efficiently (see above).

In PKC down-regulated β-cells large transients in $[Ca^{2+}]_i$, lasting for approximately 10 s, are often superimposed on the slow $[Ca^{2+}]_i$-oscillations (62). Under conditions where PKC is down-regulated, the normal inhibition of the PLC-system and as well stimulation of membrane Ca^{2+}-pumps by the enzyme should be suppressed. Consequently, an enhanced formation of $InsP_3$, in combination with reduced Ca^{2+}-efflux from the cell and maybe reduced intracellular buffering of the ion, should promote the generation of the large Ca^{2+}-transients. The $InsP_3$- and most likely also CICR-$[Ca^{2+}]_i$-spikes (see above) will activate

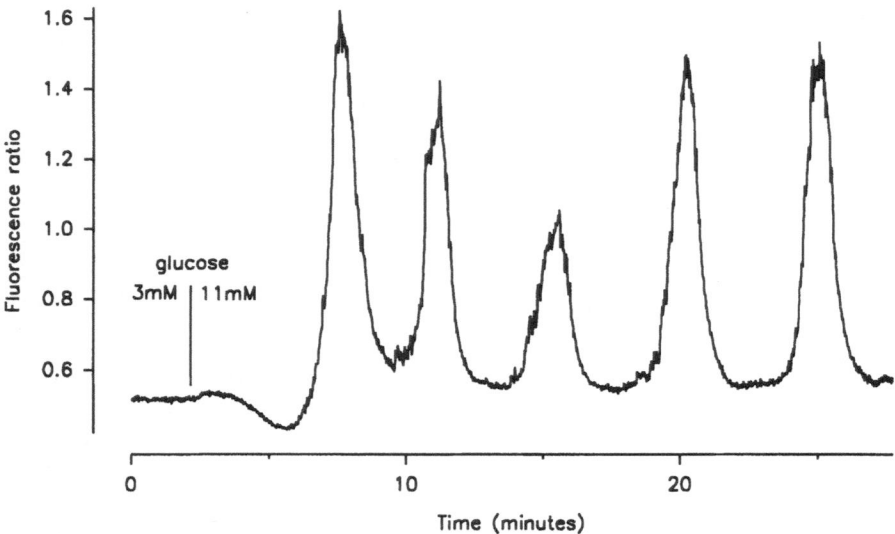

Figure 7. Glucose-induced slow oscillations in $[Ca^{2+}]_i$ in a small mouse β-cell aggregate. The cells were attached to a coverslip and loaded with the Ca^{2+}-indicator Fura-2. A monochromator based system (Spex Fluorolog-2) alternatingly excited the sample, mounted in a perifusion chamber on a epi-fluorescence microscope with light at 340 and 380 nm. Emission (selected by a 500-530 nm bandpass filter) was detected by a photon-counting photometer tube. $[Ca^{2+}]_i$ is given as the 340/380 nm fluorescence ratio.

Ca^{2+}-dependent K^+-channels (4). This leads to membrane repolarization, closure of voltage-activated Ca^{2+}-channels, a decrease in Ca^{2+}-influx and thereby a lowering in $[Ca^{2+}]_i$. Once $[Ca^{2+}]_i$ is low, the Ca^{2+}-activated K^+-channels close, the cell is depolarized, since glucose is still present in the medium, and a new cycle starts. The glucose-induced 10 s transients in $[Ca^{2+}]_i$ in PKC down-regulated cells are dependent on extracellular Ca^{2+} (62). When the PLC-pathway is directly activated by agonists like carbamylcholine and GTP-γ-S, in non-PKC-depleted β-cells, the $[Ca^{2+}]_i$-spikes are generated even in the absence of extracellular

Ca^{2+} (4). Noteworthy is, that whole pancreatic islets demonstrate both slow and fast oscillations in $[Ca^{2+}]_i$ (64, see also Figure 8). In this case, the fast oscillations, similar in duration to the fast $[Ca^{2+}]_i$-spikes obtained in single β-cells or small β-cell aggregates subjected to PKC down-regulation (62), are superimposed on the slow ones. However, so far we have no data to convincingly demonstrate, that the fast $[Ca^{2+}]_i$-spikes obtained subsequent to PKC down-regulation have a similar molecular basis as the fast oscillations obtained in whole pancreatic islets.

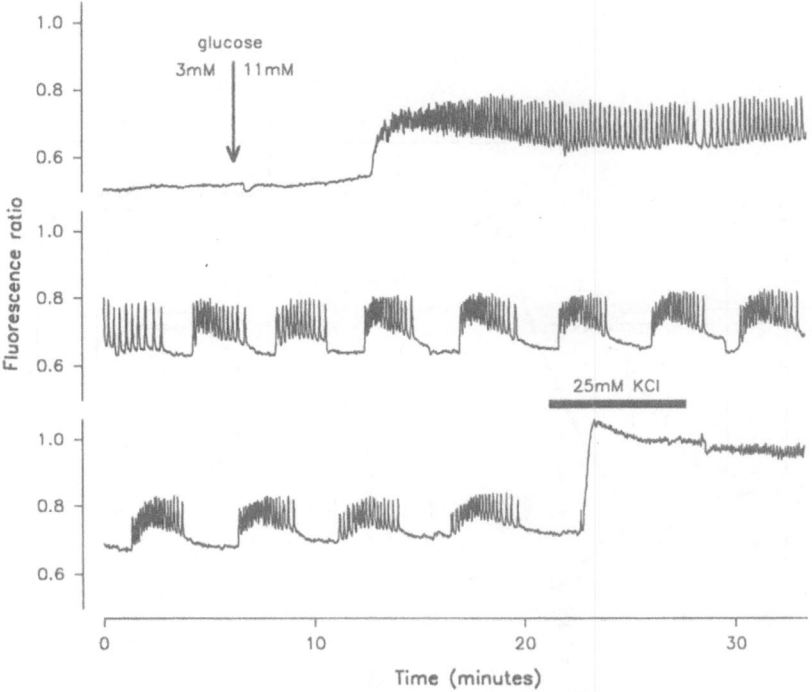

Figure 8. Glucose-induced slow and fast oscillations in $[Ca^{2+}]_i$ in a whole pancreatic islet from mouse. Experimental details as in figure 7.

It should be remembered, that β-cells situated within a pancreatic islet in vivo are continuously exposed to a variety of neuropeptides, other transmitter substances and hormones, which in addition to glucose activate the PLC-system. It is possible, that these substances remain present for a while within the pancreatic islet subsequent to its isolation and hence in combination with glucose activate the PLC-system, thereby constituting the molecular mechanism underlying activation of the fast $[Ca^{2+}]_i$-oscillations. This may also explain why the fast oscillations are not obtained in isolated single β-cells or small ß-cell aggregates, under normal conditions. Moreover, this supports the concept of a complex interplay between Ca^{2+}-influx through voltage-activated L-type Ca^{2+}-channels and Ca^{2+}-release from intracellular stores, in the regulation of the glucose-induced fast oscillations in $[Ca^{2+}]_i$ in the β-cell (see model proposed in Figure 9). This model also suggests that the β-cell is polarized, not only in terms of secretory granules, but also voltage-gated Ca^{2+}-channels, ER and Ca^{2+}-activated K^+-channels.

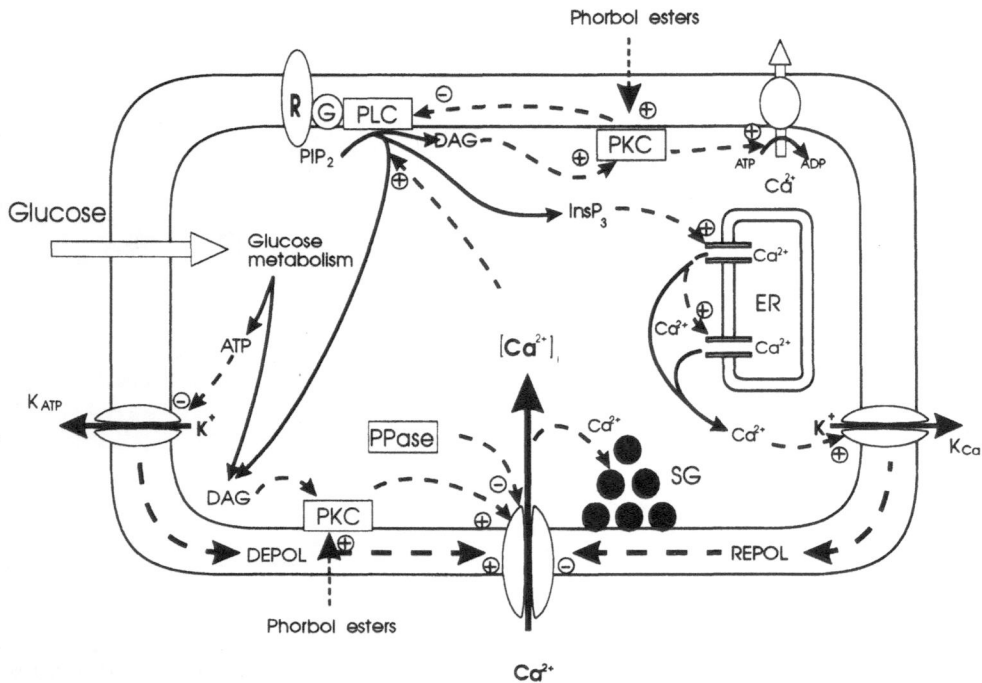

Figure 9. A model demonstrating how Ca^{2+}-influx through voltage-activated Ca^{2+}-channels and Ca^{2+}-release from intracellular stores may interact in the molecular regulation of the fast $[Ca^{2+}]_i$-oscillations. Glucose metabolism leads to the formation of ATP and DAG. ATP closes the ATP-regulated K^+-channels (K_{ATP}), resulting in depolarization, opening of voltage-activated Ca^{2+}-channels and increase in $[Ca^{2+}]_i$. The increase in $[Ca^{2+}]_i$ may activate the PLC-system. The PLC-system is also activated by various receptor agonists, resulting in the formation of $InsP_3$ and DAG. Whereas DAG activates PKC, an enzyme interacting both with voltage-activated Ca^{2+}-channels and the plasma membrane Ca^{2+}-pump, $InsP_3$ mobilizes intracellularly bound Ca^{2+}, which may promote CICR. Ca^{2+} originating from intracellular stores activates a low conductance Ca^{2+}-activated K^+-channel (K_{Ca}), resulting in repolarization of the plasma membrane and closure of voltage-activated Ca^{2+}-channels. Interestingly, the activity of voltage-gated Ca^{2+}-channels can also be modulated by serine/threonine protein phosphatases (PPase). Note also that the proposed model suggests that the β-cell is polarized, not only in terms of secretory granules (SG), but also voltage-gated Ca^{2+}-channels, ER and K_{Ca}. This means that Ca^{2+}-influx through voltage-gated Ca^{2+}-channels will take place in close contact to SG, thus initiating exocytosis. Ca^{2+} released from intracellular stores will preferentially activate K_{Ca}, resulting in repolarization of the cell and thereby closure of the voltage-gated Ca^{2+}-channels.

Whereas the model described above may offer an explanation to the molecular mechanisms regulating the glucose-induced fast oscillations in $[Ca^{2+}]_i$, we have reasons to believe that the glucose-induced slow oscillations in $[Ca^{2+}]_i$ are regulated by other means. In this context, we have preliminary data (Kindmark and Berggren, unpublished observations) which are compatible with the concept (65), that these oscillations are metabolically driven and hence reflect oscillations in for example the cytoplasmic ATP/ADP-concentration ratio. Interestingly, oscillations in $[Ca^{2+}]_i$ and NADH-fluorescence, with a strikingly similar frequency, have recently been observed in pancreatic β-cells (66).

ACKNOWLEDGEMENTS

Own research described in the present study was supported by the Swedish Medical Research Council (19X-00034, 04X-09890 and 04X-09891), the Bank of Sweden Tercentenary Foundation, the Swedish Diabetes Association, the Nordic Insulin Foundation, Fredrik and Ingrid Thurings Foundation, the Swedish Hoechst Diabetes Research Foundation, Magnus Bergvalls Foundation, Novo Nordisk A/S, Torsten and Ragnar Söderbergs Foundations and Funds of the Karolinska Institute.

References

1. Prentki, M. and Matschinsky, F.M. Ca^{2+}, cAMP and phospholipid-derived messengers in coupling mechanisms of insulin secretion. Physiol Rev 1987, 67: 1185-1248

2. Efendic, S., Kindmark, H. and Berggren, P-O. Mechanisms involved in the regulation of the insulin secretory process. J Internal Med 1991, 229: 9-22

3. Berggren, P-O., Rorsman, P., Efendic, S., Östenson, C-G., Flatt, P.R., Nilsson, T., Arkhammar, P. and Juntti-Berggren, L. Mechanisms of action of entero-insular hormones, islet peptides and neural input on the insulin secretory process. In Nutrient Regulation of Insulin Secretion (P.R Flatt ed.), Portland Press London, 1992, 289-318

4. Ämmälä, C., Larsson, O., Berggren, P-O., Bokvist, K., Juntti-Berggren, L., Kindmark, H. and Rorsman, P. Inositol trisphosphate-dependent periodic activation of a Ca^{2+}-activated K^+-conductance in glucose-stimulated pancreatic β-cells. Nature 1991, 353: 849-852

5. Islam, Md.S., Rorsman, P. and Berggren, P-O. Ca^{2+}-induced Ca^{2+}-release in insulin-secreting cells. FEBS Lett 1992, 296: 287-291

6. Catterall, W.A. Structure and function of voltage-dependent ion channels. Science 1988, 242:50-61

7. Tsien, R.W., Ellinor, P.T. and Horne, W.A. Molecular diversity of voltage-dependent Ca²⁺ channels. Trends in Pharmac Sci 1991, 12:349-354

8. Nowycky, M.C., Aaron P.F. and Tsien R.W. Three types of neuronal calcium channel with different calcium agonist sensitivity. Nature 1985, 316:440-443

9. Dean, P.M. and Matthews, C.K. Electrical activity in pancreatic islet cells. Nature 1968, 219:389-390

10. Dean, P.M. and Matthews, C.K. Glucose-induced electrical activity in pancreatic islet cells. J Physiol (London) 1970, 210:255-264

11. Meissner, H.P. and Schmelz, H. Membrane potential of beta cells in pancreatic islets. Pflügers Arch 1974, 351:195-206

12. Ashcroft, F.M. and Rorsman, P. Electrophysiology of the pancreatic β-cell. Prog. Biophys. molec. Biol. 1991, 54:87-143

13. Velasco, J.M. Calcium channels in insulin-secreting RINm5F cell line. J Physiol (London) 1987, 398:15P

14. Ashcroft, F.M., Kelly, R.P. and Smith, P.A. Two types of Ca channel in rat pancreatic ß-cells. Pfleugers Arch 1990, 415:504-506

15. Shenolikar, S. and Nairn, A.C. Protein phosphatases: Recent progress. In Advances in Second Messenger and Phosphoprotein Research. Vol 23 (P. Greengard and J.A. Robinsson eds.), Raven Press, Ltd, New York 1991, 1-121

16. Chad, J.E. and Eckert, R. An enzymatic mechanism for calcium current inactivation in dialyzed Helix neurons. J Physiol 1986, 378:31-51

17. Rorsman, P., Ashcroft, F.M and Trube, G. Single Ca channel currents in mouse pancreatic ß-cells. Pflügers Arch 1988, 412:597-603

18. Henquin, J.C. and Meissner, H.P. Cyclic adenosine monophosphate differently affects the response of mouse pancreatic ß-cells to various amino acids. J Physiol 1986, 381:77-93

19. Di Virgilio, F., Pozzan, T., Wollheim, C.B. Vicentini, L.M. and Meldolesi, J. Tumor promoter phorbol myristate acetate inhibits Ca²⁺ influx through voltage-dependent Ca²⁺ channels in two secretory cell lines, PC 12 and RINm5F. J Biol Chem 1986, 261:32-35

20. Yada, T., Russo, L.L. and Sharp, G.W.G. Phorbol ester-stimulated secretion by RINm5F insulinoma cells is linked with membrane depolarization and an increase in cytosolic free Ca^{2+} concentration. J Biol Chem 1989, 264:2455-2462

21. Rorsman, P., Arkhammar, P and Berggren, P-O. Voltage-activated Na^+ currents and their suppression by phorbol ester in clonal insulin-producing RINm5F cells. Am J Physiol 1986, 251:C912-919

22. Smith, P.A., Rorsman, P. and Ashcroft, F.M. Modulation of dihydropyridine-sensitive Ca^{2+} channels by glucose metabolism in mouse pancreatic ß-cells. Nature 1989, 342:550-553

23. Velasco, J.M., Petersen, J.U.H. and Petersen, O.H. Single channel Ba^{2+} currents in insulin secreting cells are activated by glyceraldehyde stimulation. FEBS Lett 1988, 213:366-370

24. Velasco, J.M. and Petersen, O.H. The effects of a cell-permeable diacylglycerol analogue on single Ca^{2+} (Ba^{2+}) channel currents in the insulin-secreting line RINm5F. Q J exp Physiol 1989, 74:367-370

25. Bialojan, C. and Takai, A. Inhibitory effect of a marine-spong toxin, ocadaic acid, on protein phosphatases. Specificity and kinetics. Biochem J 1988, 256:283-290

26. Berridge, M. J. Rapid accumulation of inositol trisphosphate reveals that agonists hydrolyse polyphosphoinositides instead of phosphatidylinositol. Biochem. J. 1983, 212: 849-858

27. Streb, H., Irvine, R. F., Berridge, M. J., and Schulz, I. Release of Ca^{2+} from a nonmitochondrial intracellular store in pancreatic acinar cells by inositol-1,4,5-trisphosphate. Nature 1983, 306: 67-69

28. Prentki, M., Biden, T. J., Janjic, D., Irvine, R. F., Berridge, M. J., and Wollheim, C. B. Rapid mobilization of Ca^{2+} from rat insulinoma microsomes by inositol-1,4,5-trisphosphate. Nature 1984, 309: 562-564

29. Biden, T. J., Prentki, M., Irvine, R. F., Berridge, M. J., and Wollheim, C. B. Inositol 1,4,5-trisphosphate mobilizes intracellular Ca^{2+} from permeabilized insulin-secreting cells. Biochem. J. 1984, 223: 467-473

30. Nilsson, T., Arkhammar, P., Hallberg, A., Hellman, B. and Berggren, P.-O. Characterization of the inositol 1,4,5-trisphosphate-induced Ca^{2+} release in pancreatic ß-cells. Biochem. J. 1987, 248: 329-336

31. Muallem, S., Pandol, S. J. and Beeker, T. J. Hormone-evoked calcium release from intracellular stores is a quantal process. J. Biol. Chem. 1989, 264: 205-212

32. Meyer, T., and Stryer, L. Transient Ca^{2+} release induced by successive increments of inositol 1,4,5-trisphosphate. Proc. Natl. Acad. Sci. USA. 1990, 87: 3841-3845

33. Supattapone, S., Worley, P. F., Baraban, J. M., and Snyder, S. H. Solubilization, purification, and characterization of an inositol trisphosphate receptor. J. Biol. Chem. 1988, 263:1530-1534

34. Furuichi, T., Yoshikawa, S., Miyawaki, A., Wad, K., Maeda, N., and Mikoshiba, K. Primary structure and functional expression of the inositol 1,4,5-trisphosphate-binding protein P_{400}. Nature 1989, 342: 32-38

35. Supattapone, S., Danoff, S. K., Theibert, A., Joseph, S. K., Steiner, J. and Snyder, S. H. Cyclic AMP-dependent phosphorylation of a brain inositol trisphosphate receptor decreases its release of calcium. Proc. Natl. Acad. Sci. USA. 1988, 85: 8747-8750

36. Nilsson, T., Zwiller, J., Boynton, A. L., and Berggren, P.-O. Heparin inhibits IP_3-induced Ca^{2+} release in permeabilized pancreatic ß-cells. FEBS lett. 1988, 229: 211-214

37. Worley, P. F., Baraban, J. M., Supattapone, S., Wilson, V. S., and Snyder, S. H. Characterization of inositol trisphosphate receptor binding in brain: regulation by pH and calcium. J. Biol. Chem. 1987, 262:12132-12136

38. Ferris, C. D., Huganir, R. L., Snyder, S. H. Inositol trisphosphate receptor: phosphorylation by protein kinase C and calcium calmodulin-dependent protein kinases in reconstituted lipid vesicles. Proc. Natl. Acad. Sci. USA. 1990, 87:2147-2151

39. Juntti-Berggren, L., Arkhammar, P., Nilsson, T., Rorsman, P., and Berggren, P.-O. Glucose-induced increase in cytoplasmic pH in pancreatic ß-cells is mediated by Na^+/H^+ exchange, an effect not dependent on protein kinase C. J. Biol. Chem. 1991, 266:23537-23541

40. Ferris, C. D., Huganir, R. L., Bredt, D. S., Cameron, A. M., and Snyder, S. H. Calcium flux mediated by purified inositol 1,4,5-trisphosphate receptor in reconstituted lipid vesicles is allosterically regulated by adenine nucleotides. Proc. Natl. Acad. Sci. USA. 1991, 88: 2232-2235

41. Jean, T., and Klee, C. B. Calcium modulation of inositol 1,4,5-trisphosphate-induced Ca^{2+} release from neuroblastoma x glioma hybrid (NG 108-15) microsomes. J. Biol. Chem. 1986, 261:16414-16420

42. Willems, P. H. G. M., De Jong, M. D., de Pont, J. J. H.H. M., and Van Os, C. H. Inhibition of inositol 1,4,5-trisphosphate-induced Ca^{2+} release in permeabilized pancreatic acinar cells by hormonal and phorbol ester pretreatment. J. Biol. Chem. 1990, 265:681-687

43. Bezprovanny, I., Watras, J., and Ehrlich, B. E. Bell-shaped Ca^{2+} response curves of Ins(1,4,5)P_3- and calcium-gated channels from endoplasmic reticulum of cerebellum. Nature. 1991, 351: 751-754

44. Irvine, R. F. 'Quantal' Ca^{2+} release and the control of Ca^{2+} entry by inositol phosphates - a possible mechanism. FEBS Lett. 1990, 263: 5-9

45. Arkhammar, P., Nilsson, T. and Berggren, PO. Stimulation of insulin release by phorbol ester 12-O-tertradecanoylphorbol 13-acetate in the clonal cell line RINm5F depite a lowering of the free cytoplasmic Ca^{2+} concentration. Biochim Biophys Acta 1986, 887:236-241

46. Hill, T. D., Dean, N. M., and Boynton, A. L. Inositol 1,3,4,5-tetrakisphosphate induces Ca^{2+} sequestration in rat liver cells. Science. 1988, 242: 1176-1178

47. Islam, M. S., Nilsson, T., Rorsman, P., and Berggren, P.-O. Interaction with the inositol 1,4,5-trisphosphate receptor promotes Ca^{2+} sequestration in insulin-secreting cells. FEBS Lett. 1991, 288: 27-29

48. Dunlop, M. E., and Larkins, R. G. GTP- and inositol 1,4,5-trisphosphate-induced release of $^{45}Ca^{2+}$ from a membrane store co-localized with pancreatic-islet-cell plasma membrane. Biochem. J. 1988, 253: 67-72

49. Wolf, B. A., Florholmen, J., Colca, J. R., and McDaniel, M. L. GTP mobilization of Ca^{2+} from the endoplasmic reticulum of islets. Biochem. J. 1987, 242:137-141

50. Endo, M. Calcium release from the sarcoplasmic reticulum. Physiol. Rev. 1977, 57: 71-108

51. Kuba, K. Release of Ca^{2+} ions linked to the activation of potassium conductance in a caffeine-treated sympathetic neurone. J. Gen. Physiol. 1980, 298:251-269

52. Cheek, T. R., Barry, V. A., Berridge, M. J., and Missiaen, L. Bovine adrenal chromaffin cells contain an inositol 1,4,5-trisphosphate-insensitive but caffeine-sensitive Ca^{2+} store that can be regulated by intraluminal free Ca^{2+}. Biochem. J. 1991, 275:697-701

53. Dehlinger-Kremer, M., Zeuzem, S. and Schulz, I. Interaction of caffeine-, IP_3- and vanadate sensitive Ca^{2+} pools in acinar cells of the exocrine pancreas. J. Membr. Biol. 1991, 119: 85-100

54. Galione, A., Lee, H. C., and Busa, W. B. Ca^{2+}-induced Ca^{2+} release in sea urchin egg homogenates: modulation by cyclic ADP-ribose. Science 1991, 253:1143-1146

55. Koshiyama, H., Lee, H. C., and Tashijan, A. H.Jr. Novel mechanisms of intracellular calcium release in pituitary cells. J. Biol. Chem. 1991, 266:16985:16988

56. Islam, M. S., Rorsman, P. and Berggren, P.-O. Ca^{2+}-induced Ca^{2+} release in insulin-secreting cells. FEBS Lett. 1992, 296:287-291

57. Trimm, J. L., Salama, G. and Abramson, J. J. Sulfhydryl oxidation induces rapid calcium release from sarcoplasmic reticulum vesicles. J. Biol. Chem. 1986, 261:16092-16098

58. Zaidi, N. F., Lagenaur, C. F., Abramson, J. J., Pessah, I. and Salama, G. Reactive disulfides trigger Ca^{2+} release from sarcoplasmic reticulum via an oxidation reaction. J. Biol. Chem. 1989, 264: 21725-21736

59. Swann, K. Thimerosal causes calcium oscillations and sensitizes calcium-induced calcium release in unfertilized hamster eggs. FEBS Lett. 1991, 278:175-178

60. Berridge, M.J. and Irvine, R.F. Inositol phosphates and cell signalling. Nature 1989, 341: 197-205

61. Berggren, P-O., Arkhammar, P., Bokvist, K., Efendic, S., Islam, MdS., Juntti-Berggren, L., Kindmark, H., Köhler, M., Larsson, O., Nilsson, T., Rorsman, P. and Ämmälä, C. Intracellular Ca^{2+}-stores and Ca^{2+}-oscillations in insulin secreting cells. 20th Karolinska Institute Nobel Conference on Calcium Signalling 1992, 83-86

62. Kindmark, H., Köhler, M., Efendic, S., Rorsman, P., Larsson, O. and Berggren, P-O. Protein kinase C activity affects glucose-induced oscillations in cytoplasmic free Ca^{2+} in the pancreatic β-cell. FEBS Lett. 1992, 303:85-90

63. Grapengiesser, E., Gylfe, E. and Hellman, B. Three types of cytoplasmic Ca^{2+}-oscillations in stimulated pancreatic β-cells. Arch Biochem Biophys. 1989, 268:404-407

64. Valdeolmilos, M., Santos, RM., Contreras, D., Soria, B. and Rosario, L.M. Glucose-induced oscillations of intracellular Ca^{2+}-concentration resembling bursting electrical activity in single mouse islets of Langerhans. FEBS Lett. 1989, 259:19-23

65. Longo, E.A., Tornheim, K., Deeney, J.T., Varnum, B.A., Tillotson, D., Prentki, M. and Corkey, B.E. Oscillations in cytosolic free Ca^{2+}, oxygen consumption and insulin secretion in glucose-stimulated rat pancreatic islets. J Biol Chem 1991, 266:9314-9319

66. Pralong ,W-F., Gjinovci, A. & Wollheim, C.B. Ca^{2+}-modulation of redox state in single β-cells exposed to nutrients. Diabetes 1991, 40:317

THE ß-CELL SULFONYLUREA RECEPTOR

Stephen J.H. Ashcroft[1], Ichiro Niki[1], Sue Kenna[1], Ling Weng[1], Jackie Skeer[2], Barbara Coles[2] and Frances M. Ashcroft[2]

[1]Nuffield Department of Clinical Biochemistry
John Radcliffe Hospital, Headington, Oxford OX3 9DU
2 University Laboratory of Physiology
Parks Road, Oxford OX1 3PT
U.K.

INTRODUCTION

It is now 50 years since the accidental observations which indicated that a sulfonamide, p-amino-benzene-sulfamido-isopropyl-thiodiazole (2254 RP), could induce hypoglycaemia (Janbon et al., 1942). This response was shown to be due to stimulation of insulin secretion and led to the use of this class of sulfonamides (called sulfonylureas) to treat diabetes (Loubatières, 1955; Bertram et al., 1955; Franke et al., 1955). Loubatières (1944) also found that a related sulfonamide, 3-methyl-7-chloro-1,2,4-benzothiadiazine-1,1-dioxide (diazoxide) elicited hyperglycaemia, via inhibition of insulin secretion.

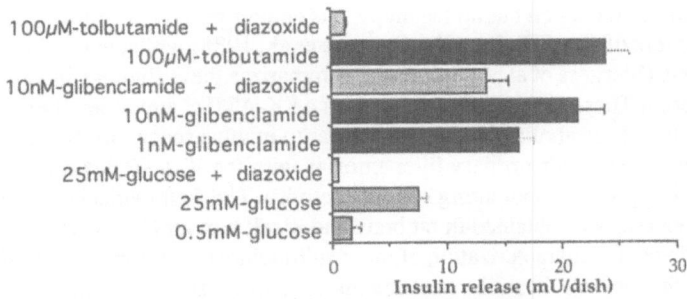

Figure 1. Inhibition by 300µM-diazoxide of insulin secretion from MIN6 ß-cells stimulated by glucose or sulfonylureas. Cells were incubated in Petri dishes for 2.5h in DMEM medium containing the additions shown.

New Concepts in the Pathogenesis of NIDDM, Edited by
C. G. Östenson *et al.*, Plenum Press, New York, 1993

Fig 1 illustrates the effects of sulfonylureas, diazoxide and glucose on insulin secretion from a cloned ß-cell line MIN6. Clearly diazoxide inhibits insulin release induced both by sulfonylureas and by glucose. A major advance in understanding in recent years has been the realization that, in fact, all these agents have a common site of action in the ß-cell, namely the ATP-sensitive K-channel (K-ATP channel), first identified in cardiac muscle (Noma, 1983), but subsequently shown to occur in many tissues and play a variety of physiological roles (Ashcroft and Ashcroft, 1990). These channels are inhibited by intracellular ATP (for review see Ashcroft, 1988) and in the ß-cell couple glucose metabolism to membrane potential. Thus in response to an increase in blood glucose concentration there is increased glucose metabolism within the ß-cell and a rise in the intracellular ratio of [ATP]/[ADP] which closes K-ATP channels. The ensuing depolarisation opens voltage-sensitive Ca^{2+}-channels and the consequent influx of Ca^{2+} ions triggers insulin release (for review see Ashcroft and Ashcroft, 1989; Ashcroft and Rorsman, 1989). Fig. 2 demonstrates the dependence of ß-cell K-ATP channel activity, measured by an isotope flux method, on intracellular ATP concentration and shows the marked modulation of K-ATP channel activity caused by changes in intracellular ATP in the millimolar range.

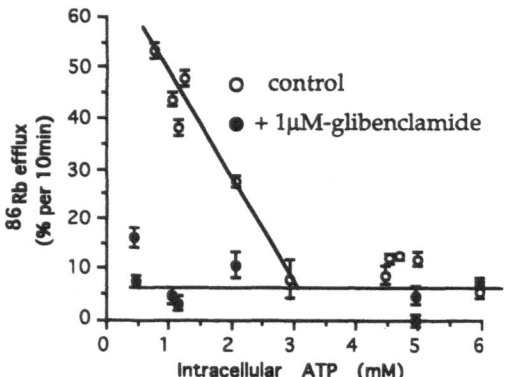

Figure 2. Intracellular ATP levels were varied by exposure to different concentrations of oligomycin in HIT T15 ß-cells loaded with [86]Rb. K-ATP channel activity was measured as glibenclamide-inhibitable ^{86}Rb-efflux. The data indicate that changes in intracellular ATP in the millimolar range modulate ß-cell K-ATP activity (data from Niki et al., 1989).

Application of the patch clamp technique has demonstrated that sulfonylureas are specific and potent blockers of K-ATP channels (Sturgess et al., 1985) whilst diazoxide causes opening of these channels (Sturgess et al., 1988). Fig. 2 summarizes these findings in a model for control of insulin secretion. These actions of sulfonamides on K-ATP channels are direct in that they are independent of ß-cell metabolism and do not appear to involve second messengers. However it is still not clear whether their primary interaction is with the K-ATP channel itself or involves binding to another protein modulating channel activity. The first evidence for specific binding sites for sulfonylureas was obtained in rat brain and ß-cell tumour (Kaubisch et al., 1982; Geisen et al., 1985). Molecular characterization of these sulfonylurea receptors will clarify their relation to the K-ATP channel: we describe here results of current studies on the nature of the ß-cell sulfonylurea receptor.

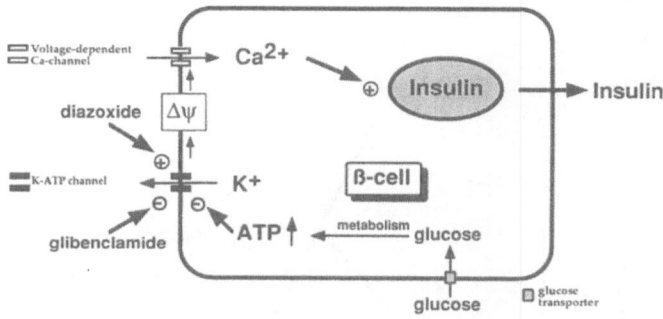

Figure 3. The central role of K-ATP channels in the regulation of insulin secretion

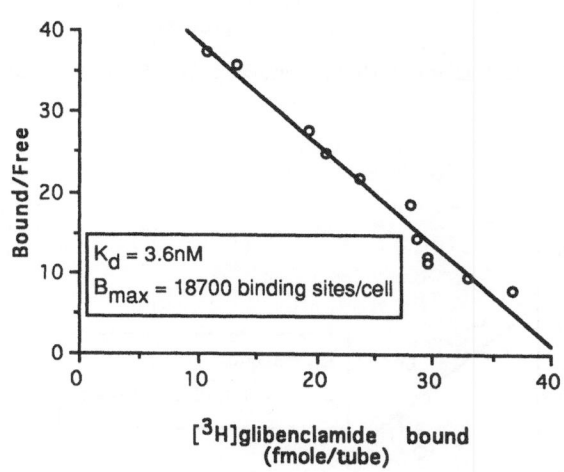

Figure 4. Scatchard plot for the binding of [3H]-glibenclamide to intact HIT T15 ß-cells (from Niki et al., 1990)

PANCREATIC ß-CELLS CONTAIN HIGH AFFINITY SULFONYLUREA BINDING SITES

Fig. 4 shows that intact HIT T15 ß-cells contain a single class of [3H]-glibenclamide-binding sites with a K_d of 3.6nM and a B_{max} of 18700 binding sites per cell. An additional class of low affinity binding sites can be seen when [3H]-glibenclamide binding to HIT T15 ß-cell membranes is studied (Fig. 5). However the significance of these low affinity sites is unclear. For purification of the receptor, it is important to be able to solubilize the receptor with retention of specific binding properties. Fig. 6 shows that the high affinity ß-cell sulfonylurea receptor can be solubilized in an active form with the detergent CHAPS.

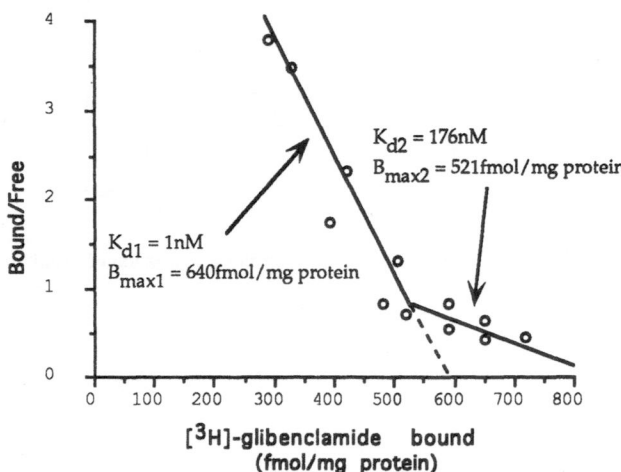

Figure 5. Scatchard plot for the binding of [^{3}H]-glibenclamide to HIT T15 ß-cell membranes (data from Niki et al., 1989)

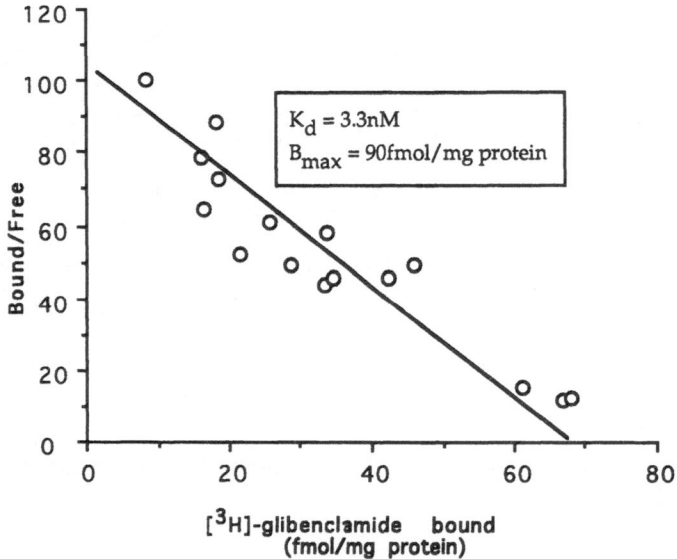

Figure 6. Scatchard plot for the binding of [^{3}H]-glibenclamide to the HIT T15 ß-cell sulfonylurea receptor solubilized with CHAPS (data from Niki et al., 1991)

The specificity of the solubilized receptor is shown Fig. 7. Clearly displacement of [3H]-glibenclamide binding by unlabelled sulfonylureas parallels their known potencies in blocking K-ATP channels and stimulating insulin release.

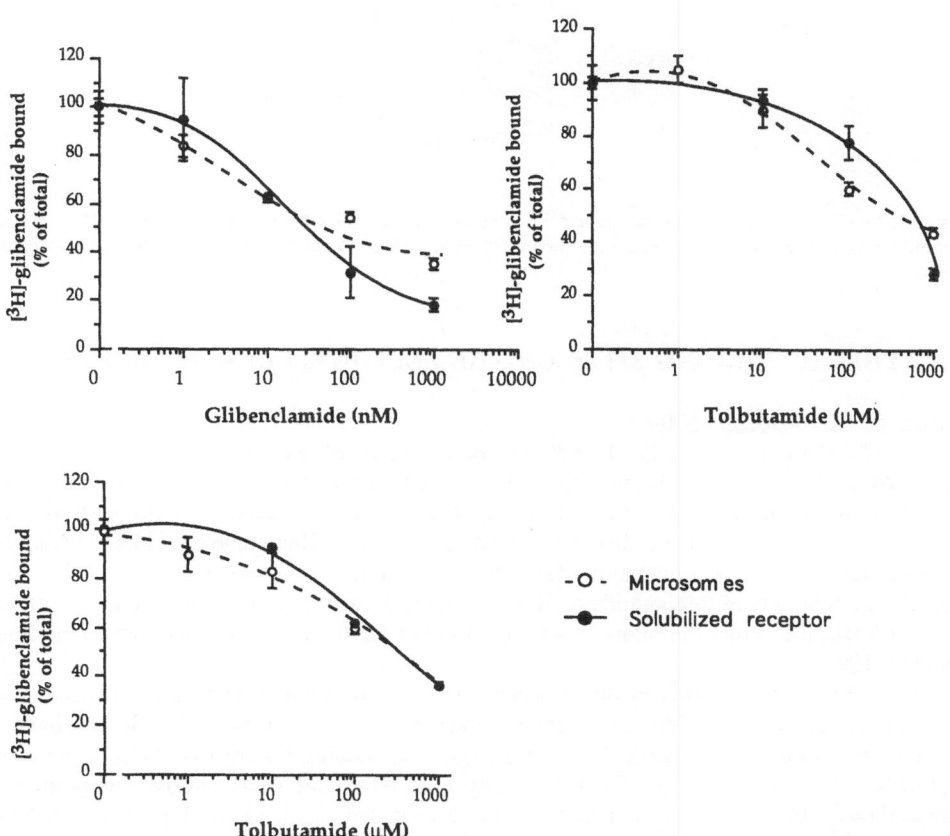

Figure 7. Specificity of the HIT T15 ß-cell sulfonylurea receptor is retained after solubilization with CHAPS (data from Niki et al., 1991)

ENZYMIC MODIFICATION OF THE SULFONYLUREA RECEPTOR

Modification of the membrane by enzymatic treatment can give information about receptor structure. Fig. 8 shows that specific binding of [3H]-glibenclamide to MIN6 ß-cell membranes is markedly and dose-dependently inhibited by exposure to proteases or lipases. Inhibition by treatment with phospholipase C is particularly effective in decreasing [3H]-glibenclamide binding. These data suggest that the ß-cell sulfonylurea receptor contains a protein component associated with phospholipid.

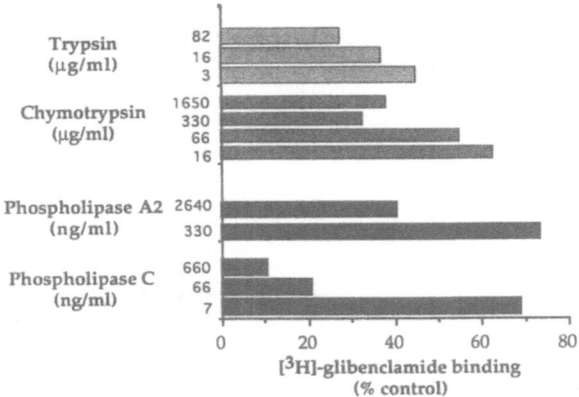

Figure 8. Modification of [³H]-glibenclamide binding to MIN6 ß-cell membranes by exposure to enzymes. Membranes were treated with the enzymes shown for 30min prior to measurement of sulfonylurea-binding activity.

LOCATION OF BINDING SITES ON THE SULFONYLUREA RECEPTOR

Sulfonylurea-Binding Site

K-ATP channels are blocked by tolbutamide regardless of whether the drug was applied to the inside or the outside of the membrane which suggests that tolbutamide dissolves in the membrane and then diffuses laterally to the blocking site. The location of the sulfonylurea receptor is not defined by these observations, however, since tolbutamide could be reaching an extracellular receptor via diffusion through the membrane. However studies on the pH dependence of effects of tolbutamide on K-ATP channels suggested that the unionised form of the drug forms the actual inhibitory ligand by virtue of its ability to enter the membrane lipid (Findlay, 1992).

On the other hand, other evidence argues that the sulfonylurea-binding site is accessible from the extracellular milieu. First, tolbutamide and several other sulfonylureas have a distribution volume in pancreatic islets only slightly exceeding that of extracellular markers (Hellman et al., 1971; Sehlin, 1973). Secondly both ADP and ADP-agarose competitively displaced [³H]-glibenclamide from intact HIT T15 ß-cells (Niki et al, 1990) This effect of ADP is accompanied by increased intracellular Ca²⁺ and insulin secretion. Thirdly the existence of an endogenous peptide ligand (endosulfine) for the sulfonylurea receptor which displaces [³H]-glibenclamide and stimulates insulin secretion (Virsolvy-Vergine et al., 1992) also suggests that the [³H]-glibenclamide-binding site is extracellular.

ATP-Binding Site(s). It has been shown (De Weille et al., 1992) that fluorescein derivatives which have been used to label the nucleotide-binding sites in various ATPases have both activatory and inhibitory effects on K-ATP channel activity in HIT T15 ß-cells and inhibit binding of [³H]-glibenclamide to HIT cell membranes. These data suggest that there are interactions between the sulfonylurea-binding site and one or more nucleotide-binding sites on the sulfonylurea receptor. We have further studied the interaction of one such fluorescein derivative, Rose Bengal, with the ß-cell sulfonylurea receptor. Fig. 9 shows that Rose Bengal inhibits [³H]-glibenclamide binding to MIN6 cell membranes with a K_i of 162nM.

The inhibition is non-competitive, consistent with Rose Bengal binding to a different site than the [³H]-glibenclamide. Further support for this conclusion is the relative inability of Rose Bengal to displace [³H]-glibenclamide from intact MIN6 cells shown in Fig. 10.

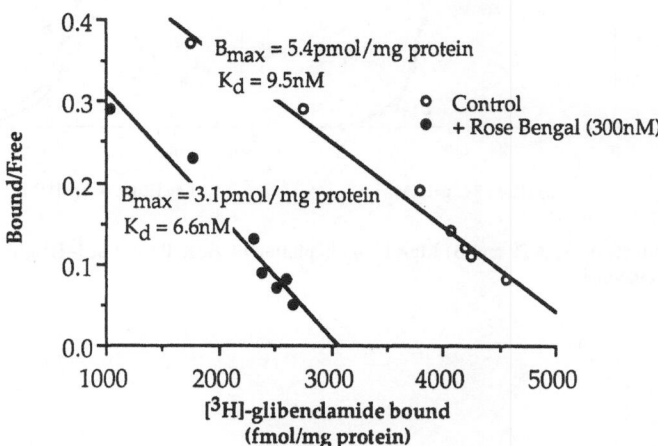

Figure 9. Scatchard plots for the inhibitory effect of Rose Bengal on [^{3}H]-glibenclamide binding to MIN6 ß-cell membranes.

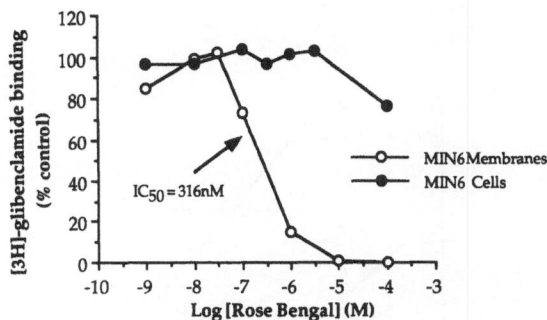

Figure 10. Rose Bengal inhibits [^{3}H]-glibenclamide binding to MIN6 ß-cell membranes but not to intact MIN6 ß-cells

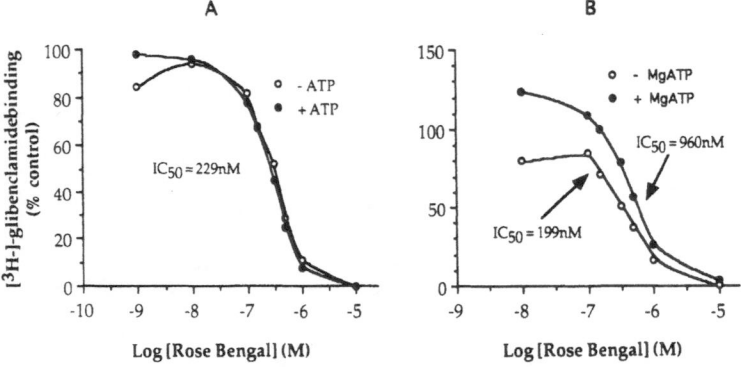

Figure 11. The effects of A) ATP and B) MgATP on inhibition by Rose Bengal of [3H]-glibenclamide binding to MIN6 ß-cell membranes

However, if Rose Bengal is binding to the site at which ATP binds to inhibit the K-ATP channel, it would be expected that ATP should block the effect of Rose Bengal.

Fig. 11A shows that this is not the case, suggesting an alternative mode of action of Rose Bengal. Since, on the other hand, MgATP modified the inhibition by Rose Bengal (Fig. 11B) it is possible that the action of Rose Bengal on [3H]-glibenclamide binding is indirect and is mediated by an effect on protein phosphorylation (see below). We suggest that caution must be used in interpreting the effects of fluorescein derivatives on the sulfonylurea receptor.

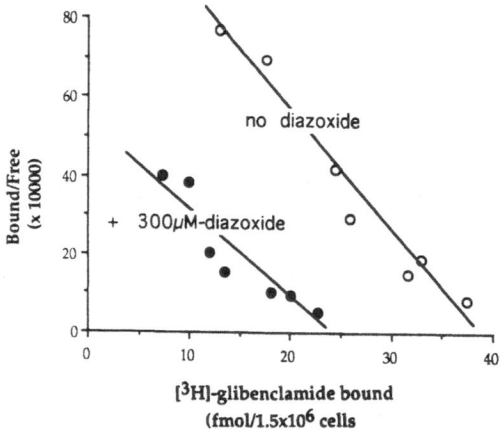

Figure 12. Scatchard plots for the inhibition by diazoxide of [3H]-glibenclamide binding to intact HIT T15 ß-cells (data from Niki and Ashcroft, 1991).

Diazoxide-Binding Site. As shown in Fig 12, diazoxide inhibits binding of [3H]-glibenclamide to intact ß-cells non-competitively, suggesting that the diazoxide site is distinct from that for glibenclamide. However the interaction of diazoxide with the sulfonylurea receptor is critically dependent on phosphorylation of the receptor and is considered further in the next section.

THE BINDING PROPERTIES OF THE ß-CELL SULFONYLUREA RECEPTOR CAN BE MODULATED BY PHOSPHORYLATION

Several observations suggest that phosphorylation of the ß-cell sulfonylurea receptor modifies its binding properties. Thus MgATP (but not ATP4-) produces inhibition of [3H]-glibenclamide binding (Fig. 13).

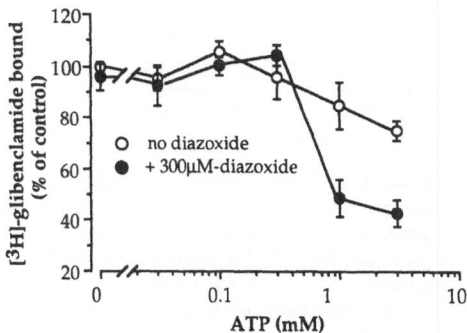

Figure 13. The effects of MgATP and diazoxide on [3H]-glibenclamide binding to HIT T15 ß-cell membranes (data of Niki and Ashcroft, 1991).

The ability of diazoxide to decrease [3H]-glibenclamide binding to intact HIT T15 ß-cells was less pronounced after ATP depletion (Niki et al., 1991). With isolated ß-cell membrane preparations, diazoxide alone was unable to inhibit [3H]-glibenclamide binding (Schwanstecher et al., 1991; Niki et al., 1989) but when MgATP is also present diazoxide is able to displace [3H]-glibenclamide from HIT ß-cell membranes (Fig 12). This effect of MgATP is reproduced by ATPγS but not by non-hydrolysable ATP analogues (AMP-PNP or AMP-PCP) or by ATP in the absence of Mg2+ (Niki and Ashcroft, 1989; Schwanstecher et al., 1991). The presence of MgATP was also necessary for displacement of [3H]-glibenclamide from HIT cell and rat cortex membranes by pinacidil, an effect which was sustained after solubilization of the membranes (Schwanstecher et al., 1992).

Does diazoxide directly inhibit [3H]-glibenclamide binding to the receptor by binding to the phosphorylated receptor or does it modulate a kinase or phosphatase? This question was first studied by examining the time course of the effects of MgATP and diazoxide on [3H]-

glibenclamide binding. HIT cell membranes contain high activity of ATPase and, under the conditions of our binding assays, added 100μM-MgATP is completely hydrolysed after 15min. However, as shown in Fig. 14, diazoxide is still effective when added at 15min i.e. after the ATP had been hydrolysed; diazoxide therefore is not acting on a kinase. After a further 15min, at which time it is assumed that the sulfonylurea receptor has become dephosphorylated, diazoxide is no longer effective in inhibiting [3H]-glibenclamide binding. These data suggest that diazoxide may interact specifically with a phosphorylated form of the sulfonylurea receptor.

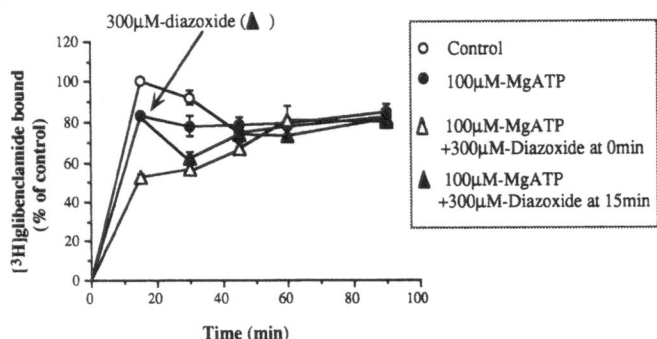

Figure 14. Effects of MgATP and diazoxide on the time-course of [3H]-glibenclamide binding to HIT T15 ß-cell membranes. HIT cell membranes were incubated with [3H]-glibenclamide for the times shown. Measurements of ATPase activity showed that the concentration of ATP in the medium was reduced essentially to zero after 15min. However the inhibitory effect of MgATP and MgATP plus diazoxide persisted for up to 45min. When diazoxide was added to samples already incubated with MgATP for 15min (i.e. after ATP had been hydrolysed) the sulfonamide reduced binding at the 30min time-point to the level found in samples incubated with diazoxide plus MgATP from time zero. Thus the effect of diazoxide is not exerted via a protein kinase.

We then tested the effects of phosphatase inhibitors on sulfonylurea binding. NaF, a non-specific inhibitor of protein phosphatases, augmented the inhibitory effects of both MgATP and diazoxide on [3H]-glibenclamide binding, consistent with the involvement of protein phosphorylation in the inhibition of [3H]-glibenclamide binding by MgATP and diazoxide. However diazoxide still inhibited the binding even in the presence of NaF and had no effect on dephosphorylation of HIT cell membrane proteins favouring the idea that diazoxide binds directly to the sulfonylurea receptor.

It is possible that the phosphorylation of the sulfonylurea receptor required for diazoxide to modulate [3H]-glibenclamide binding to HIT cell membranes may occur at two different sites, since we find that (i) inhibition of [3H]-glibenclamide binding by MgATP disappears more rapidly than inhibition by diazoxide in the presence of MgATP; and (ii) ATP-γS itself has only a small inhibitory effect on binding but permits sustained inhibition by diazoxide.

The view that both protein phosphorylation and dephosphorylation of the sulfonylurea receptor may occur in ß-cell membrane preparations requires that these preparations contain endogenous kinase and phosphatase activities. Since the sulfonylurea receptor has not been

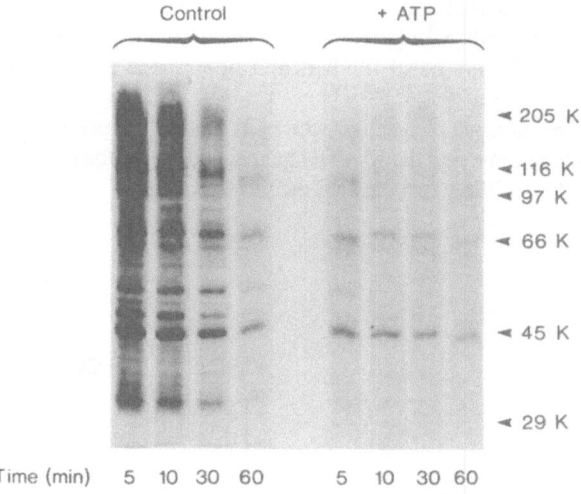

Figure 15. Phosphorylation and dephosphorylation of HIT T15 ß-cell membrane proteins. Membranes were incubated with [γ-^{32}P]-ATP for up to 60min (first four lanes marked 'Control'). Unlabelled ATP was then added to samples incubated for 60min and incubation continued for up to a further 60min (second four lanes marked '+ ATP'). Maximal phosphorylation occurred within 5min and dephosphorylation clearly occurred subsequently.

characterized at the molecular level, is not currently possible to demonstrate phosphorylation of the receptor itself . However, endogenous protein kinase and protein phosphatase activities are clearly present in ß-cell membranes.

Fig. 15 demonstrates that HIT cell membranes rapidly incorporate ^{32}P from [γ-^{32}P]-ATP into many endogenous proteins. Incorporation is maximal after 5min and declines after that. Since no cyclic AMP or Ca^{2+} was present in these incubations the kinase activity responsible is distinct from protein kinase A, protein kinase C or Ca^{2+}/calmodulin-dependent protein kinase. This conclusion is reinforced by the lack of effect of inhibitors of protein kinase A or C on phosphorylation of endogenous proteins. Phosphatase inhibitors inhibited dephosphorylation of several phosphoproteins - NaF and vanadate preserved label in 45, 50, 54 and 57kDa species; vanadate also inhibited dephosphorylation of a 115kDa band. Microcystin reduced dephosphorylation of 45, 50, 54, 60, 68 and 135 (but not 115kDa) species. Ca^{2+} potentiated dephosphorylation of a 45kDa species but did not affect other bands.

A MODEL FOR THE SULFONYLUREA RECEPTOR/K-ATP CHANNEL

Comparison of the present findings with those from electrophysiological studies may illuminate the relationship between the sulfonylurea receptor and the K-ATP channel. Diazoxide is able to open K-ATP channels only in the presence of ATP (Dunne et al, 1987; Sturgess et al, 1988; Zünckler et al, 1988). Two pieces of evidence suggest that the effect of diazoxide may involve protein phosphorylation of the channel or associated protein: i) in the presence of non-hydrolysable ATP analogues the activating effects of diazoxide were lost (Dunne, 1989; Kozlowski et al., 1989); ii) in the absence of Mg^{2+}, diazoxide became an inhibitor of K-ATP channels (Kozlowski et al., 1989). There is also evidence that other K-ATP channel openers (pinacidil, nicorandil, cromakalim and RP 49356) require MgATP to open ß-cell K-ATP channels but become inhibitory in the absence of ATP (Dunne, 1990; Dunne et al, 1990). Thus protein phosphorylation may be of fundamental importance for the function of the K-ATP channel.

A model of the sulfonylurea receptor which fits most of the present experimental findings is given in Fig. 16. We assume that the K-ATP channel and the sulfonylurea receptor form a single functional unit but do not imply that they are the same protein.

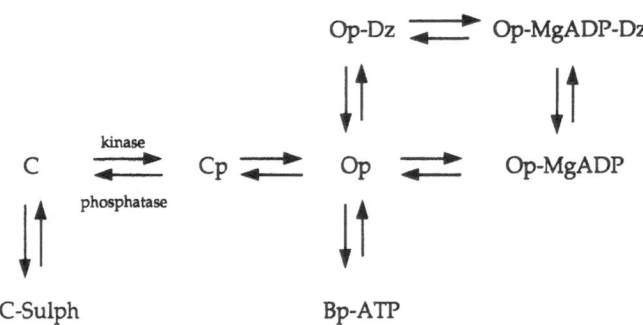

Figure 16. A model for the ß-cell sulfonylurea receptor/K-ATP channel

Kinetic analysis indicates that ß-cell K-ATP channels may exist in at least one open state and two closed states (Ashcroft et al., 1988). The channel openings occur in bursts; the channel flickers between the open state (Op) and a short closed state (Cp) during the burst, and enters a much longer closed state(s) (C) between bursts of openings.

The rundown of K-ATP channel activity which occurs upon formation of an inside-out membrane patch can be restored by brief exposure of the intracellular membrane surface to MgATP (Ohno-Shosaku et al., 1987). Since Mg-free solutions and non-hydrolysable ATP analogues are ineffective, restoration of channel activity may involve protein phosphorylation. As rundown is associated with an increase in the long closed state, we suggest that in this state (C) the channel is dephosphorylated and that phosphorylation results in the channel spending an increased time in the shorter closed state, Cp. We also assume that the channel remains phosphorylated in the open state, Op.

ATP clearly binds to the open state of the channel and inhibits it: we define this ATP-blocked state as Bp-ATP. Other nucleotides may also bind to this state and inhibit channel activity but, as they are less effective, they must do so with lower affinity. In pancreatic ß-cells (Ashcroft and Kakei, 1989), the free ion ATP^{4-} is more effective than MgATP. Although the model shows the channel entering the ATP-blocked state from the open state, the data are equally consistent with the channel entering the ATP-blocked state from the phosphorylated closed state, Cp.

Since electrophysiological studies also indicate that both diazoxide (Sturgess et al., 1988; Kozlowski et al., 1989; Trube et al, 1986) and MgADP (Kakei et al., 1986;Bokvist et al., 1991) can increase channel activity we assume that these agents bind to and stabilise the open state (Op-

Dz and Op-MgADP, respectively). By contrast, sulfonylureas inhibit channel activity by increasing the long closed time; we therefore suggest that sulfonylureas bind preferentially to the dephosphorylated closed state, C.

This model derived from electrophysiological data can account for most of the data on sulfonylurea binding. In the absence of MgATP it may be assumed that the channels are mainly in the dephosphorylated closed state (C.) Sulfonylureas bind to this state (C), and therefore ATP would not be expected to influence binding, as is found experimentally. In the presence of MgATP, some channels will be phosphorylated to the open state (Op) and may therefore enter the blocked state Bp-ATP. The consequent reduction in the availability of the state C receptor would account for the inhibitory effect of MgATP on sulfonylurea binding. The requirement for MgATP for diazoxide to reduce sulfonylurea binding is explained by the fact that diazoxide preferentially binds to the open phosphorylated state (Op).

There are some findings which this model does not easily account for. First, MgADP can decrease sulfonylurea binding (Niki et al, 1990). Our model can only explain this finding if MgADP is converted to MgATP by endogenous transphosphorylating enzymes, as suggested by Schwanstecher et al (1991). Secondly, the model cannot account for the ability of fluorescein derivatives to displace glibenclamide binding and to reactivate rundown K-ATP channels in the absence of MgATP (De Weille et al., 1992). However, as discussed above, it is not established that these agents interact directly with the ATP-binding site: an indirect action of fluorescein derivatives to modulate the phosphorylation state of the receptor could explain their effects on both sulfonylurea binding and channel activity. Finally, we have not attempted to account for the finding that in pancreatic ß-cells intracellular MgADP increases the ability of tolbutamide to inhibit K-ATP channels (Zünkler et al., 1988). It may be necessary to refine the model by incorporating an additional nucleotide-binding site.

Acknowledgements

Our own studies have been supported by grants from the Medical Research Council, the British Diabetic Association, the EP Abraham Fund, the Medical Research Fund and Glaxo Research. We thank Professor J.-I. Miyazaki for MIN6 ß-cells and Professor A.E. Boyd III for HIT T15 ß-cells.

REFERENCES

Ashcroft, F.M., 1988, Adenosine triphosphate-sensitive K+-channels, *Ann. Rev. Neurosci.* 11: 97-118.

Ashcroft, F.M, Ashcroft, S.J.H. and Harrison, D.E., 1988, Properties of single potassium channels modulated by glucose in rat pancreatic ß-cells, *J. Physiol.* 400: 501-527.

Ashcroft, F. M. and Kakei, M., 1989, ATP-sensitive K+ channels in rat pancreatic beta-cells: Modulation by ATP and Mg2+ ions, *J. Physiol.* 416: 349-367.

Ashcroft, S. J. H. and Ashcroft, F. M., 1990, Properties and functions of ATP-sensitive K-channels, *Cellular Signalling* 2: 197-214.

Ashcroft, S. J. H. and Ashcroft, F.M, 1989, The role of the ATP-sensitive K-channel in stimulus-response coupling in the pancreatic ß-cell, *in* "Hormones and Cell Regulation No.14," Eds J.Nunez, J.E.Dumont. Colloque INSERM/J.Libbey Eurotext Ltd. 198: 99-103.

Ashcroft, F.M. and Rorsman, P, 1989, Electrophysiology of the pancreatic ß-cell, *Prog. Biophys. Molec. Biol.* 54: 87-143.

Bertram, F., Bendfeldt, E. and Otto, H., 1955, Über ein wirksames perorales Antidiabeticum (BZ 55), *Deutsch Med. Woschenschrift* : 1455.

Bokvist, K., Ämmälä, C., Ashcroft, F.M., Bergrren, P.-O., Larsson, O. and Rorsman, P., 1991, Separate processes mediate nucleotide-induced inhibition and stimulation of the ATP-regulated K+ channels in mouse pancreatic ß-cells, *Proc. Roy. Soc. B*: 243: 139-144.

De Weille, J. R., Müller, M., and Lazdunski, M., 1992, Activation and inhibition of ATP-sensitive K+ channels by fluorescein derivatives, *J.Biol.Chem.* 267: 4557-4563

Dunne M.J., 1990, Effects of pinacidil, RP 49356 and nicorandil on ATP-sensitive potassium channels in insulin-secreting cells, *Br J Pharmacol*. 99: 487-492.

Dunne, M.J., Aspinall,R.J., and Petersen, O.H., 1990, The effects of cromakalim on ATP-sensitive potassium channels in insulin-secreting cells, *Br. J. Pharmacol*. 99: 169-175.

Dunne, M.J., Ilott, M.C. and Petersen. O.H., 1987, Interactions of diazoxide, tolbutamide and ATP4- on nucleotide-dependent K+ channels in an insulin-secreting cell line, *J. Membr. Biol*. 99: 215-224.

Findlay, I., 1992, Effects of pH upon the inhibition by sulphonylurea drugs of ATP-sensitive K+ channels in cardiac muscle *J.Pharmacol.Exp.Ther*. 262: 71-79.

Franke, H. and Fuchs, J., 1955, Ein neues antidiabetisches Prinzip, *Deutsch Med. Woschenschrift* 80: 1449.

Geisen, K., Hitzel, V., Ökomonopoulos, R., Pünter, J., Weyer, R., and Summ, H. D., 1985, Inhibition of [3H]-glibenclamide binding to sulfonylurea receptors by oral antidiabetic agents. *Arzeim. Forsch* 35: 707-712.

Hellman, B., Sehlin, J., and Täljedal, I.-B., 1971, The pancreatic ß-cell recognition of insulin secretagogues. II. Site of action of tolbutamide. *Biochem. Biophys. Res. Commun*. 45: 1384-88.

Janbon, M., Chapal, J., Vedel, A., and Schaap, J., 1942, Accidents hypoglycémiques graves par un sulfamidothiazol (VK 57 ou 2254 RP), *Montpellier méd*. 21-22: 441

Kakei, M. Kelly ,R.P., Ashcroft,S.J.H., and Ashcroft,F.M., 1986, The ATP-sensitivity of K+ channels in rat pancreatic B-cells is modulated by ADP, *FEBS Lett* 208: 63-66.

Kaubisch, N. Hammer ,R., Wollheim, C.B., Renold, A.E., and Offord,R.E., 1982, Specific receptors for sulfonylureas in brain and in a B-cell tumor of the rat. *Biochem.Pharmacol*. 31: 1171-1174.

Kozlowski, R.Z., Hales, C.N., and Ashford, M.L.J., 1989, Dual effects of diazoxide on ATP-K+ currents recorded from an insulin-secreting cell line, *Br. J. Pharmacol*. 97: 1039-1050.

Loubatières, A., 1944 Relations entre la structure moléculaire et l'activité hypoglycémiante des aminobenzène-sulfamido-alkylthiodiazols, *Comptes Rendus Soc. Biol. (Paris)* 138: 830.

Loubatières, A., 1955, Effets chez l'homme diabétique du p-amino-benzène-sulfamido-isopropyl-thiodiazol, *Montpellier méd*. 48: 618.

Niki, I. and Ashcroft, S. J. H., 1991, Possible involvement of protein phosphorylation in the regulation of the sulphonylurea receptor of a pancreatic beta-cell line, HIT T15, *Biochim.Biophys.Acta Mol.Cell Res*. 1133: 95-101.

Niki, I., Ashcroft, F. M. and Ashcroft, S. J. H., (1989), The dependence on intracellular ATP concentration of ATP-sensitive K-channels and of Na-K-ATPase in intact HIT-T15 ß-cells, *FEBS Lett*. 257: 361-364.

Niki, I. Kelly , R.P., Ashcroft, S.J.H., and Ashcroft, F.M., 1989, ATP-sensitive K-channels in HIT T15 ß-cells studies by patch-clamp methods, 86Rb efflux and glibenclamide binding, *Pflügers Arch*. 415: 47-55.

Niki, I., Nicks, J.L., and Ashcroft, S.J.H., 1990, The beta-cell glibenclamide receptor is an ADP-binding protein, *Biochem J* 268: 713-718.

Niki, I., Welsh, M., Berggren, P.-O., Hubbard, P., and Ashcroft, S. J. H., 1991, Characterization of the solubilized glibenclamide receptor in a hamster pancreatic beta-cell line, HIT T15, *Biochem.J*. 277: 619-624.

Noma, A., 1983, ATP-regulated K+ channels in cardiac muscle, *Nature* 305: 147-8.

Ohno-Shosaku, T., Zünckler, B., and Trube, G., 1987, Dual effects of ATP on K+ currents of mouse pancreatic ß-cells, *Pflügers Arch* 408: 133-138.

Schwanstecher, M., Brandt, C., Behrends, S., Schaupp, U., and Panten, U., 1992, Effect of MgATP on pinacidil-induced displacement of glibenclamide from the sulphonylurea receptor in a pancreatic beta-cell line and rat cerebral cortex *Br.J.Pharmacol.* 106: 295-301.

Schwanstecher, M., Löser, S., Rietze, I., and Panten, U., 1991, Phosphate and thiophosphate group donating adenine and guanine nucleotides inhibit glibenclamide binding to membranes from pancreatic islets, *Naunyn-Schmiedeberg's Arch. Pharmacol.* 343: 83-89.

Sehlin, J., 1973, Evidence for specific binding of tolbutamide to the plasma membrane of the pancreatic ß-cells, *Acta Diabetol. Lat.* 10: 1052-1060.

Sturgess, N. C., Ashford, M.L.I., Cook, D.L., and Hales, C.N., 1985, The sulphonylurea receptor may be an ATP-sensitive potassium channel, *Lancet* 2: 474-475.

Sturgess, N.C., Kozlowski, R.Z., Carrington, C.A., Hales, C.N., and Ashford, M.L.J., 1988, Effects of sulphonylureas and diazoxide on insulin secretion and nucleotide-sensitive channels in an insulin-secreting cell line, *Br.J. Pharmacol.* 9583-94; 1988.

Trube, G., Rorsman, P., and Ohno-Shosaku, T., 1986, Opposite effects of tolbutamide and diazoxide on the ATP-dependent K+ channel in mouse pancreatic beta-cells, *Pflügers Arch.* 407: 493-499.

Virsolvy-Vergine, A., Leray,, H., Kuroki, S., Lupo, B., Dufour, N., and Bataille, D., 1992, Endosulfine, an endogenous peptidic ligand for the sulfonylurea receptor: purification and partial characterization from ovine brain, *Proc. Natl. Acad. Sci.* 89: 6629-6623.

Zünckler, B.J., Lenzen, S., Manner, K., Panten, U., and Trube, G., 1988, Concentration-dependent effects of tolbutamide, meglitinide, glipizide, glibenclamide and diazoxide on ATP-regulated K+ currents in pancreatic B-cells, *Naunyn-Schmiedeberg's Arch. Pharmacol.* 337: 225-230.

Zünkler, B.J., Lins, S., Ohno-shosaku, T., Trube, G., and Panten, U., 1988, Cytosolic ADP enhances the sensitivity to tolbutamide of ATP-dependent K+ channels from pancreatic ß-cells, *FEBS Lett.* 239: 241-244.

REGULATION OF GLUCOSE TRANSPORTERS AND THE Na/K-ATPase BY INSULIN IN SKELETAL MUSCLE

Harinder S Hundal and Amira Klip

Division of Cell Biology
The Hospital For Sick Children
Toronto, Ontario
Canada

INTRODUCTION

In addition to its functional role of conferring bodily movement, skeletal muscle, by virtue of its total body mass also plays a key role affecting whole body metabolism. Skeletal muscle represents the chief storage site for protein (in the form of free amino acid pools and contractile muscle protein) (Rennie, 1985) and inorganic ions, notably potassium (Bergstrom et al, 1981). In the post-prandial state muscle represents the principal tissue responsible for insulin-stimulated glucose utilization (for oxidative metabolism and/or glycogen synthesis) (DeFronzo et al, 1981) and is a significant contributor in the inter-organ flow of carbon and nitrogen (in the form of muscle alanine and glutamine efflux) (Hundal, 1991). The ability to store or exclude organic and inorganic nutrients against a concentration gradient is maintained through the activity of specific membrane transport proteins as well as the selective permeability properties of the membrane. In skeletal muscle, a primary insulin target, the activation of glucose transport and Na/K transport (mediated by the Na pump or its enzymic equivalent the Na/K-ATPase) represent two of the best documented responses to the hormone (Klip et al, 1987; Klip and Paquet, 1990; Hirshman et al, 1990; Marette et al, 1992a; Erlij and Grinstein, 1976; Clausen and Kohn, 1977). The present chapter reviews work from our laboratory which has focused on the molecular basis by which insulin activates glucose transport and the Na/K-ATPase in skeletal muscle and the possible implications that regulation of these membrane processes may have during non-insulin dependent diabetes mellitus (NIDDM).

New Concepts in the Pathogenesis of NIDDM, Edited by
C. G. Östenson *et al.*, Plenum Press, New York, 1993

GLUCOSE TRANSPORTERS IN SKELETAL MUSCLE AND THEIR ACUTE REGULATION BY INSULIN

Measurements of arterio-venous glucose differences during euglycemic clamp studies in humans, indicate that in the presence of fasting or basal circulating insulin levels skeletal muscle normally represents a minor site of whole body glucose uptake (DeFronzo et al, 1981; Baron et al, 1988). However, during hyperglycemia or insulin treatment following an intravenous glucose load, skeletal muscle becomes the predominant tissue involved in glucose uptake reflecting its important role in limiting a rise in circulating glucose (Baron et al, 1988). This capacity to increase glucose uptake from the circulation does not result in intramuscular glucose accumulation suggesting (i) that the hexose is channelled directly into oxidative and/or storage (as glycogen) pathways and (ii) that the transport of glucose across the muscle membrane must be rate limiting for metabolism signifying that its regulation is likely to have important consequences for glucose utilization in the fed and fasted states (Baron et al, 1988; Daniel et al, 1975).

Except for the active (energy requiring) glucose uptake that occurs in the luminal membranes of the small intestine and the proximal tubule of the kidney, glucose transfer across the cell membrane of mammalian cells occurs by a facilitative diffusion mechanism (Bell et al, 1990; Gould and Bell, 1990). The latter is mediated by a family of structurally related transport proteins that are the products of distinct genes. These transporter proteins have been assigned GLUT1 through to GLUT5 (based on the order in which encoding cDNA sequences were identified) and are expressed in a tissue specific manner (Table 1).

GLUT4 or the "insulin-regulatable glucose transporter" is expressed exclusively in insulin-sensitive tissues, namely fat and skeletal muscle (Bell et al, 1990; Gould and Bell, 1990), and is thought to be largely responsible for mediating the stimulated uptake of glucose in these tissues in response to insulin (Zorzano et al, 1989; Klip and Paquet,

Table 1. Facilitative glucose transporters in mammalian tissues

Transporter	Sites of expression
GLUT1	Erythrocyte (human only), kidney, brain, placenta
GLUT2	Liver, small intestine, ß-cell, kidney
GLUT3	Brain, kidney, placenta
GLUT4	Skeletal muscle, fat, heart
GLUT5	Small intestine, spermatazoa and skeletal muscle (human)

1990). Over a decade ago Cushman & Wardzala (1980) and Suzuki & Kono (1980) independently proposed that in fat cells the insulin-dependent increase in glucose uptake was mediated through the recruitment of glucose transporters to the plasma membrane from an intracellular storage site. An increase in transporter number in the membrane was initially detected as a gain in the number of D-glucose protectable binding of cytochalasin B (CB, a fungal metabolite that specifically interacts with the cytoplasmic glucose binding domain of the transporter). This rise in plasma membrane glucose transporter content was paralleled by a concomitant fall in the number of CB binding sites in the light-microsoms (here referred to as the intracellular membranes) isolated from fractionated adipocytes; these observations formed the basis of the translocation hypothesis. This hypothesis was subsequently tested immunologically using monoclonal and ployclonal antibodies that recognised separately GLUT4 and GLUT1 glucose transporters in adipocytes (Zorzano et al, 1989). In the basal state (ie. in the absence of insulin) GLUT4 protein was predominantly localized in the light microsomes (accounting for nearly 90% of the transporters in this fraction). After treatment of adipocytes with insulin these microsomes were largely depleted of GLUT4. The loss of GLUT4 from the intracellular fraction is recovered in the plasma membrane of insulin treated adipocytes. Adipocytes also express the GLUT1 glucose transporter, but in contrast to GLUT4 this isoform represents a much smaller fraction of the total cellular glucose transporter population (as judged by CB photolabelling) (Zorzano et al, 1989) and is more homogeneously distributed within the adipocyte. In response to insulin intracellular GLUT1 also appears to translocate to the plasma membrane but its gain in the plasma membrane constitutes only a small fraction of the recruited glucose transporters (Zorzano et al, 1989).

It remained unknown whether the translocation of glucose transporters observed in response to insulin in the adipocyte also formed the molecular basis by which insulin stimulated glucose transport in skeletal muscle. A major reason for this was the lack of an appropriate fractionation procedure that would allow the isolation of fractions enriched with specific membranes from skeletal muscle that were relatively free of contamination with membranes originating from other cellular compartments or organelles. By using conventional differential centrifugation techniques, enzyme-free low-salt sucrose buffers and sucrose gradient fractionation our laboratory has, over the course of the past five years, successfully developed a protocol allowing the isolation of plasma membranes (PM), sarcoplasmic reticulum (SR) and internal muscle membranes (IM) (Klip et al, 1987; Douen et al, 1989; Douen et al, 1991). We have extensively characterized the membranes isolated by density separation on discontinous sucrose gradients (25%, 30% and 35% sucrose wt/wt) both enzymatically and immunologically using specific membrane markers allowing an assessment of purity and contamination of each membrane fraction (Table 2). Based on the increased activites of 5' nucleotidase and p-nitrophenylphospahatase (PM markers) in membranes recovered on top of 25% sucrose this fraction was considered to be enriched with PM relative to the muscle homogenate. Membranes banding atop 30% sucrose are considered to represent a mixture of PM and membranes of intracellular origin that are not completely resolved on the sucrose gradient (Ramlal et al, 1989). The 35% sucrose fraction contains predominantly intracellular membranes (IM) (but which are largely of non-SR origin) since there is no detectable enrichment of PM markers in this fraction and because it was depleted of SR Ca^{2+}-ATPase activity (which is mostly recovered in the membrane fraction isolated in the gradient pellet).

The availability of isoform-specific antibodies to the GLUT1 and GLUT4 glucose transporters allowed us to detect their presence immunologically in the PM and IM enriched fractions. Moreover, prior treatment of rats with insulin *in vivo* has allowed us to assess the effect of the hormone on the distribution of glucose transporters in isolated muscle membranes after subcellular fractionation and to perform parallel studies of D-glucose protectable CB binding. By Western blot analyses we have found that the PM fraction isolated from muscles of non-insulin treated rats contained modest levels of GLUT1 but nearly two fold more GLUT4 protein, whereas the IM fraction contained only trace amounts of GLUT1 but represented a sizable intracellular store of GLUT4 transporters (containing over 7 fold more GLUT4 in this fraction than in the PM based on overall recoveries) (Klip and Marette, 1992). In response to rapid insulin treatment *in vivo* we observed a marked increment in GLUT4 abundance in the PM fraction accompanied by

Table 2. Characterization of membrane fractions isolated from rat hindlimb skeletal muscle.

	25% sucrose (PM)	35% sucrose (IM)	Transverse Tubules	Pellet (SR containing)	Triads
GLUT 4	++	+++	++	+	-
GLUT 1	+++	-	-	-	-
α_1 (PM marker)	+++	-	+++	-	?
α_2	+++	++	?	-	?
β_1	++	++	?	++	?
β_2	++	++	?	-	?
DHP receptor (TT marker)	-	-	+++	-	+
Ryanodine receptor (cisternal SR marker)	-	+	++	+++	+++
Calsequestrin (cisternal SR marker)	-	+	-	++	++
Ca^{2+}-ATPase (longitudinal SR marker)	-	+	-	+++	++

Symbols (+ and -) denote relative enrichments for equal amounts of protein in the indicated membrane fractions. α_1, α_2, β_1, and β_2 represent the various isoforms of the Na/K-ATPase. DHP (Dihydropyridine receptor). Table modified from Marette et al (1992b)

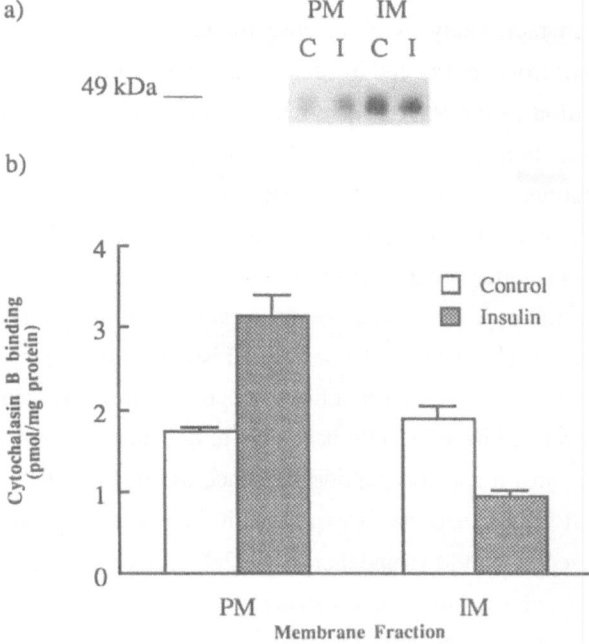

Figure 1. (a) Representative Western blot showing GLUT4 distribution in equal amounts of PM and IM prepared from red (type I) skeletal muscle of control (C) or insulin (I) treated rats (adapted from Marette et al, 1992a) (b) Effect of insulin treatment on D-glucose protectable cytochalasin B binding in PM and IM fractions of rat skeletal muscle. Bars represent mean ± SE for 3 to 4 preparations.

a concomitant decrease in the amount of immunoreactive GLUT4 in the IM fraction (figure 1); GLUT1 abundance in the PM remained unaffected by the acute insulin treatment (Douen et al, 1990). Interestingly, these immunological results confirmed earlier observations regarding the pattern of CB binding after insulin treatment which showed that the number of D-glucose protectable CB binding sites in the PM increases with an associated decrease in the IM fraction (Klip et al, 1987).

The exact identity of the organelle or store from which GLUT4 molecules are recruited in response to insulin remains controversial and a topic of intense current interest. One morphological study (at the electron microscope level using immunogold labelling of GLUT4 molecules) has suggested that the internal pool in human muscle appears to include the triad region (comprised of the terminal cisternae of the SR and transverse tubules) (Friedman et al, 1991), whereas another study in rat muscle has claimed that this pool is probably sub-sarcolemmal being closely associated with the PM and the trans-Golgi network (Rodnick et al, 1992). Our own biochemical observations tend to rule out the involvement of the triad region in intracellular GLUT4 transporter storage since the GLUT4 rich IM fraction does not contain any detectable amounts of the dihydropyridine receptor (a marker for the transverse tubules) and as indicated above is depleted of SR Ca^{2+}-ATPase activity. In fact, analyses of glucose transporter content in purified transverse tubule membranes from control and insulin treated skeletal muscle indicate that these membranes (in addition to the PM) also receive a significant proportion of the intracellular GLUT4 proteins in response to insulin (Marette et al, 1992a). These results suggest that GLUT4 transporters are stored in a non-SR unique organelle and migrate to both the PM and the transverse tubules in response to insulin (Marette et al, 1992a).

Are other glucose transporter isoforms expressed in skeletal muscle? GLUT2 (the liver type transporter) does not appear to be expressed in rat skeletal muscle and we are unable to presently confirm whether the GLUT3 or GLUT5 isoforms are expressed in rat muscle since the cDNA and isoform-specific antisera reactive with rat tissues are not currently available. However, we have recently been able to demonstrate, using an anti-GLUT5 antibody raised against a specific peptide sequence to the COOH-terminus of human GLUT5, that the GLUT5 transporter is expressed in human skeletal muscle and that it is specifically localized to the PM (Hundal et al, 1992a). The exact role of GLUT5 in human skeletal muscle remains unknown but oocyte expression studies have led to the suggestion that it may function as a specific fructose transporter (Burant et al, 1992). Owing to ethical reasons and practical limitations we presently do not know whether expression of this isoform in human skeletal muscle is regulated by insulin but the cloning of a rat GLUT5 cDNA and isolation of an anti-rat GLUT5 antibody may enable this question to be addressed more fully in animal studies in the future.

GLUCOSE UPTAKE AND GLUCOSE TRANSPORTERS IN SKELETAL MUSCLE DURING NIDDM

Type II (non-insulin dependent) diabetes mellitus is characterized by insulin resistance in peripheral tissues such as fat and skeletal muscle (DeFronzo, 1988). Although the molecular basis of muscle resistance to insulin action is currently not well understood, defects in hormone signalling at the level of insulin-receptor binding (and in its associated kinase activity) in skeletal muscle, fat and liver has been suggested to be a contributing factor in the development of NIDDM (see Klip and Leiter, 1990; Kahn, 1992). Moreover, it is likely that the insulin resistance of skeletal muscle during NIDDM may in itself contribute significantly towards the hyperglycemic state through a substantial reduction in insulin-stimulated muscle glucose uptake (DeFronzo et al, 1985; DeFronzo, 1988). The latter may arise through a dysfunction of skeletal muscle glucose transport, transporter expression and localization which may play a contributory role in the pathogenesis of NIDDM (Kahn, 1992). Unfortunately, there are no suitable animal models of NIDDM to test this proposition and available information from human studies is at present both scant and controversial. In adipocytes isolated from NIDDM patients the total GLUT4 transporter content is lower in both the PM and in the intracellular light microsomes possibly as a result of pre-translational regulation since GLUT4 mRNA is also decreased (Garvey et al, 1991). Studies in skeletal muscle show that total GLUT4 protein content does not appear to be affected by either NIDDM or obesity (Pedersen et al, 1990), a finding that is inconsistent with the observed decrease in insulin stimulated muscle glucose uptake *in vivo*. One possible explanation for this discrepancy may be that all biochemical studies of human muscle to date have utilized crude unfractionated muscle membranes; incorporation of GLUT4 protein into the plasma membrane cannot be assessed by this method. A defect in this process may still account for the decrease in muscle glucose uptake during NIDDM. Interestingly in streptozotocin (STZ)-diabetic rats (a model of Type-I diabetes) chronic insulin lack (ie 7 days post-STZ injection) results in decreased incorporation of GLUT4 glucose transporters in the PM after insulin treatment despite a comparable loss of GLUT4 molecules from the IM fraction in both control and diabetic rats (Klip et al, 1990). Whether a similar scenario also exists in muscle of NIDDM patients may be more easily resolved with the advent of appropriate biochemical fractionation techniques of small quantities of human skeletal muscle obtained by biopsy, required to assess the subcellular abundance of GLUT4 transporters in skeletal muscle.

REGULATION OF THE Na/K-ATPase IN SKELETAL MUSCLE BY INSULIN

Insulin has long been known to be a physiological regulator of Na/K homeostasis in skeletal muscle (Zierler and Rabinowitz, 1964; Clausen and Kohn, 1977; Moore, 1983). The significance of this hormonal regulation has become increasingly more apparent over the last two decades and it is now well recognised that it is crucial for the maintenance of normal cytosolic concentrations of Na and K. In the presence of insulin, muscle cells hyperpolarize in spite of stimulation by the hormone of various Na-coupled transport processes (such as the uptake of certain amino acids (Hundal, 1991), Na/H exchange (Moore, 1981; Grinstein et al, 1989) and inorganic phosphate uptake (Clausen, 1985)). Early studies in isolated rat muscles demonstrated that insulin stimulated K influx and Na efflux and that the hormone's effect on these two processes was abolished when the cardiac glycoside ouabain was present (Clausen and Kohn, 1977). The latter finding suggested that insulin activated Na/K transport by stimulating the activity of the Na pump (Na/K-ATPase), which is directly inhibited by ouabain. Yet despite the large body of information documenting the stimulation of the Na/K transport activity by insulin in several cell systems (Erlij and Grinstein, 1976; Moore, 1983; Lytton, 1985; McGill and Guidotti, 1991; Rosic et al, 1985), the molecular basis of this regulation in skeletal muscle has remained elusive. Using the frog sartorius muscle preparation Erlij and Grinstein (1976) proposed that the ability of insulin to activate Na/K transport lay in its capacity to modulate the number of Na pumps (by "unmasking" dormant pump units) at the plasma membrane based on the measured increase in the number of specific [^{3}H]-ouabain binding sites after exposure of muscle to insulin (since ouabain binds only to the active confirmation of the pump). Using an isolated rat muscle preparation Clausen & Hansen (1977) performed time course measurements of [^{3}H]-ouabain binding under equilibrium conditions in the absence and presence of insulin and suggested that the apparent increase in [^{3}H]-ouabain binding sites could alternatively be interpreted as a modification in the ability of existing Na pumps in the membrane to take on a configuration that would allow them to bind ouabain. Subsequent sudies using the BC$_3$H1 muscle cell line and adipocytes led to the proposal of additional hypotheses that a rise in cytosolic Na concentration, occuring as a result of activation of Na/H exchange modifies the pumps affinity for its substrates (ie. ATP, Na and K) and respectively may form the basis by which the hormone stimulates Na/K transport (Rosic et al, 1985; Lytton, 1985; Resh et al, 1980). However, the finding that inhibition of Na/H exchange by amiloride does not prevent insulin stimulation of ouabain-sensitive K uptake in rat soleus muscle implies that, in skeletal muscle, factors other than a rise in intracellular Na play a role in mediating the insulin stimulation of the pump (Weil et al, 1991). Moreover, the recent observations of Omatsu-Kanbe & Kitasato (1990) that insulin diminshes the Na/K-ATPase activity of an intracellular "light" membrane fraction prepared from amphibian muscle and elevates it in the isolated PM also rules out the proposition that mass-action or allosteric effects of intracellular Na are directly responsible for the insulin activation of the pump.

Structurally the Na/K-ATPase is a dimeric protein consisting of a catalytic α subunit (Mr 112 kDa) and a glycosylated ß subunit (Mr 35 kDa for the native protein). Association of one α subunit with one ß subunit constitutes the minimum requirement for a functional unit. Three isoforms of the α subunit (α1, α2 and α3) have been identified which, in rodents, have different sensitivites to ouabain and are expressed in a tissue specific manner (for review see Sweadner, 1989). α2 and α3 have about 100-times higher sensitivity to the glycoside than α1. Hetereogeneity in the ß subunit also exists (ß1, ß2 and ß3) but their tissue distribution and knowledge about which ß isoform pairs up with which α subunit in different tissues is presently poorly defined. The availability of isoform-specific antibodies to the various α and ß isoforms of the Na-pump has allowed us to apply the methodological approach described in the previous section for detecting the presence of different glucose transportes in muscle, to identify which α and ß subunits of the Na/K-ATPase are expressed in skeletal muscle and assess their cellular localization (Hundal et al, 1992b). This strategy has led us to propose that insulin causes recruitment to the plasma membrane of specific subunits of the Na/K-ATPase in a manner analogous to that of the GLUT4 glucose transporter in muscle.

Only the α1 and α2 subunits were detected immunologically in rat skeletal muscle membranes. We found no significant reaction to α3 by Western blots when anti-α3 antibody was used to probe isolated muscle membranes indicating that this isoform is not expressed in rat skeletal muscle. Moreover, we have also recently observed that neural tissue does not contribute towards the observed $\alpha 2$ signal on Western blots since a sciatic nerve extract showed no detectable reaction with anti-$\alpha 2$ antibody (Hundal & Klip, unpublished work). PM prepared from skeletal muscle of control rats were found to be endowed with the $\alpha 1$ subunit of the Na/K-ATPase and its abundance was unaffected in PM prepared from skeletal muscle of acutely insulin-treated rats. Immunologically, the $\alpha 1$ subunit was barely detectable in the IM fraction and insulin had no effect on its abundance in this fraction. Consistent with these observations, the content of the $\alpha 1$ subunit was also markedly lower in crude membranes which had not been subjected to sucrose gradient fractionation. In contrast, the $\alpha 2$ subunit was abundant in both the PM and the IM fraction consistent with the suggestion that it may be stored in an intracellular compartment prior to its transfer to the PM. When we probed muscle membranes from insulin-treated animals with the anti-$\alpha 2$ antibody the amount of the $\alpha 2$ subunit in the PM was found to be significantly elevated whereas it fell in the IM fraction. When one considers the total protein yield of each fraction it can be calculated that the net gain in $\alpha 2$ in the PM matches the amount lost in the IM, suggesting that insulin causes the $\alpha 2$ subunit to translocate from the IM fraction to the PM (figure 2). The increased abundance of $\alpha 2$ in the PM is unlikely to have arisen from regulation of its biosynthesis or degradation since probing of crude unfractionated membranes revealed that insulin had no effect on the net amount of $\alpha 2$ in skeletal muscle.

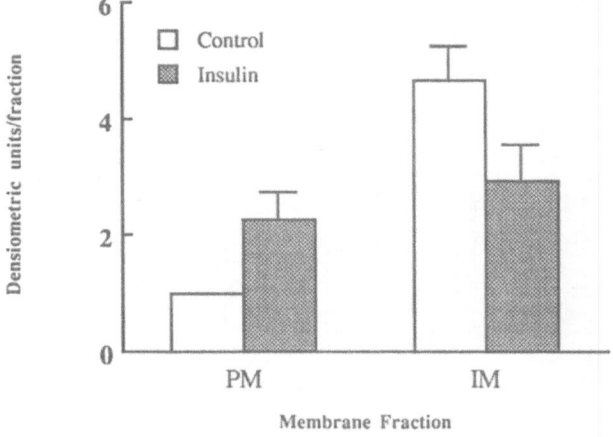

Figure 2. Densiometric scanning data of Western blot analyses showing the effect of insulin on the distribution of the $\alpha 2$ subunit in isolated PM and IM membrane fractions in rat skeletal muscle. Bars represent $\alpha 2$ subunit recovery in each fraction expressed in arbitary units/g of muscle, relative to control PM. Results are mean $\pm$ SE for five to eight muscle preparations (adapted from Hundal et al, 1992b).

The gain in α2 subunits in the muscle PM suggests that this isoform is responsible for the insulin-stimulated Na-pump activity. This is indeed supported by the observation that insulin augmented specifically the component of Na efflux of high sensitivity to ouabain (Lytton et al, 1985). In preparations of other rat tissues such as isolated adipocytes and synaptosomes, activation of the pump by insulin has also been proposed to be largely mediated through the α2 isoform based on the higher sensitivity to ouabain of ATPase activity after insulin action (Resh et al, 1980; Brodsky, 1990).

Given that ß subunit association to the α-subunit is required for full enzymatic activity we next examined whether insulin also caused a parallel subcellular movement of ß subunits in skeletal muscle. The PM-enriched fraction was found to contain both ß1 and ß2 subunits and these isoforms were also detected in the intracellular cognate fraction containing α2. When muscle membranes prepared from skeletal muscle of insulin-treated rats were probed with anti-ß1 and anti-ß2 antibodies the content of the ß1 subunit, but not that of the ß2 subunit, was elevated in the PM. Provocatively, the observed gain in ß1 in the PM after insulin treatment is not derived from the same intracellular compartment as the α2 subunit but appeared to occur at the expense of ß1 subunits from the "heavier" fraction recovered at the bottom of the sucrose gradient, which contains part of the sarcoplasmic reticulum . In response to insulin treatment the amount of ß1 subunit in this fraction was noted to fall by nearly half and was calculated to be sufficient to account for the net gain in ß1 in the PM (figure 3).

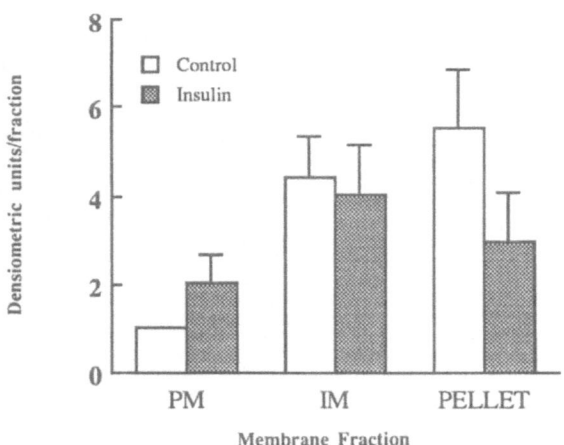

Figure 3. Densiometric scanning data of Western blot analyses showing the effect of insulin on the distribution of the ß1 subunit in PM , IM and in membranes isolated in the gradient pellet upon rat skeletal muscle fractionation. Bars represent the the amount of ß1 subunit recovered in each fraction expressed in arbitary units/g of muscle, relative to control PM. The results are presented as mean ± SE of four muscle preparations (adapted from Hundal et al, 1992b)

The movement of both α2 and ß1 from different cellular sources suggests that these two isoforms come togeather at some point either during their passage to the PM or within the membrane itself and that they are likely to constitute the insulin responsive species. Therefore, it is pertinent to ask whether the translocation of the α2ß1 dimer to the PM after insulin treatment contributes towards the increase in functional Na/K-ATPase activity. We are presently addressing this question in collaboration with Dr Rhoda Blostein (Montreal, Canada) by assessing Na/K-ATPase activity (measured as K-stimulated p-nitrophenyl phosphatase) in muscle membranes prepared from skeletal muscle of control and insulin-treated rats. Our preliminary observations indicate that the activity attributable to the the high affinity ouabain component (ie the α2 subunit) is reduced in the IM fraction whilst this component is slightly increased in the PM after insulin treatment.

We currently do not know whether a subunit resembling ß3 is also expressed in mammalian tissues but it is likely that the expression of multiple isoforms must confer some specific function within the tissue. As yet, the exact functional role of the different ß subunits remains unknown but these have been implicated in assembly and export to the PM of mature Na/K-ATPase molecules (Geering, 1991). In skeletal muscle the ß1 and ß2 subunits that are present in the same fraction as the α2 may function as chaperone proteins of the catalytic subunit conferring structural integrity during intracellular storage. The ß1 subunit in the SR-associated membrane fraction, that translocates to the PM in response to insulin, may bestow upon the α subunit the ability to become catalytically active and functionally responsive to substrates such as Na which may, in turn, exert additional regulatory effects on the Na/K-ATPase. Intriguingly, it has been suggested in a preliminary communication that α subunits may reach the PM of yeast cells but becomes enzymatically active only when coupled to ß subunits (Liu & Guidotti, 1992). This suggests the provocative possibility that α2 subunits in skeletal muscle may translocate to the PM independently of ß1 subunits and that final assembly of the functional dimer occurs in the PM. This possibility however remains to be tested experimentally.

EFFECTS OF DIABETES ON THE Na/K-ATPase IN SKELETAL MUSCLE

The importance of insulin as an activator of the Na pump is best illustrated by the observation that in streptozotocin (STZ)-induced diabetes in the rat there is an increase in intramuscular Na associated with the drug-induced hypoinsulineamia (Moore et al, 1983). The increase in muscle Na during diabetes has largely been thought to result from a reduction in Na/K-ATPase activity but it is plausible that acute changes in Na/K-ATPase subunit expression and cellular distribution may contribute to the rise in intracellular Na. Recent data on the regulation of pump subunits in diabetes were reported for STZ-treated rats (Nishida et al, 1992) where Na/K-ATPase activity and α2 mRNA in skeletal muscle were assessed 2 and 14 days post-STZ injection. As had been observed previously (Moore et al, 1983) enzyme activity fell in crude muscle membranes prepared from diabetic muscle 2 days post-injection, and showed partial recovery at day 14 but remained lower than that measured in control muscle. The fall in activity could not be attributed to changes in $\alpha2$ subunit mRNA expression which was unaffected on day 2 and was found to be increased on day 14. Surprisingly, $\alpha2$ and $\beta1$ mRNA were differentially regulated, expression of the latter being diminished 14 days post-STZ treatment (Nishida et al, 1992). The study did not comphensively address the effects of the various systemic alterations that occur in diabetes consequently the fall in ATPase activity was suggested to arise simply from the notorious hypoinsulinemic effect.

How could diabetes alter the Na pump? The cellular events that lead to the translocation of $\alpha2$ and $\beta1$ from separate intramuscular pools after acute insulin treatment remain presently unknown. However, a defect in insulin signalling may block subunit translocation with obvious consequences for Na/K homeostasis during the diabetic state. We have examined this possibility in preliminary work investigating the effect of STZ-diabetes on α and β protein levels and the ability of insulin to translocate the $\alpha2$ and $\beta1$ subunits to the PM from their respective cellular compartments. Firstly, our observations have revealed that 2 days after STZ injection rat skeletal muscle augments its content of $\alpha1$, $\alpha2$, $\beta1$ and $\beta2$ proteins. The increase is detectable in crude unfractionated membranes prepared from diabetic muscle suggesting increased biosynthesis or stability of subunits rather than a mere subcellular redistribution of the various isoforms. The signal that prompts elevated synthesis or stability is presently not known but it may include a rise in intracellular Na (Moore et al, 1983) which probably occurs through the reduced activation of the Na pump during the hypoinsulinemia that prevails after STZ-induced pancreatic β-cell lysis. Secondly, probing membrane fractions prepared from skeletal muscle of 2-day diabetic rats and diabetic rats treated acutely with insulin (30 min prior to sacrifice) revealed that the insulin-induced translocation of $\alpha2$ and $\beta1$ subunits was lost in muscle of diabetic rats. This finding suggested that the observed increase in subunit protein content may be a compensatory mechanism which comes into play to counteract the loss in insulin-induced recruitment of specific subunits to the membrane caused by insulin deficiency. Moreover, the finding that acute insulin provision does not re-install the translocation of $\alpha2\beta1$ subunits implicates the participation of other cellular mediators, which are either defective or missing, in diabetic muscle. In contrast, the mobilization of intracellular GLUT4 glucose transporter to the PM in response to insulin in 2-day diabetic muscle is not affected, suggesting that some components of the insulin signalling pathway that regulate the cellular distribution of the pump subunits may differ from those which are involved in GLUT4 translocation. The finding that GLUT4 and the $\alpha2$ subunit are regulated differently in diabetic muscle with respect to acute insulin treatment suggests that these two molecules may reside on different vesicular membranes despite co-localising in the same sucrose gradient fraction (ie. the 35% sucrose fraction representing the IM).

In summary our work shows that in skeletal muscle, insulin causes the recruitment to the plasma membrane of the GLUT4 glucose transporter and the α2 and ß1 subunits of the Na pump. We propose that this is likely to represent the principal means by which the hormone stimulates the uptake of glucose and Na/K transport in skeletal muscle when circulating insulin levels are normally elevated (e.g., after a meal). Incorporation of these protein components into the muscle surface membrane occurs at the expense of transporters/subunits which are stored in unique intracellular compartments whose identity presently remains to be defined. The signals which emanate from the insulin receptor that result in the observed translocation of GLUT4 and the pump subunits also remain unknown but they may serve as potential regulatory sites during insulin deficiency or tissue resistance to insulin. In this regard it will be particularly important to establish a cell culture system displaying translocation of GLUT4 and the α2 and ß1 subunits of the Na pump in which it would be more amenable to study the signalling pathway.

ACKNOWLEDGMENTS

We are grateful to André Marette, Jeanne Richardson, Jeffrey Pessin and Toolsie Ramlal for participation in some of the studies described in this chapter. This work was supported by a grant from the MRC (Canada) to AK. H.S.H. is an International Human Frontier Science Program Fellow.

REFERENCES

1. Baron, A.D., Brechtel, G., Wallace, P. and Edelman, S.V. 1988, Rates and tissue sites of non-insulin and insulin mediated glucose uptake in humans. *Am J Physiol*, 255:E769.

2. Bell, G.I., Kayano, T., Buse, J.B., Burant, C.F., Takeda, J., Lin, D., Fukomoto, H. and Seino, S. 1990, Molecular biology of mammalian glucose transporters. *Diabetes Care*, 13:198.

3. Bergstrom, J., Furst, P., Holmstrom, B.O., Vinnars, E., Askanazi, J., Elwyn, D.H., Michelson, C.B.and Kinney, J.M. 1981, Influence of injury and nutrition on muscle water and electrolytes. *Ann Surg*, 193:810.

4. Brodsky, J.L. 1990, Insulin activation of brain Na/K-ATPase is mediated by α2-form of enzyme. *Am J Physiol*, 258:C812.

5. Burant, C.F., Takeda, J., BrotLaroche, E., Bell, G.I. and Davidson, N.O. 1992, Fructose transporter in human spermatozoa and small intestine is GLUT5. *J Biol Chem*, 267:14523.

6. Clausen, T. 1985, The significance of the effects of insulin on Na,K-transport in muscle cells. In: *Regulatory Peptides*, pp. 59-67. Elsevier Science, Netherlands.

7. Clausen, T. and Hansen, O. 1977, Active Na-K transport and the rate of ouabain binding. The effect of insulin and other stimuli on skeletal muscle and adipocytes. *J Physiol*, 270:415430.

8. Clausen, T. and Kohn, P.G. 1977, The effect of insulin on the transport of sodium and potassium in rat soleus muscle. *J Physiol*, 254:19.

9. Cushman, S.W. and Wardzala, L.J. 1980, Potential mechanism of insulin action on glucose transport in the isolated rat adipose cell. *J Biol Chem*, 225:4758.

10. Daniel, P.M., Love, E.R. and Pratt, O.E. 1975, Insulin-stimulated entry of glucose into muscle in vivo as a major factor in the regulation of blood glucose. *J Physiol*, 247:273.

11. DeFronzo, R.A., Jacot, E., Jequier, E., Maeder, E., Wahren, J. and Felber, J.P. 1981, The effect of insulin on the disposal of intravenous glucose: Results from indirect calorimetry and hepatic and femoral venous catheterization. *Diabetes*, 30:1000.

12. DeFronzo, R.A., Gunnarsson, R., Bjorkman, O. and Wahren, J. 1985, Effect of insulin on peripheral and splanchnic glucose metabolism in non-insulin dependent (Type II) diabetes mellitus. *J Clin Invest*, 76:149.

13. DeFronzo, R.A. 1988, The triumvirate: ß-cell, muscle, liver: a collusion responsible for NIDDM. *Diabetes*, 37:667.

14. Douen, A., Burdett, E., Ramlal, T., Rastogi, S., Vranic, M. and Klip, A. 1991, Characterization of glucose transporter enriched membranes from rat skeletal muscle: Assessment of endothelial cell contamination and the presence of sarcoplasmic reticulum and transverse tubules. *Endocrinology*, 128:611.

15. Douen, A.G., Ramlal, T., Klip, A., Young, D.A., Cartee, G.D. and Holloszy, J.O. 1989, Exercise-induced increase in glucose transporters in plasma membranes of rat skeletal muscle. *Endocrinology*, 124:449.

16. Douen, A.G., Ramlal, T., Rastogi, S., Bilan, P.J., Cartee, G.D., Vranic, M., Holloszy, J.O. and Klip, A. 1990, Exercise induces recruitment of the "insulin-responsive glucose transporter". *J Biol Chem*, 265:13427.

17. Erlij, D. and Grinstein, S. 1976, The number of sodium ion pumping sites in skeletal muscle and its modification by insulin. *J Physiol*, 259:13.

18. Friedman, J.E., Dudek, R.W., Whitehead, D.L., Downes, D.L., Frisell, W.R., Caro, J.F. and Dohm, L. 1991, Immunolocalization of glucose transporter GLUT4 within human skeletal muscle. *Diabetes*, 40:150.

19. Garvey, W.T., Maianu, L., Huecksteadt, T.P., Birnbaum, M.J., Molina, J.M. and Ciaraldi, T.P. 1991, Pretranslational supression of a glucose transporter protein causes insulin resistance in adipocytes from patients with non-insulin dependent diabetes mellitus and obesity. *J Clin Invest*, 87:1072.

20. Geering, K. 1991, The functional role of the ß-subunit in the maturation and intracellular transport of Na,K-ATPase. *FEBS Letters*, 285:189.

21. Gould, G.W. and Bell, G.I. 1990, Facilitative glucose transporters: an expanding family. *TIBS*, 15:18.

22. Grinstein, S., Rotin, D. and Mason, M.J. 1989, Na/H exchange and growth factor-induced cytosolic pH changes. Role of cellular proliferation. *Biochim Biophys Acta*, 988:73.

23. Hirshman, M.F., Goodyear, L.J., Wardzala, L.J., Horton, E.D. and Horton, E.S. 1990, Identification of an intracellular pool of glucose transporters from basal and insulin stimulated rat skeletal muscle. *J Biol Chem*, 265:987.

24. Hundal, H.S. 1991, Role of membrane transport in the regulation of skeletal muscle glutamine turnover. *Clinical Nutrition*, 10 (Suppl):33.

25. Hundal, H.S., Ahmed, A., Guma, A., Mitsumoto, Y., Marette, A., Rennie, M.J. and Klip, A. 1992a, Biochemical and immunocytochemical localization of the "GLUT5 glucose transporter" in human skeletal muscle. *Biochem J*, 286:348.

26. Hundal, H.S., Marette, A., Mitsumoto, Y., Ramlal, T., Blostein, R. and Klip, A. 1992b, Insulin induces translocation of the α2 and β1 subunits of the Na/K-ATPase from intracellular compartments to the plasma membrane in mammalian skeletal muscle. *J Biol Chem*, 267: 5040.

27. Kahn, B.B. 1992, Alterations in glucose transporter expression and function in diabetes: mechanisms for insulin resistance. *J Cell Biochem*, 48:122.

28. Klip, A., Ramlal, T., Young, D.A. and Holloszy, J.O. 1987, Insulin induced translocation of glucose transporters in rat hindlimb muscles. *FEBS Letters*, 224:224.

29. Klip, A. and Leiter, L. 1990, Cellular mechanism of action of Metformin. *Diabetes Care*, 13:696.

30. Klip, A., Ramlal, T., Bilan, P.J., Cartee, G.D., Gulve, E.A. and Holloszy, J.O. 1990, Recruitment of GLUT-4 glucose transporters by insulin in diabetic rat skeletal muscle. *Biochem Biophys Res Comm*, 172:728

31. Klip, A. and Marette, A. 1992, Acute and chronic signals controlling glucose transport in skeletal muscle. *J Cell Biochem*, 48:51.

32. Klip, A. and Paquet, M. 1990, Glucose transport and glucose transporters in muscle and their metabolic regulation. *Diabetes Care*, 13:228.

33. Liu, J.Y. and Guidotti, G. 1992, In vitro assembly of the Na/K-ATPase from subunits. *J Gen Physiol*, 100:135.(Abstract)

34. Lytton, J. 1985, Insulin affects the sodium affinity of rat adipocyte (Na,K)-ATPase. *J Biol Chem*, 260:10075.

35. Lytton, J., Lin, J.C. and Guidotti, G. 1985, Identification of two molecular forms of (Na,K)-ATPase in rat adipocytes: relation to insulin stimulation of the enzyme. *J Biol Chem*, 260: 1177.

36. Marette, A., Burdett, E., Douen, A.G., Vranic, M. and Klip, A. 1992a, Insulin stimulates the translocation of GLUT4 glucose transporters from a unique intracellular organelle to both the plasma membrane and transverse tubules in rat skeletal muscle. *Diabetes* (In press)

37. Marette, A., Hundal, H.S. and Klip, A. (1992b): Regulation of glucose transporter proteins in skeletal muscle. In: *Diabetes Mellitus and Exercise*, edited by J. Devlin, E.S. Horton & M. Vranic, pp. 27-43. Smith-Johnson, London.

38. McGill, D.L and Guidotti, G. 1991, Insulin stimulates both the α1 and α2 isoforms of the rat adipocyte (Na,K) ATPase. *J Biol Chem*, 266:15824.

39. Moore, R.D. 1981, Stimulation of Na-H exchange by insulin. *Biophys J*, 33:203.

40. Moore, R.D. 1983, Effects of insulin upon ion transport. *Biochim Biophys Acta*, 737:1.

41. Moore, R.D., Munford, J.W. and Pillsworth, T.J. 1983, Effects of streptozotocin diabetes and fasting on intracellular Na and ATP in rat soleus muscle. *J Physiol*, 338:277.

42. Nishida, K., Ohara, T., Johnson, J., Wallner, J.S., Wilk, J., Sherman, N., Kawakami, K., Sussman, K. and Draznin, B. 1992, Na/K-ATPase activity and its αII subunit gene expression in rat skeletal muscle: Influence of diabetes, fasting and refeeding. *Metabolism*, 41:56.

43. Omatsu-Kanbe, M. and Kitasato, H. 1990, Insulin stimulates the translocation of Na/K-dependent

ATPase molecules from intracellular stores to the plasma membrane in frog skeletal muscle. *Biochem J*, 272:727.

44. Pedersen, O., Bak, J.F., Anderson, P.H., Lund, S., Moller, D.E., Flier, J.S. and Kahn, B.B. 1990, Evidence against altered expression of GLUT1 or GLUT4 in skeletal muscle of patients with obesity or NIDDM. *Diabetes*, 39:865.

45. Ramlal, T., Rastogi, S., Vranic, M. and Klip, A. 1989, Decrease in glucose transporter number in skeletal muscle of mildly diabetic (Streptozotocin-Treated) rats. *Endocrinology*, 125:890.

46. Rennie, M.J. 1985, Muscle protein turnover and the muscle wasting due to injury and disease. *Brit Med Bull*, 41:257.

47. Resh, M.D., Nemenoff, R.A. and Guidotti, G. 1980, Insulin stimulation of (Na,K)-Adenosine Triphosphatase-dependent ^{86}Rb$^+$ uptake in Rat adipocytes. *J Biol Chem*, 255:10938.

48. Rodnick, K.J., Slot, J.W., Studelska, D.R., Hanpeter, D.E., Robinson, L.J., Geuze, H.J. and James, D.E. 1992, Immunocytochemical and biochemical studies of GLUT4 in rat skeletal muscle. *J Biol Chem*, 267:6278.

49. Rosic, N.K., Standaert, M.L. and Pollet, R.J. 1985, The mechanism of insulin stimulation of (Na,K)-ATPase transport activity in muscle. *J Biol Chem*, 260:6206.

50. Suzuki, K. and Kono, T. 1980, Evidence that insulin causes translocation of glucose transport activity to the plasma membrane from an intracellular storage site. *Proc Natl Acad Sci*, 77:2542.

51. Sweadner, K.J. 1989, Isozymes of the Na/K-ATPase. *Biochim Biophys Acta*, 988:185.

52. Weil, E., Sasson, S. and Gutman, Y. 1991, Mechanism of insulin-induced activation of Na-K-ATPase in isolated rat soleus muscle. *Am J Physiol*, 261:C224.

53. Zierler, K.L. and Rabinowitz, D. 1964, Effect of very small concentrations of insulin on forearm metabolism. Persistence of its action on potassium and free fatty acids without its effect on glucose. *J Clin Invest*, 43:950.

54. Zorzano, A., Wilkinson, W., Kotliar, N., Thoidis, G., Wadzinkski, B.E., Ruoho, A.E. and Pilch, P.F. 1989, Insulin-regulated glucose uptake in rat adipocytes is mediated by two transporter isoforms present in at least two vesicle populations. *J Biol Chem*, 264:12358.

INSULIN RECEPTOR: ASPECTS OF ITS STRUCTURE AND FUNCTION

Cecil C. Yip

Banting and Best Department of Medical Research
University of Toronto
Toronto, Ont. Canada M5G 1L6

INTRODUCTION

Insulin receptors are composed of two 130-kDa α subunits and two 90-kDa ß subunits which are linked by disulfide bonds forming a heterotetrameric structure (see reviews by Czech[1], and Goldfine[2]). The two subunits are synthesized as a single-chain precursor protein encoded by 22 exons[3]. The α subunit, extracellularly located, is anchored to the cell membrane by disulfides to the ß subunit which contains a putative trans-membrane domain[4]. Photoaffinity labeling with light-sensitive insulin photoprobes[5] and chemical crosslinking[6] have established that insulin binds to the extracellular α subunit. Insulin binding causes the autophosphorylation of the insulin receptor ß subunit, the latent tyrosine kinase activity of which is thus stimulated as a consequence. The activation of insulin receptor kinase activity initiates a cascade of intra-cellular protein phosphorylation which regulates the cellular metabolic and mitogenic activity in response to insulin (see reviews by Goldfine[2] and by Rosen[7]). Studies of the effect on the cellular response to insulin in mutant receptors in which the intracellular ß subunit has been modified by site-mutagenesis or sequence deletion have demonstrated the domain related multifunctional nature of the insulin receptor (see review by Olefsky[8]).

In the study of the structure and function of the insulin receptor we are seeking answers to three questions:

1. Where is the insulin-binding site or domain on the receptor subunit?

2. How does the binding of insulin lead to the autophosphorylation of the ß subunit and the stimulation of tyrosine kinase activity?

3. What are the intracellular components that directly interact with the ß subunit?

My presentation will focus on the first two questions. Regarding the third question, many attempts have been made to identify cellular substrate(s) of the insulin receptor kinase. Recently a cytosolic phosphoprotein of 180 kDa has been identified as an endogenous substrate of the insulin receptor and given the name insulin receptor substrate 1 (IRS-1)[9]. IRS-1, phosphorylated by the insulin-activated receptor, may interact with the SH2 domains of the 85-kDa regulatory subunit of the enzyme phosphatidyinositol 3-kinase, thus providing a coupling point in the pathway of signal transduction initiated by insulin binding.

THE INSULIN-BINDING DOMAIN

Applying the technique of photoaffinity we in 1978 were first to report on the identification of the insulin receptor[10]. We have since continued to use this techniques to study the insulin-binding domain of the insulin receptor. Two different photoreactive derivatives of insulin have been prepared and used in my laboratory: a non-cleavable derivative, $N^{\varepsilon B29}$-monoazidobenzoyl [^{125}I]-iodoinsulin (B29-[^{125}I]-MABI), and a cleavable derivative, N-[4-[(4'-azido-3'-[^{125}I]-iodophenyl)-azo]benzoyl]-(3-aminopropyl) insulin ([^{125}I]AZAP-insulin), from which insulin can be released by azo cleavage after crosslinking. Since both derivatives retain nearly full receptor binding and biological activity, and since both derivatives are derivatized at Lys-B29 located in the receptor-binding domain of insulin, it is reasonable to expect that insulin is crosslinked to the ligand-binding domain in receptor photo-labeled with these derivatives. Figure 1 graphically illustrates the photolabeling of the receptor α subunit by these derivatives.

When insulin receptors were photoaffinity labeled with [^{125}I]-AZAP-insulin and its labeled α subunit isolated and digested with the enzyme endoproteinase Glu-C, we obtained a labeled fragment of 23 kDa[11] which can be further digested with trypsin to a labeled fragment of less than 3 kDa. Based on several lines of indirect evidence we suggested at that time that the 23-kDa fragment originated from the sequence 205-316 in the cysteine-rich domain of the receptor α subunit, the cysteine-rich domain being encoded by exon 3 of the insulin receptor gene[3]. The origin of the 23-kDa fragment was later confirmed when we showed that the fragment was precipitated by an antiserum raised against the amino acid sequence 241-251 of the receptor α subunit[12]. These observations therefore suggest that insulin binds to the cysteine-rich domain of the insulin receptor subunit. The amino acid sequence 241-251 containing CPPPYYHFQDW was of particular interest since it could provide the required hydrophobic residues to interact with those present in the receptor-binding domain in the C-terminal of the B chain of insulin[13]. To test this hypothesis we mutated this sequence to CPRRYYDFQDW and expressed the mutant receptor in rat hepatoma cells[14]. Cells expressing the mutant receptor showed an increased binding affinity for insulin and an increased sensitivity to insulin. The positive effect obtained with the mutant receptor strongly indicates that the mutations have directly affected the ligand-binding site. In order to further characterize the insulin-binding site we studied the inter-action of the insulin receptor with the antiserum against the amino acid sequence 241-251. We found that the antiserum, though able to immunoprecipitate the denatured receptor α subunit, did not recognize the native insulin receptor, and did not inhibit insulin binding[12]. Interestingly, receptors photolabeled with the cleavable photoprobe, AZAP-insulin, were recognized by the antiserum only after insulin had been cleaved off, suggesting that the sequence 241-251 is blocked by bound insulin. These observations lead us to suggest that the insulin-binding domain containing the sequence 241-251 exists as a crevice which is accessible to insulin but not to large protein molecules like antibody, and that insulin binding induces a conformational change in the binding domain such that it becomes accessible to large protein molecules (Figure 2).

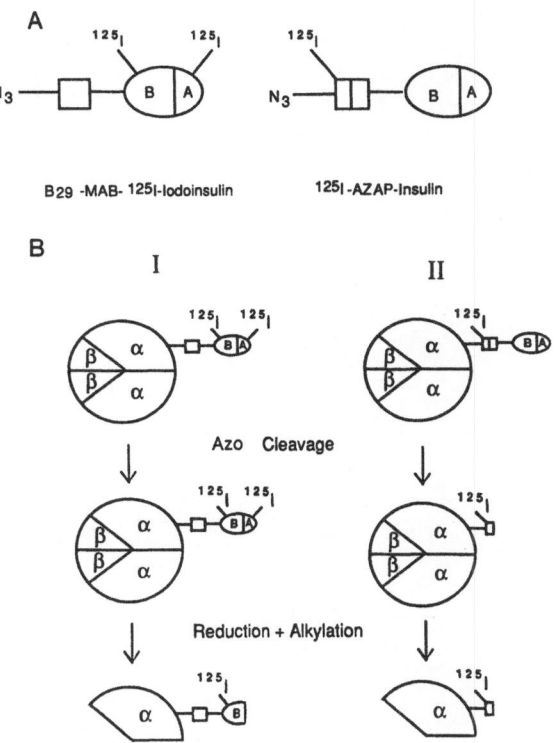

Figure 1. Schematic representation of photoaffinity labeling of the insulin receptor with either a non-cleavable (B29-MAB-^{125}I-iodoinsulin) or a cleavable photoreactive insulin (^{125}I-AZAP-Insulin) derivative (From Yip et al.[12]).

The structural and sequence homology between insulin receptor and insulin-like growth factor I (IGF-I) receptor together with their ligand binding specificity has provided an useful experimental approach to study the ligand-binding domain of these receptors. Chimeric receptors have been produced by domain exchange, and their ligand binding specificity and affinity studied[15-18]. Results of these studies have lead to the suggestion that the insulin-binding domain is encoded by exon 2 and exon 3 (see review by Yip[19]). It appears that the cysteine-rich region, encoded by exon 3, is a binding domain for both insulin and IGF-I whereas the specificity of ligand binding involves exon 2 and perhaps other parts of the receptor.

ACTIVATION OF THE RECEPTOR

Insulin binding of the insulin receptor leads to the activation of the receptor as a tyrosine kinase which is preceded by receptor autophosphorylation of tyrosine residues of the receptor ß subunit. Although the sites of tyrosine phosphorylation have been extensively studied and identified[20-22], the mechanism through which insulin activates this process remains unknown. Since the receptor subunits

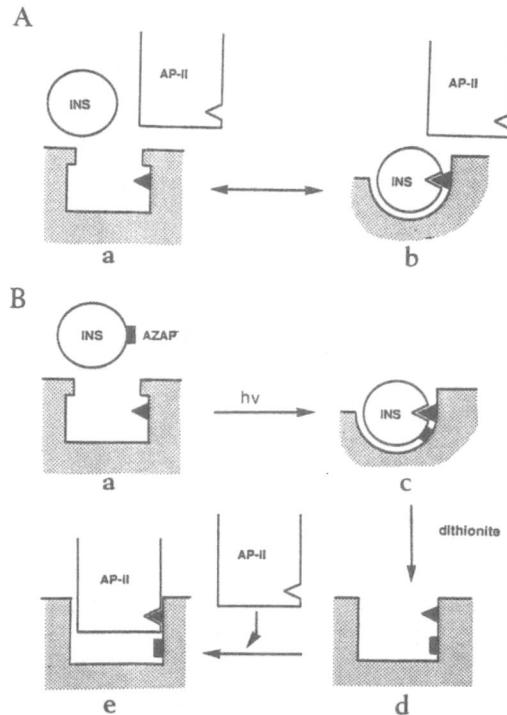

Figure 2. Schematic representation of the binding of insulin to the ligand-binding domain which is not accessible to the antibody AP-II raised against part of the ligand-binding domain. Insulin binding caused a conformational change of the domain so that it became accessible to the antibody AP-II which could now bind to its epitope after the removal of insulin

are linked by di-sulfides to form the heterotetrameric structure, the effects of disulfide reducing reagents and thiol alkylating reagents on receptor autophosphorylation and kinase activity were investigated in several studies[23-29]. Results of these studies support the view that receptor thiols and disulfides are involved in receptor activation by insulin. However, the reagent of interest was added to receptor preparations in these studies. We felt that if indeed receptor thiols or disulfides were involved, the procedures used to obtain the receptor preparation might have already altered these functional groups. Therefore we[30] prepared liver plasma membranes with or without the addition of the sulfhydryl alkylating reagent iodoacetamide in the homogenization buffer. Membranes were solubilized in Triton. Insulin receptors solubilized from the two types of membranes were compared for their insulin-stimulated receptor autophosphorylation and kinase activity. As shown in Figure 3A, the addition of iodoacetamide to the homogenization buffer greatly enhanced both the basal and insulin-stimulated autophosphorylation of insulin receptors from mouse and rat livers. The increased autophosphorylation was found to occur on tyrosine residues. Tryptic phosphopeptide analysis showed rapid phosphorylation of juxtamembrane tyrosines (Engl and Yip, manuscript in preparation). The enhancement of receptor autophosphorylation was accompanied by a similar change of receptor kinase activity measured as the phosphorylation of poly(Glu-Tyr) (Figure 3B). Iodoacetamide produced similar effects when human placental tissues were used as the source of insulin receptors (Engl and Yip, unpublished observations). It is important to point out that the addition of iodoacetamide subsequent to tissue homogenization was without effect. Indeed, as previously observed by others, a slight inhibitory effect was obtained. The observed enhancement of insulin activation could be the result of an increase in insulin binding. However, Scatchard analysis of insulin binding by receptors from either membrane preparation showed no difference in binding affinity and binding capacity. Alternately, the effect may reflect a change in the requirement for ATP. We therefore studied the effect of ATP concentration of insulin-stimulated receptor autophosphorylation obtained in a short incubation period of 2 minutes. Lineweaver-Burk analysis of the data showed a K_m of 50 μM ATP for receptors from membrane prepared with iodoacetamide, compared with 143 μM for receptors from control membranes. In addition, the V_{max} was increased from 157 fmol/min for the control to 555 fmol/min for receptors from membranes prepared with the reagent. Thus, iodoacetamide added to the homogenization buffer produced a significant positive effect on the interaction between the insulin receptor and ATP resulting in a decrease in the K_m and an increase in the V_{max} for ATP. We have carried out several control experiments to rule out the possibility that iodoacetamide produced these effects by inhibiting the activity of enzymes such as proteases which could degrade the receptor ß subunit or phosphatases which could dephosphorylate phosphotyrosines. Furthermore, we[30] found that, like iodoacetamide, two other sulfhydryl alkylating reagents, namely iodoacetate and p-chloro-mercuriphenyl sulfonate, produced a similar enhancing effect. Therefore it is most likely that iodoacetamide functioned as a thiol alkylating reagent in these experiments. Considered together these observations strongly suggest an important role of thiols/disulfide in the activation of insulin receptor by insulin.

How insulin activates receptor autophosphorylation and hence stimulates the expression of receptor kinase activity is not known. It is possible that insulin binding releases the conformational constraints imposed by the receptor α subunit on the ß subunit[31,32]. The conformational change induced by insulin may involve the participation of receptor thiols and disulfides in intrareceptor thiol/disulfide exchange reaction. Thiol/disulfide exchange reaction has been found to modify reversibly the activity of several enzymes (see references 33 for review). If such exchange reaction is required to activate the insulin receptor, experimental conditions favoring the preservation of receptor disulfides involved in the exchange would be expected to produce an enhancing effect. Accordingly we propose that iodoacetamide acted to preserve these putative reactive receptor thiols or disulfides during membrane preparation. Since the enhancing effect of iodoacetamide was obtained only when it was present during tissue homogenization, it must have reacted with thiols released from tissue disruption. We suggest that, in the absence of the sulfhydryl alkylating reagent, the thiol(s) released during homogenization would react with the reactive disulfide(s) of the insulin receptor. As a consequence, the putative intrareceptor thiol/disulfide exchange reaction cannot occur

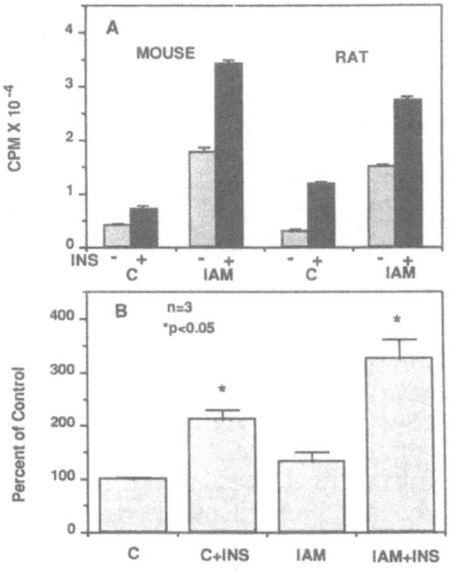

Figure 3. A. Insulin stimulated autophosphorylation of insulin receptor from mouse or rat liver plasma membranes prepared in the absence (C) or presence (IAM) of iodoacetamide. B. Insulin (+INS) stimulation of receptor tyrosine kinase activity of insulin receptor from rat liver plasma membranes prepared in the absence (C) or presence (IAM) of iodoacetamide. (From Li *et al.*[30]).

following insulin binding. Iodoacetamide alkylated the thiol(s) released during homogenization, thus preventing it from reacting with the receptor. In this context it has been observed that the reducing reagent dithiothreitol (DTT) at low concentrations added directly to standard preparations (i.e. without iodoacetamide) of insulin receptor stimulated receptor autophosphorylation and kinase activity[23-27]. In these experiments mild reduction of the insulin receptor by DTT might have regenerated the crucial reactive thiol(s) required for the productive intrareceptor thiol/disulfide exchange reaction.

There are 47 cysteine residues in the αß dimer of the insulin receptor, 37 of which are in the extracellular α subunit. It is not known how many and which ones of these residues exist as disulfides. However, it is surprising that the two αß dimers are linked by only 3 disulfides in the holoreceptor, and that there is evidently only one free thiol which is found in the ß subunit[34]. If receptor activation by insulin requires intrareceptor thiol/disulfide exchange as we propose, this free thiol of the ß subunit may participate in the exchange reaction.

PROSPECTS

In this presentation I have touched on two important aspects of the structure and function of the insulin receptor: where insulin binds, and how it activates the receptor. Current data generated from photoaffinity labeling and chimeric receptors strongly indicate that the insulin-binding domain is encoded by exon 2 and exon 3 of the insulin receptor gene. However, it is possible that other parts of the receptor may also be involved. The precise structure of this domain remains unknown and its ulitmate elucidation will require the X-ray crystallographic study of the holoreceptor and its complex with insulin. Nevertheless, the fact that the insulin receptor is structurally made up of two dimers composed of αß subunits has raised the question of the number of insulin-binding sites on the receptor, i.e. receptor valency of binding. This question can be answered by the mathematical analysis of binding data by means of Scatchard plots. The difficulty with this approach is the need to have a highly purified receptor preparation with known concentration. This has been further complicated by the curvilinear Scatchard plots that have been obtained. We have taken a direct approach to answer this question. We took advantage of two observations that: a) IGF-I can bind, though apparently with low affinity, to the insulin receptor, and b) antibody to insulin or to IGF-I can respectively immunoprecipitate insulin receptor photoaffinity labeled with insulin or IGF-I. Insulin receptors were photoaffinity labeled with radioactive insulin photoprobe (B29-[^{125}I]-MABI) in the presence or absence of non-radioactive IGF-I photoprobe, and immunoprecipitated with antiserum to IGF-I. It was expected that if there was only one ligand-binding site on the insulin receptor no radioactively labeled insulin receptors would be precipitated by the antiserum to IGF-I regardless of the presence of IGF-I photoprobe during photolabeling. On the other hand, if there were two binding sites on the receptor, radioactively labeled insulin receptors would be precipitated by the antiserum only when IGF-I photoprobe was presence during photolabeling, which was what we found[35]. Therefore we have concluded that the insulin receptor is a bivalent molecule, i.e. there are two ligand-binding sites on each receptor molecule. This conclusion in turn raises an important and interesting question of whether the insulin receptor requires the occupancy of both binding sites by insulin to achieve maximal activity.

Conformational changes are implicit to and underlie the mechanism of activation of macromolecules. In the case of the activation of the insulin receptor by insulin, we visualize that insulin binding to the ligand-binding domain of the receptor initiates a local conformational change. Since the insulin receptor is bivalent, the occupation of the binding domain by two insulin molecules, assuming that there is only one binding domain or pocket, and the avidity of insulin to dimerize would be an important component to induce this local change. In this context it is relevant to note the very low biological activity of guinea pig insulin which, in contrast to other mammalian insulins, does not dimerize[36]. The local conformational change could propagate across the membrane bilayer. In the process crucial reactive thiol(s) and disulfide(s) would be brought into appropriate proximity for thiol/disulfide exchange giving rise to further conformational changes to promote the transfer of phosphate from bound ATP to tyrosine residues. The involvement of thiol/disulfide exchange in the activation of the insulin receptor raises the possiblity that the activity of the insulin receptor in response to insulin binding may be regulated by cellula redox potential which in turn is affected by cellular metabolism. Indeed the *in vivo* activity of the mouse Ltk transmembrane protein tyrosine kinase has been found to be markedly enhanced by alkylating and thiol-oxidizing agents[37]. This consideration may be relevant to our attempt to understand the complexity of the pathogenesis of insulin resistance in NIDDM.

REFERENCES

1. M.P. Czech, The nature and regulation of the insulin receptor: Structure and function, *Annu. Rev. Physiol.* 47:357 (1985).
2. I.D. Goldfine, The insulin receptor: Molecular biology and transmembrane signaling, *Endocr. Rev.* 8:235 (1990).
3. S. Seino, M. Seino, S. Nishi, and G.I. Bell, Structure of the human insulin receptor gene and characterization of its promoter, *Proc. Natl. Acad. Sci. U.S.A.* 86:114 (1989).
4. A. Ullrich, J.R. Bell, E.Y. Chen, R. Herrera, L.M. Petruzzelli, J.T. Dull, A. Grey, L. Coussens, Y.-C. Liao, M. Tsubokawa, M. Mason, P.H. Seeburg, C. Grunfeld, O.M. Rosen, and J. Ramachandra, Human insulin receptor and its relationship to the tyrosine kinase family of oncogenes, *Nature* 313:756 (1985).
5. C.C. Yip, C.W.T. Yeung, and M.L. Moule, Photoaffinity labeling of insulin receptor proteins of liver plasma membrane preparations, *Biochem.* 19:70 (1980).
6. P.F. Pilch, and M.P. Czech, The subunit structure of the high affinity insulin receptor, *J. Biol. Chem.* 255:1722 (1980).
7. O.M. Rosen, After insulin binds, *Science* 237:1452 (1987).
8. J.M. Olefsky, The insulin receptor: A multifunctional protein, *Diabetes* 39:1009 (1990).
9. X.J. Sun, P. Rothenberg, C.R. Kahn, J.M. Backer, E. Araki, P.A. Wilden, D.A. Cahill, B.J. Goldstein, and M.F. White, Structure of the insulin receptor substrate IRS-1 defines a unique signal transduction protein, *Nature* 342:73 (1991).
10. C.C. Yip, C.W.T. Yeung, and M.L. Moule, Photoaffinity labelling of insulin receptor of rat adipocyte plasma membrane, *J. Biol. Chem.* 253:1743 (1978).
11. C.C. Yip, H. Hsu, R.G. Patel, D.M. Hawley, B.A. Maddux, and I.D. Goldfine, Localization of the insulin-binding site to the cysteine-

rich region of the insulin receptor α-subunit, *Biochem. Biophys. Res. Commun.* 157:321 (1988).

12. C.C. Yip, C. Grunfeld, and I.D. Goldfine, Identification and characterization of the ligand-binding domain of the insulin receptor: Insulin-induced conformational changes demonstrated by the use of an anti-peptide antiserum against the amino acid sequence 242-251 of the alpha subunit, *Biochem.* 30:695 (1991).

13. R.G. Mirmira, S.H. Nakagawa, and H.S. Tager, Importance of the character and configuration of residues B24, B25, and B26 in insulin-receptor interactions, *J. Biol. Chem.* 266:1428 (1991).

14. R. Rafaeloff, R. Patel, C.C. Yip, I.D. Goldfine, and D.M. Hawley, Mutation of the high cysteine region of the human insulin receptor α-subunit increases insulin receptor binding affinity and transmembrane signalling, *J. Biol. Chem.* 264:15900 (1989).

15. T.A. Gustafson, and W.J. Rutter, The cysteine-rich domains of the insulin and insulin-like growth factor I receptors are primary determinants of hormone binding specificity: Evidence from receptor chimeras, *J. Biol. Chem.* 265:18663 (1990).

16. A.S. Andersen, T. Kjeldsen, F.C. Wiberg, P.M. Christensen, J.S. Rasmussen, K. Norris, K.B. Moller, and N.P.H. Moller, Changing the insulin receptor to possess insulin-like growth factor I ligand specificity, *Biochem.* 29:7363 (1990).

17. T. Kjeldsen, A.S. Andersen, F.C. Wiberg, J.S. Rasmussen, L. Schaffer, P. Balschmidt, K.B. Moller, and N.P.H. Moller, The ligand specificities of the insulin receptor and the insulin-like growth factor I receptor reside in different regions of a common binding site, *Proc. Natl. Acad. Sci. U.S.A.* 88:4404 (1991).

18. B. Zhang, and R.A. Roth, Binding properties of chimeric insulin receptors containing the cysteine-rich domain of either the insulin-like growth factor I receptor or the insulin receptor related receptor, *Biochem.* 30:5113 (1991).

19. C.C. Yip, The insulin-binding domain of insulin receptor is encoded by exon 2 and exon 3, *J. Cell. Biochem.* 48:19 (1992).

20. C.K. Chou, T.J. Dull, D.S. Russell, R. Gherzi, D. Lebwohl, A. Ullrich, and O.M. Rosen, Human insulin receptors mutated at the ATP-binding site lack protein tyrosine kinase activity and fail to mediate postreceptor effects of insulin, *J. Biol. Chem.* 262:1842 (1987).

21. H.E. Tornqvist, M.W. Pierce, A.R. Frackelton, R.A. Nemenoff, and J. Avruch, Identification of insulin receptor tyrosine residues autophosphorylated *in vitro*, *J. Biol. Chem.* 262:10212 (1987).

22. M.F. White, S.E. Shoelson, H. Keutmann, and C.R. Kahn, A cascade of tyrosine autophosphorylation in the ß-subunit activates the phosphotransferase of the insulin receptor, *J. Biol. Chem.* 262:2969 (1987).

23. M.A. Shia, J.R. Rubin, and P.F. Pilch, The insulin receptor protein kinase: Physicochemical requirements for activity, *J. Biol. Chem.* 263:6822 (1988).

24. L.M. Petruzzelli, R. Herrera, and O.M. Rosen, Insulin receptor is an insulin-dependent tyrosine protein kinase: Copurification of insulin-binding activity and protein kinase activity to homogeneity from human placenta, *Proc. Natl. Acad. Sci. U.S.A.* 81:3327 (1984).

25. Y. Fujita-Yamaguchi, and S. Kathuria, The monomeric αß form of the insulin receptor exhibits much higher insulin-dependent tyrosine-specific protein kinase activity than the intact $\alpha_2\beta_2$ form

of the receptor, *Proc. Natl. Acad. Sci. U.S.A.* 82:6095 (1985).

26. L.J. Sweet, P.A. Wilden, and J.E. Pessin, Dithiothreitol activation of the insulin receptor/kinase does not involve subunit dissociation of the native $\alpha_2\beta_2$ insulin receptor subunit complex, *Biochem.* 25:7068 (1986).

27. L.J. Pike, A.T. Eakes, and E.G. Krebs, Characterization of affinity-purified insulin receptor/kinase: Effects of dithiothreitol on receptor/kinase function, *J. Biol. Chem.* 261:3782 (1986).

28. L.J. Pike, E.A. Kuenzel, J.E. Casnellie, and E.G. Krebs, A comparison of the insulin- and epidermal growth factor-stimulated protein kinases from human placenta, *J. Biol. Chem.* 259:9913 (1984).

29. P.A. Wilden, and J.E. Pessin, Differential sensitivity of the insulin-receptor kinase to thiol and oxidizing agents in the absence and presence of insulin, *Biochem. J.* 245:325 (1987).

30. C. Li, M.L. Moule, and C.C. Yip, Insulin receptors prepared with iodoacetamide show enhanced autophosphorylation and receptor kinase activity, *J. Biol. Chem.* 266:7051 (1991).

31. S. Tamura, Y. Fujita-Yamaguchi, and J. Larner, Insulin-like effect of trypsin on the phosphorylation of rat adipocyte insulin receptor, *J. Biol. Chem.* 258:14749 (1983).

32. L. Ellis, D.O. Morgan, E. Clauser, R.A. Roth, and W.J. Rutter, A membrane-anchored cytoplasmic domain of the human insulin receptor mediates a constitutively elevated insulin-independent uptake of 2-deoxyglucose, *Mol. Endocrinol.* 1:15 (1987).

33. H. Gilbert, Redox control of enzyme activities by thiol/disulfide exchange, *Methods Enzymo.* 107:330 (1984).

34. F.M. Finn, K.D., Ridge, and K. Hofmann, Labile disulfide bonds in human placental insulin receptor, *Proc. Natl. Acad. Sci. U.S.A.* 87:419 (1990).

35. C.C. Yip, and E. Jack, Insulin receptors are bivalent as demonstrated by photoaffinity labeling, *J. Biol. Chem.* 267:13131 (1992).

36. A.E. Zimmerman, M.L. Moule, and C.C. Yip, Guinea pig insulin: II. Biological activity, *J. Biol. Chem.* 249:4026 (1974).

37. A.R. Bauskin, I. Alkalay, and Y. Ben-Neriah, Redox regulation of a protein tyrosine kinase in the endoplasmic reticulum, *Cell* 66:685 (1991).

THE DIABETOGENES CONCEPT OF NIDDM

Pierre De Meyts

Hagedorn Research Institute
Niels Steensens Vej 6
DK 2820 Gentofte
Denmark

INTRODUCTION

It is not very often that I am asked to lecture about the pathogenesis of NIDDM, and I even imagine that many of my distinguished colleagues assume that I don't have an original opinion on the matter. I felt therefore very privileged to be asked to present my views on NIDDM in this Symposium honoring Professor Rolf Luft.

I first presented the concept that I will develop in this brief assay in 1985 at a Diabetes course organized by Joe Larner at the University of Virginia. In the discussion following one of Ron Kahn's always excellent lectures, I argued that the failure to reach a consensus on a unique cause for NIDDM, or in other words to identify what is today called a unique "candidate gene", may be due to an underestimation of how heterogeneous the disease actually is. I proposed that, in much the same way as "cancer" is now thought to represent a compound perturbation of cellular growth resulting from the cumulative activation of multiple oncogenes, which are mutated molecular components of the normal growth control mechanism, "NIDDM" may also be a complex syndrome resulting from the cumulative failure of a number of gene products involved in key steps of the glucose homeostasis biochemical machinery. I proposed to call such genes "diabetogenes" to stress the parallelism with the oncogenes concept. The main thrust of the diabetogenes concept is that different combinations of diabetogenes may be prevalent in different patient subgroups or even families, making the identification of the primary defects by traditional approaches unlikely to succeed.

I offered the same view in some more detail at a Symposium that I organized in Pasadena in June 1986, in honor of Rachmiel Levine's 75th birthday, shortly after moving to the City of Hope.

This is my first attempt to put some of these notions in writing. It is prompted by a strong feeling that several years later, despite substantial progress in the molecular biology of insulin secretion and action and glucose metabolism, we are pretty much in the same deadlock with respect to what comes first in NIDDM, i.e. beta cell failure or insulin resistance, and that much of our problem in identifying the primary defect in NIDDM stems from a primary defect in NIDDM research, i.e. some degree of intellectual reductionism and empiricism, as schematized in the cartoons shown in Figures 1 and 2.

New Concepts in the Pathogenesis of NIDDM, Edited by
C. G. Östenson *et al.*, Plenum Press, New York, 1993

There have been a number of enlightened balanced overviews of NIDDM pathogenesis in recent years; the reader should not expect to find such a traditionally objective and documented approach here, but rather a spirited and provocative critique aimed at stimulating discussion. I have chosen to focus on a critical evaluation of the role of insulin receptors in NIDDM, the small segment of the field that I know better, as a paradigm to demonstrate the failure of reductionist approaches in explaining NIDDM.

Figure 1. Current views on the pathogenesis of NIDDM.

Figure 2. [1]Substitute your favorite molecule: the insulin receptor, the insulin receptor tyrosine kinase, IRS 1, GLUT 2, GLUT 4, hexokinase II, glucokinase, protein tyrosine phosphatase, glycogen synthetase, amylin (to be continued).

BETA CELL DEFECT VERSUS INSULIN RESISTANCE

The debate as to whether NIDDM is primarily due to a beta cell defect or to peripheral insulin resistance, or both, and in which order these two defects appear, has been raging for decades. Definitive but contradictory statements abound:

- "Thus, as believed several decades ago, the entire spectrum of carbohydrate intolerance in diabetes mellitus appears to be characterized by varying degrees of relative insulin deficiency[1]".
- "In many patients with type II diabetes, insulin deficiency does not exist[2]".
- "In type II diabetes, insulin resistance is certainly a major pathogenic factor, if not the most important one[3]".
- "It is unlikely that insulin insensitivity is a primary derangement in type II diabetes, and that low insulin response would represent a secondary phenomenon[4]".
- "Insulin resistance is almost uniformly present in patients with any degree of clinically detectable glucose intolerance[5]".
- "...NIDDM is a heterogeneous disorder characterized by impaired insulin secretion that is usually but not invariably accompanied by insulin resistance in both hepatic and extrahepatic tissues[6]".

For a more detailed and documented discussion of the issues, I refer the reader to several recent reviews on the pathogenesis of NIDDM, usually flavored towards either the beta cell defect or peripheral resistance preponderance theory[4,7-10], as well as to other chapters in this book.

ROLE OF INSULIN RECEPTORS IN THE INSULIN RESISTANCE OF NIDDM

In both genetic and acquired forms of obesity in humans and rodents, increased concentrations of circulating insulin have been found to be associated with decreased insulin binding to various tissues, due to a decreased concentration of insulin receptors in target cells[11-15]. The receptor decrease was reversed upon dietary manipulations that corrected the hyperinsulinemia. An inverse correlation between ambient insulin levels and cellular receptor concentrations could also be demonstrated directly in cell culture[16]. These and other influential studies established the concept that became known as "receptor downregulation", which has become an essential paradigm in endocrine regulation. Receptor internalization by ligand-induced endocytosis is thought to play a major role in receptor downregulation[17].

Whether the generally accepted role of hyperinsulinemia and receptor downregulation in the insulin resistance of obesity can be uncritically extended to NIDDM is the issue on which I would like now to focus.

Olefsky and Reaven were the first to extend the findings of an inverse correlation between fasting insulin levels and cellular receptor concentrations to a group of nonobese patients with NIDDM[18], and others reported similar findings usually in obese patients with NIDDM (e.g. ref. 19). The universality of this correlation has become textbook dogma, as examplified in a figure from a review by Ron Kahn in Joslin's Diabetes Mellitus[20] reproduced here as Fig. 3, emphasizing the parallelism between Olefsky and Reaven's data[18] and the NIH study[15].

It is not surprising therefore to find the following definition of NIDDM in one of the currently popular Biochemistry textbooks[21]: "NIDDM usually afflicts adults over the age of 40 and appears to be determined by genetic factors affecting both insulin production and insulin receptor proteins".

I see, however, two problems with the Joslin textbook's interpretation of Fig. 3. First, the caption of the Figure mentions "thin (left) and obese (right) patients **with**

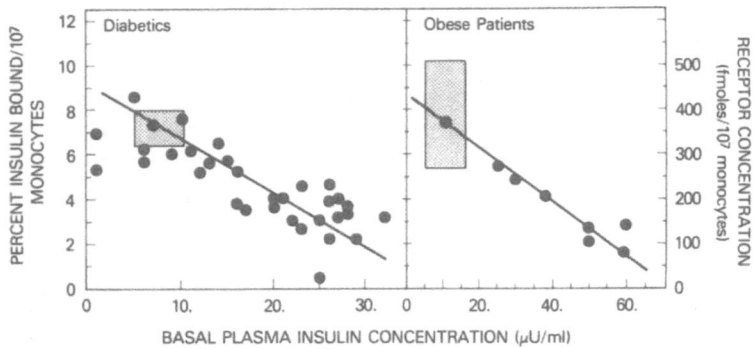

Figure 3. The caption in Joslin's Diabetes Mellitus reads: "Insulin binding to circulating monocytes in thin (left) and obese (right) patients with NIDDM. (Redrawn from refs. 18 and 15)". Used with permission.

NIDDM". In fact, only two out of eight patients in the NIH study[15] had NIDDM and one IGT, as shown in their glucose tolerance test (Fig. 1 of ref. 15); interestingly, the word "diabetes" was not even mentioned in that paper, except once when referring to Olefsky and Reaven's patients[18], and once in reference to Chinese hamsters (p. 1133).

Interestingly, the two patients with the diabetic glucose tolerance test also had the smallest insulin response to glucose (patient E had the highest fasting insulin but showed no insulin increase after glucose). One may therefore argue that even in that small number of patients, while there is indeed an inverse correlation between receptor number and fasting insulinemia, it is the failure of the beta cell to sustain an enhanced insulin response after challenge that precipitates NIDDM. A bundle of other data suggests that the transition from decreased oral tolerance to manifest diabetes is mainly precipitated by progressive impairment of insulin release[4] (see below).

The second problem I have with Fig. 3 is that the parallelism shown is somewhat artificial; the range of fasting insulins on the horizontal axis in the NIH study covers 60 μU/ml, which is not unexpected in massive obesity (the ideal body weights ranged from 199 to 288%), but the range of fasting insulins in Olefsky and Reaven's patients covers only 30 μU/ml (I doubt that such high values would be common in European thin NIDDM patients). Therefore, if one reports the "downregulation line" from Olefsky and Reaven's patients on the same scale as the NIH patients (Fig. 4), it becomes apparent that Olefsky and Reaven's patients markedly downregulated their receptors over a range of insulin concentrations in which the NIH patients still demonstrated binding within the normal range. Does this mean that the patients on the left are hypersensitive to insulin as compared to the patients on the right? But if the patients on the right are more resistant to downregulation, should they then not be more sensitive to insulin if downregulation explains resistance? Without trying to solve these paradoxes, a minimal conclusion from such analysis is that there is a marked heterogeneity in the sensitivity of the two groups with respect to receptor downregulation, and it follows that one should be careful in extrapolating findings from obesity to NIDDM, and from one subgroup of NIDDM patients to other subgroups, especially of other ethnic or geographic origins.

I want to make clear that I am not focusing on these two particular studies in order to single out two groups of excellent investigators; Ron Kahn's review[20] in fact emphasized the complexity of NIDDM pathogenesis and the difficulty in identifying the primary defect, while I actually wrote myself some years ago a "reductionist" review stressing the role of receptor downregulation in NIDDM[22]!

My scepticism on the prevalence of receptor downregulation in NIDDM started when, in collaboration with the group of André Lambert at the University of Louvain in Belgium, we tried to select mildly diabetic patients with fasting hyperinsulinemia for a clinical study on gliclazide. It had been reported at that time by several groups that sulfonylureas corrected the receptor decrease observed in NIDDM patients[23], and that sulfonylureas increased receptor levels in animals as well as in cell culture[24, 25]. We

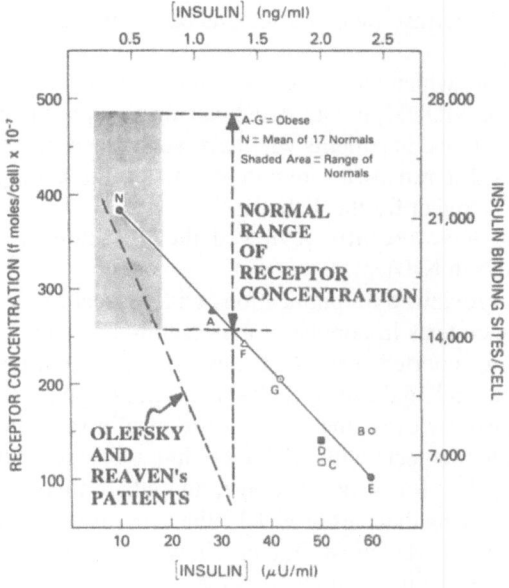

Figure 4. The regression line of monocyte insulin binding data from ref. 18 has been superimposed on the data from the NIH study (ref. 15). Modified from ref. 15.

simply failed to find **hyperinsulinemic** nonobese Belgian NIDDM patients, and selected a group of mild diabetics with fasting insulin levels as close as possible to nondiabetic controls. Gliclazide proved effective in normalizing glycemia in such patients, but the receptor concentration (on erythrocytes) was unchanged, meaning it was normal both before and after treatment[26]. It is not surprizing that we did not find receptor downregulation since there was no fasting hyperinsulinemia.

Subsequently, we[27] and others[28,29] failed to document effects of gliclazide or other sulfonylureas on insulin receptor concentrations *in vitro* or after *in vivo* administration to animals, and the consensus appears to be today that if there is a peripheral effect of sulfonylureas, it is at a postreceptor step.

We also failed to find evidence for receptor downregulation in NIDDM in a large scale study in Maltese subjects in collaboration with Joseph J. Hoet, Zdenko Skrabalo and the W.H.O (unpublished data). Malta has a high incidence of obese NIDDM. We compared fasting insulin levels and erythrocyte receptor concentrations in about 300 subjects with either normal glucose tolerance, IGT or NIDDM, and found comparable mean receptor levels on erythrocytes in all three groups. There was no difference in mean fasting insulin between nondiabetics and NIDDM patients and a very modest elevation in fasting insulin in those with IGT.

In agreement with such results, there has also been little evidence for decreased receptor concentration in other groups of NIDDM patients such as the Pima Indians[30].

This leads me to question the apparently generally accepted concept that associates NIDDM with hyperinsulinemia. First, authors who mention hyperinsulinemia often neglect to specify whether they mean fasting hyperinsulinemia, postprandial hyperinsulinemia, or hyperinsulinemia after glucose challenge, and altogether "hyper" with respect to what.

Yalow and Berson, which have often been quoted as being the first to demonstrate the hyperinsulinemia of NIDDM, in fact stated that "fasting plasma insulin concentrations in early maturity onset diabetic patients who have never been treated with insulin and in non diabetic subjects **did not differ markedly**, while the average integrated insulin concentration was 26% higher for the diabetics[31] ".

Several authors have recently reviewed the literature on insulin secretion in response to oral glucose in NIDDM.

De Fronzo, who reviewed 32 publications with respect to plasma insulin response during glucose tolerance tests in nonobese non-insulin-dependent diabetic subjects with fasting hyperglycemia, reported that 27 of these studies found normal fasting insulin, while 11 found normal, and 16 decreased, plasma insulin response to glucose[7].

Temple et al. also reviewed the literature on insulin secretion in response to oral glucose in NIDDM and subjects with IGT, including only those studies where controls and patients were matched for weight and age, and where there were sufficient data to define the subjects as having diabetes or IGT by the currently accepted diagnostic criteria. They found that in NIDDM, "most studies show diminished early and late insulin secretion in all but the very mildest diabetic subjects, which is contrary to what has often been stated in the past."

Owens reviewed the plasma insulin responses of NIDDM patients to OGTT and IVGTT in 28 publications; fasting plasma insulin was increased in only 22% of 959 patients, and early insulin response was decreased in 78% (D.R. Owens, personal communication).

My conclusion from these studies is that **the majority of patients with NIDDM do not present with fasting hyperinsulinemia and receptor downregulation**. Moreover, as pointed out by Porte [9], the confounding effect of obesity as an independent variable affecting basal and stimulated insulin levels should be taken into consideration, and several authors have pointed out that the "normalcy" of an insulin level should be defined with respect to the prevalent glycemia: a "normal" insulin level in the face of hyperglycemia clearly indicates a defective beta cell function. Moreover, the contribution of increased proinsulin and partially proteolyzed derivates should be taken in consideration in the evaluation of "hyperinsulinemia" [9, 32].

The above arguments are not meant to completely dismiss a role for the insulin receptor in the pathogenesis of NIDDM. Nearly 50 different mutations in the insulin receptor gene have now been reported in various syndromes associated with severe insulin resistance (for review, see ref. 33). It is clear however that they represent only a minute fraction of NIDDM cases. Moreover, one is struck in such cases, like in the case of the even more rare mutations in the insulin molecule[34], by the variable penetrance of the mutations in causing diabetes (as opposed to insulin resistance). These data suggest that an isolated mutation in one of the components of the complex biochemical machinery regulating blood glucose, while increasing the predisposition to NIDDM, may not alone be sufficient to precipitate overt diabetes.

The insulin receptor has recently made a comeback as a potential etiologic factor in NIDDM with studies showing an abnormal expression of the two alternatively spliced isoforms of the insulin receptor in muscle tissue from NIDDM patients. While muscle of nondiabetics contain predominantly the receptor form that does not contain the 12 residues encoded by exon 11, muscle from NIDDM patients was reported to contain also variable amounts of the longer receptor form[35-37]. Two other groups have however failed to confirm these findings[38-39].

The above discussion should make it clear that I am not any more a firm believer in the importance of the association "hyperinsulinemia - receptor downregulation" in the pathogenesis of NIDDM.

Multiple studies have, on the other hand, stressed the importance of alterations in early phase insulin release in the progression towards glucose intolerance and NIDDM[9,40-42], (D.R Owens, I. Ismaïl, S. Luzio, and A. Vølund, to be submitted).

The interpretation of alterations in beta cell responsiveness and in peripheral sensitivity to insulin in molecular terms is still in its infancy , although the search for "candidate genes" is proceeding at full steam.

Mutations in the glucokinase gene have now been documented in a substantial number of families with maturity-onset diabetes of the young, a form of NIDDM characterized by monogenic autosomal transmission and early age of onset (for review, see ref. 43).

Other potentially relevant factors that I lack both space and expertise to discuss are impaired fetal growth[44], islet amyloid polypeptide or amylin[9], glucose toxicity and the fact that islet cell function and insulin sensitivity do not appear to be independent variables but are related by a hyperbolic relationship[9].

The above discussion, while it may shed little light on the pathogenesis of NIDDM, may help leading to the concept that no single cause is likely to provide a satisfactory explanation for the progression from normal glucose tolerance to IGT to NIDDM, and give support to the thesis that NIDDM is a multifactorial and polygenic disease. In other words, the problem as I see it is not "the beta cell versus insulin resistance", or even the conspiracy of a "triumvirate[45]"; I am proposing that NIDDM is in fact far more heterogeneous than has been previously estimated.

THE DIABETOGENES CONCEPT

The premise behind this concept is that the system that maintains glucose homeostasis (Fig. 5) is sturdy and redundant, and can compensate so that normoglycemia is not likely to be affected by single gene defects. For example, the decreased potency of a mutant insulin is usually almost exactly compensated by an increased half-life due to a slow clearance probably explained by a decreased receptor-mediated endocytosis in the liver. Therefore, it is only when the accumulation of a number of defective gene products reaches a certain threshold that glucose homeostasis will collapse. By analogy with the pathogenesis of cancer, I have proposed to call **diabetogenes** the candidate genes potentially involved in NIDDM. The analogy with cancer (and probably other multifactorial and polygenic diseases) is shown as follows.

Table 1

	CANCER	DIABETES
Initiation	**Oncogenes**	**Diabetogenes**
Promotion	**Carcinogens**	**Diabetogenic Factors** (Nutrition, obesity, environmental toxins, lack of exercise, ...).

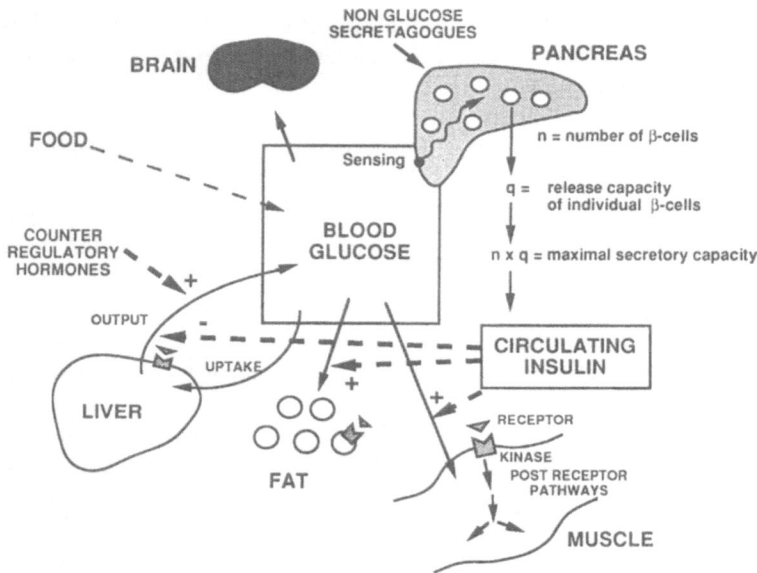

Figure 5. The complex regulation of glucose homeostasis.

A similar analogy between diabetes, cancer and other multifactorial diseases has been proposed in Granner and O'Brien's excellent recent review of the molecular physiology and genetics of NIDDM[46].

An inspection of Fig. 5 reveals that the number of potential diabetogenes may be considerable, including (the list is not limitative):

- Proteins involved in the glucose sensing mechanism: transporter(Glut 2), glucokinase, other enzymes involved in beta cell glucose metabolism;
- Factors determining the beta cell sensitivity to nonglucose secretagogues;
- Processes controlling beta cell growth and regeneration;
- Processes involved in insulin synthesis and release;

- The insulin molecule itself;
- Factors controlling the transendothelial transport of insulin (see the chapter by Rich Bergman in this book);
- The insulin receptor and its tyrosine kinase moiety;
- Insulin receptor substrate 1,
- Other substrates of the tyrosine kinase;
- Insulin-responsive enzymes of intracellular metabolism;
- Insulin-responsive elements controlling gene expression;
- Metabolic processes controlling glucose uptake and output by the liver;
- Processes controlling the secretion of counterregulatory hormones;
- Transcription factors controlling the expression of the above mentioned diabetogenes are themselves potential diabetogenes.

A more logically organized discussion of candidate genes can be found in Granner and O'Brien's review[44]. As they also point out, according to this concept, the **sequence** of appearance of defective gene products is not as important as the **accumulation** of these defects in reaching a **threshold** above which glucose homeostasis collapses.

A related concept valid for both IDDM and NIDDM has been proposed by Aldo Rossini under the name of "tumbler hypothesis[47]". Jørn Nerup and colleagues have proposed a model of IDDM as a polygenic disorder in which no IDDM-specific gene exists, but where genetic susceptibility is rather conferred by an unfavorable combination of common alleles of normal genes, each of which controls a part of the pathogenic process[48] (J. Nerup, personal communication). These putative genes could be easily incorporated in the diabetogenes concept (as factors affecting n and/or q in Fig.5) to provide a unifying model of the pathogenesis of diabetes mellitus.

An important consequence of the diabetogenes concept is that a much better phenotypic definition of NIDDM patients will be required in order to understand the pathogenesis and to identify candidate genes. This will demand, as proposed by Granner and O'Brien[46], careful **metabolic staging**, focusing on the study of patients with IGT and subjects at risk such as first degree relatives of NIDDM patients.

Paraphrasing Granner and O'Brien[46], I agree that it is clear that "metabolic staging, in which the critical events are defined at a biochemical level, will make it easier to interpret studies regarding the pathophysiology of the disease by defining 1) the nature of the diabetogenes, 2) the number of diabetogenes; 3) the relative contribution of each diabetogene 4) the temporal relationship of each diabetogene in the progression of the disease; and 5) the contributing role of each factor to etiology, as opposed to their role in complicating an established disease".

A putative example of such metabolic staging is shown in Fig. 6, borrowed from Granner and O'Brien's review[46]. This scheme presents the initial defect as an impairment in peripheral glucose utilization (insulin resistance), but equally plausible schemes can be designed with either faulty hepatic glucose production or insulin secretion being the initial defect [46].

Given the potential complexity of the disease mechanism, one has to hope that epidemiological studies are correct in predicting that a **major** gene defect and several more minor defects contribute to the genetic basis of NIDDM[46].

I plan to develop in a more rigorous way the diabetogenes concept and its implications in a review in preparation; the purpose of this preliminary exercise is to stimulate discussion, and to encourage younger investigators to take a fresh look at NIDDM pathogenesis with an open mind and to adopt a critical attitude toward textbook dogmas.

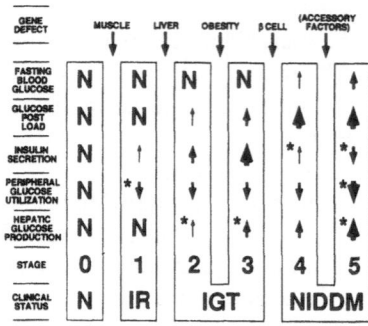

Figure 6. Example of metabolic staging in NIDDM. Metabolic stages (0-5) in progression from normal glucose homeostasis through insulin resistance (IR) and impaired glucose tolerance (IGT) to NIDDM in one patient is shown. Asterisks denote events critical to each transition. In this example, the defect in peripheral glucose utilization is the first to appear. The appearance of a second defect results in unrestrained hepatic glucose production, which increases insulin resistance. When a beta cell defect compromises insulin secretion, NIDDM occurs. From ref. 46, used with permission.

ACKNOWLEDGEMENTS

I am deeply indebted to Daryl Granner, Dan Porte, David Owens, Suad Efendic and Oluf Pedersen for stimulating discussions, and to Ron Shymko for critical reading of the manuscript. Ingeborg Drøhse provided skillful secretarial assistance, and Henrik Wengholt precious help with desktop publishing.

SELECTED REFERENCES

1. D.J. Porte, and J.D. Bagdade, Human insulin secretion: an integrated approach, *Ann. Rev. Med.* 21: 219 (1970).

2. J.R. Olefsky, and O.G. Kolterman, Mechanism of insulin resistance in obesity and noninsulin-dependent (type II) diabetes, *Am. J. Med.* 70:151 (1981).

3. C.R. Kahn, Insulin resistance: a common feature of diabetes mellitus, *N. Engl. J. Med.* 315:252 (1986).

4. S. Efendic, R. Luft, and A. Wajngot, Aspects of the pathogenesis of type 2 diabetes, *Endocrine Reviews.* 5:395 (1984).

5. R.A. De Fronzo, E. Ferrannini, V. Koivisto, New concepts in the pathogenesis and treatment of noninsulin-dependent diabetes mellitus, *Am. J. Med.* 74:1A, 52 (1983).

6. L.J. Mandarino, and J.E. Gerich, Prolonged sulfonylurea administration decreases insulin resistance and increases insulin secretion in non-insulin-dependent diabetes mellitus: evidence for improved insulin action at postreceptor site in hepatic as well as extrahepatic tissues, *Diabetes Care* 7:suppl. 1, 89 (1984).

7. R.A. De Fronzo, R.C. Bonadonna and E. Ferrannini, Pathogenesis of NIDDM: a balanced overview, *Diabetes Care* 15:318 (1992).

8. R.A. De Fronzo, Pathogenesis of type 2 (non-insulin dependent) diabetes mellitus: a balanced overview, *Diabetologia* 35:389 (1992).

9. D. Porte, Jr., Beta cells in type II diabetes mellitus, *Diabetes* 40:166 (1990).

10. S. Dinneen, J. Gerich and R. Rizza, Carbohydrate metabolism in non-insulin-dependent diabetes mellitus, *N. Engl. J. Med.* 327:707 (1992).

11. C.R. Kahn, D.M. Neville, Jr., and J. Roth, Insulin receptor interaction in the obese-hyperglycemic mouse: a model of insulin resistance, *J.Biol.Chem.* 248:244 (1973).

12. A.H. Soll, C.R. Kahn, and D.M. Neville, Jr., Insulin binding to liver plasma membranes in the obese hyperglycemic (ob/ob) mouse. Demonstration of a decreased number of functionally normal receptors, *J. Biol. Chem.* 250:4702 (1975).

13. A.H. Soll, C.R. Kahn, D.M. Neville, Jr., and J. Roth, Insulin receptor deficiency in genetic and acquired obesity. *J. Clin. Invest.* 56:769 (1975).

14. J.M. Olefsky, Decreased insulin binding to adipocytes and circulating monocytes from obese patients, *J. Clin. Invest.* 57:1165 (1976).

15. R.S. Bar, P. Gorden, J. Roth, C.R. Kahn, and P. De Meyts, Fluctuations in the affinity and concentration of insulin receptors on circulating monocytes of obese patients. Effects of starvation, refeeding and dieting, *J. Clin. Invest.* 58:1123 (1976).

16. J.R. Gavin, III, J. Roth, D.M. Neville, Jr, P. De Meyts, and D.N. Buell, Insulin-dependent regulation of insulin receptor concentrations. A direct demonstration in cell culture, *Proc. Natl. Acad. Sci. U.S.A.* 71:74 (1974).

17. J.L. Carpentier, The cell biology of the insulin receptor, *Diabetologia* 32:627 (1989).

18. J.M. Olefsky, and G.M. Reaven, Insulin binding in diabetes, relationships with plasma insulin levels and insulin sensitivity, *Diabetes* 26:680 (1977).

19. H. Beck-Nielsen, The pathogenic role of an insulin receptor defect in diabetes mellitus of obese, *Diabetes* 27:1175 (1978).

20. C.R. Kahn, Pathophysiology of Diabetes Mellitus: an Overview, in Joslin Diabetes Mellitus, Twelth Edition, A. Marble, L.P. Krall, R.F. Bradley, A.R. Christlieb, J.S. Soeldner, eds., Lea and Fibiger, Philadelphia, p. 43 (1985).

21. J.D. Rawn, Biochemistry, Neil Patterson Publishers, Burlington, North Carolina, p. 398 (1989).

22. P. De Meyts, Insulin receptors and diabetes, *Medicographia* vol. 3, 3:8 (1983).

23. J.M. Olefsky, and G.M. Reaven, Effects of sulfonylurea therapy on insulin binding to mononuclear leukocytes of diabetic patients, *Am.J.Med..* 60:89 (1976).

24. M.N. Feinglos, and H.E. Lebovitz, Sulphonylureas increase the number of insulin receptors, *Nature (London)* 276:184 (1978).

25. J.M. Prince, and J.M. Olefsky, Direct in vitro effect of a sulfonylurea to increase human fibroblast insulin receptors, *J. Clin. Invest.* 66:608 (1980).

26. E. Marchand, F. Grigorescu, M. Buysschaert, P. De Meyts, J.M.Ketelslegers, H. Brems, M.C. Nathan, and A.E. Lambert, The hypoglycemic effect of a sulfonylurea (gliclazide) in moderate type II diabetes is not accompanied by changes in insulin action and insulin binding to erythrocytes, *Molecular Physiol.* 4:83 (1983).

27. P. De Meyts, F. Grigorescu, and B. Lambert, The role of insulin receptors in the hypoglycemic effect of sulfonylureas: the status of 1982, in Rationale for Sulfonylurea Therapy, E.F. Pfeiffer, ed. Excerpta Medica, Amsterdam, p. 71 (1983).

28. R. Vigneri, V. Pezzino, K.Y. Wong, and I.D. Goldfine, Comparison of the in vitro effect of biguanides and sulfonylureas on insulin binding to its receptors in target cells, *J. Clin. Endo. Metab.* 54:95 (1982).

29. J. Dolais-Kitabgi, F. Alengrin, and P. Freychet, Sulfonylureas in vitro do not alter insulin binding and effects on aminoacid transport in hepatocytes, *Diabetologia* 24:441 (1983).

30. B.L. Nyomba,V.M. Ossowski, C. Bogardus, and D.M. Mott, Insulin-sensitive tyrosine kinase relationship with in vivo insulin action on humans, *Am. J. Physiol .* 258:E964 (1990).

31. R.S. Yalow, and S.A. Berson, Immunoassay of endogenous plasma insulin in man, *J. Clin. Invest.* 39:1157 (1960).

32. R. Temple, P.M.S. Clarck, and C.N. Hales, Measurement of insulin secretion in type 2 diabetes: problems and pitfalls, *Diabetic Medicine* 9:503 (1992).

33. S. Taylor, Molecular mechanisms of insulin resistance. Lessons from patients with mutations in the insulin receptor gene, *Diabetes* 41:1473 (1992).

34. H.S. Tager, Insulin gene mutations and abnormal products of the human insulin gene, in Hormone Resistance and Other Endocrine Paradoxes, M.P. Cohen and P.P. Foà, eds., Springer-Verlag, New York, p. 35 (1987).

35. L. Mosthaf, B. Vogt, H. Häring, and A. Ullrich, Altered expression of insulin receptor types A and B in the skeletal muscle of non-insulin-dependent diabetes mellitus patients, *Proc. Natl. Acad. Sci. U.S.A.* 88:4728 (1991).

35. G. Sesti, M.A. Marini, A.N. Tullio, A. Motemurro, P. Borboni, A. Fusco, D. Accili, and R. Lauro, Altered expression of the two naturally occuring human insulin receptor variants in isolated adipocytes of non-insulin- dependent diabetes mellitus patients, *Biochem. Biophys. Res. Commun.*. 181:1419 (1991).

37. L. Mosthaf, J. Erikson, H.U. Häring, L. Groop, E. Widen, and A. Ullrich, Insulin receptor isotype expression correlates with risk of non-insulin-dependent diabetes, *Proc. Natl. Acad. Sci. U.S.A.* 90: 2633 (1993).

38. H.Benecke, J.S. Flier, and D.E. Moller, Alternatively spliced variants of the insulin receptor protein. Expression in normal and diabetic human tissues, *J. Clin. Invest.* 89:2066 (1992).

39. T. Hansen, C. Bjorbæk, H. Vestergaard, J.F. Bak, and O. Pedersen, Alternatively spliced variants of the insulin receptor and its functional correlates in muscle from patients with type 2 diabetes and normal subjects, *Diabetologia* 35, suppl. 1:A 76, abstract 290 (1992).

40. S. Efendic, V. Grill, R. Luft and A. Wajngot, Low insulin response: a marker of prediabetes. *Adv. Exp. Med. Biol.* 246:167 (1988).

41. E. Cerasi, R. Luft, and S. Efendic, Decreased sensitivity of the pancreatic beta cells to glucose in pre-diabetic and diabetic subjects, *Diabetes* 21:224 (1972).

42. H. Yoneda, H. Ikegami, Y. Yamamoto, E. Yamato, T. Cha, Y. Kawaguchi, Y. Tahara and T. Ogihara, Analysis of early-phase insulin response in nonobese subjects with mild glucose intolerance, *Diabetes Care* 15:1517 (1992).

43. M.A. Permutt, K.C. Chu and Y. Tanizawa, Glucokinase and NIDDM. A candidate gene that paid off, *Diabetes* 41:1367 (1992).

44. K. Phipps, D.J.P. Barker, C.N. Hales, C.H.D. Fall, C. Osmond and P.M.S. Clark, Fetal growth and impaired glucose tolerance in men and women, *Diabetologia* 36:225 (1993).

45. R.A. De Fronzo, The triumvirate: beta cell, muscle, liver: a collusion responsible for NIDDM, *Diabetes* 37: 667 (1988).

46. D.K. Granner, and R.M. O'Brien, Molecular physiology and genetics of NIDDM. Importance of metabolic staging, *Diabetes Care* 15:369 (1992).

47. A.A. Rossini, J.P. Mordes, and E.S. Handler, Speculations on etiology of diabetes mellitus. Tumbler hypothesis, *Diabetes* 37:257 (1988).

48. J. Nerup, T. Mandrup-Poulsen, and J. Mølvig, The HLA-IDDM association: implications for etiology and pathogenesis of IDDM, *Diabetes/ Metabolism Reviews* 3:779 (1987).

MOLECULAR GENETICS OF NIDDM AND THE GENES FOR INSULIN AND INSULIN RECEPTOR

Holger Luthman, Ingrid Delin, Anna Glaser, Rolf Luft, Svante Norgren, and Anna Wedell

Departments of Clinical Genetics and Endocrinology
Rolf Luft Center for Diabetes Research
Karolinska Hospital
S-104 01 Stockholm

EPIDEMIOLOGY AND PHYSIOLOGY OF NIDDM

It is generally accepted that non-insulin dependent diabetes mellitus (NIDDM) is an inherited disease. This is illustrated, e. g., by the high concordance rate of NIDDM among monozygotic twins (90 - 95%) and the somewhat lower concordance in dizygotic twins (approx. 40%).[1-3] NIDDM also displays obvious familial aggregation. First degree relatives of NIDDM patients run a 40% lifetime risk of developing the disease, compared to a 10% risk in individuals without diagnosed NIDDM in first degree relatives.[4] Furthermore, the ethnic variability of the frequency of the disease, ranging from over 50% in Pima Indians and Naruans to 2-3% in the North European population, has been taken to illustrate that genetic factors are of major importance in determining the risk to develop NIDDM in response to environmental influence.[5-7] These and a number of other findings were derived from studies of populations with NIDDM, and in a few instances subjects with decreased glucose tolerance. As long as we lack an exact description regarding the genetics of common NIDDM, we must presume that the susceptibility to NIDDM is inherited in a polygenic fashion with the effect of some major genes, interactions between genes and environment, and environmental effects alone. Furthermore, the disease is probably heterogeneous, such that different combinations of susceptibility alleles can lead to overt disease in a permissive milieu.

If NIDDM is an inherited disease, there must be at least one physiological character which is inherited and, under certain circumstances, may lead to the development of manifest disease. The closer the physiological studies mirror functions of the susceptibility genes, the more accurate should they be able to identify the traits predisposing for NIDDM. For this reason, one turning point for the studies of the pathophysiology of NIDDM was the

New Concepts in the Pathogenesis of NIDDM, Edited by
C. G. Östenson *et al.*, Plenum Press, New York, 1993

realization of insulin secretion and insulin action as important factors characterizing glucose intolerance, see Cerasi and Luft.[8] In this study the importance of low insulin response in relation to the need, which is determined by the prevailing insulin sensitivity, was considered the major factor determining the liability to NIDDM. In later studies on a large number of nuclear families, using segregation and path analyses, it was clearly demonstrated that one of the factors, reflecting insulin response to glucose, was genetically determined with a genetic heritability (h^2) of 0.73. Despite that the glucose infusion test used in these studies was not developed to specifically measure the second factor, insulin sensitivity, it was found to have a genetic heritability of almost 0.4.[9] By the use of these two interrelated factors, it has been possible to obtain a physiological explanation of almost all conditions predisposing to NIDDM, e. g., conditions with decreased insulin sensitivity such as obesity, acromegaly, and physical inactivity.[10,11] Incapacity to comply with the demand for insulin (decreased relative insulin secretion) is a significant factor determining whether NIDDM will develop in a given situation.

MOLECULAR BIOLOGY OF NIDDM

In some families NIDDM segregates as a dominantly inherited monogenic disease.[12] However, so far, in the vast majority of NIDDM patients, we have to assume that the disease is the result of additive effects of several genes. Consequently, the road to pursue is to identify and investigate genes which may be considered as candidates for causing inherited susceptibility to NIDDM. In this context it is worth remembering that often in biology, we must try to understand metabolic traits with only scattered information on the participating reactions and reactants. Despite this limitation, it has been possible to understand the etiology of many diseases by studies of genes and gene products, which were identified by biochemical, cell biological, as well as physiological methods (e. g., familial hypercholesterolemia).

Table 1. List of some candidate functions involved in NIDDM susceptibility.

Hormones	Intracellular signaling	Gluconeogenesis
Neurotransmitters	Membrane transport	Krebs cycle
Receptors	Glycogen metabolism	Oxidative phosphorylation
Ion channels	Glycolysis	Lipid metabolism

In the etiology of NIDDM, we may consider as candidate susceptibility genes all genes involved in regulation of glucose homeostasis, cellular transport, and metabolism of energy. A candidate gene must not only be part of an important function within glucose homeostasis but, in addition, display genetic variation which can be associated with an increased risk of developing NIDDM. Table 1 lists candidate functions predicted to influence glucose homeostasis. Each of these functions represents one or several candidate genes. The number of candidate genes will inevitably be long at the outset of the studies. It can, however, be reduced by choosing only those candidate genes which display genetic variation of functional importance. If these variations are significantly more frequent in NIDDM, they may be regarded as important risk factors for the development of the disease. Obtaining information regarding the existence of genetic variation in these genes requires substantial experimental efforts with the techniques available. Obviously, a limitation of this approach is the fact that the list of candidate genes is compiled on the basis of our present knowledge of the genome. Today, the functions of less than 5% of all mammalian genes have been identified.

LINKAGE ANALYSIS TO IDENTIFY CHROMOSOMAL REGIONS CARRYING CANDIDATE GENES

The identification of chromosomal regions, carrying genes causing susceptibility to the monogenic forms of the disease, is made possible by studies of cosegregation of NIDDM and genetic markers covering the entire genome in large human pedigrees with apparently monogenic inheritance of NIDDM. Such chromosomal regions, encompassing hundreds of genes, are the basis for the cloning and identification of the actual gene predisposing to NIDDM. Linkage analysis has the potential to identify previously unknown genes of importance for glucose homeostasis. When known candidate genes are located in a chromosomal region to which linkage has been obtained, they must clearly be considered as the primary candidate genes in the region. This kind of identification will naturally be greatly facilitated as more genes are characterized during the advancement of the Genome Project.

A linkage study of a large family segregating dominantly inherited NIDDM has been able to identify a chromosomal region close to the adenosine deaminase gene located at the long arm of chromosome 20, as causing diabetes in the family.[13] Further studies of this, and other families, may ultimately lead to the isolation of a hitherto unknown gene with major influence on glucose homeostasis at least in this family.

An alternative genetic approach would be the utilization of animal models to identify genes causing inherited NIDDM and/or glucose intolerance.[14,15] In some animal models, NIDDM is segregating as a monogenic character, in others there is additive or polygenic inheritance of the disease as such, or of traits known to predispose to diabetes, e. g., obesity. Since a considerably increased risk of NIDDM has been observed in first degree relatives of IDDM,[4] animals with IDDM-like disease can serve as tools for identifying new candidate genes for NIDDM. The common susceptibility genes for NIDDM and IDDM are not likely to be involved in the autoimmune process characterizing IDDM, but could be any of the other genes predisposing to IDDM in these animals.

Advantages with animal models include the potential to work with genetically homogeneous inbred strains and unlimited access to several generations of affected offspring. High-resolution linkage mapping of susceptibility genes without the presence of complicating heterogeneity of the disease thus become feasible. In addition, the genetic homogeneity and constant environment achieved by the use of inbred animals, in combination with genetic typing of the entire genomes of offsprings, make it possible to map a limited set of genes with additive effects on quantitative traits influencing glucose tolerance.[16-18] Quantitative traits of importance in NIDDM are glucose and insulin levels during glucose loading. As demonstrated in studies of human diabetes, understanding such traits have been instrumental in dissection of the pathophysiology of NIDDM. Moreover, when chromosomal regions causing diabetes in animals have been mapped, the homologous regions in the human genome can be identified by synteny mapping. Genetic markers for these homologous regions can subsequently be used to investigate whether these regions are important also for the etiology of human NIDDM by sib-pair analysis and population association studies. However, it must be kept in mind that susceptibility genes in animals are not necessarily applicable to the common forms of NIDDM in man.

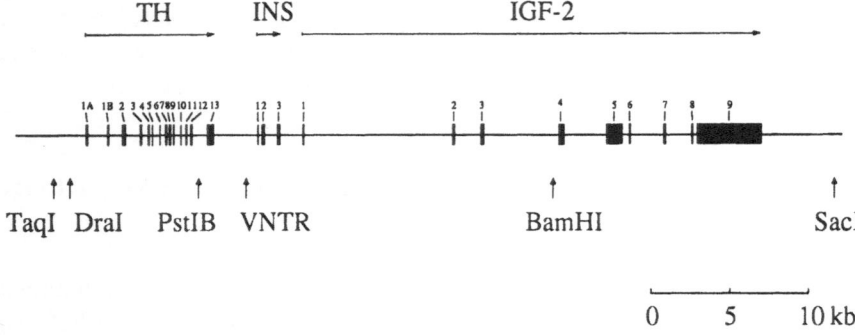

Figure 1. Human tyrosine hydroxylase (TH)/insulin (INS)/insulin-like growth factor 2 (IGF2) chromosomal region located on chromosome 11p15.5. The filled boxes denote exons with their corresponding numbers shown above, and the arrows indicate the direction of transcription. The locations and names of the used polymorphic markers are shown below.

STUDIES OF CANDIDATE GENES WITH SPECIAL INTEREST TO NIDDM

Insulin Gene Region

The first candidate gene accessible for molecular genetic studies in diabetes was the insulin gene. Consequently, many studies have been dealing with the influence of this gene on NIDDM. So far, five different rare mutations of the insulin gene have been identified.[19,20] Three of the mutations were located in the mature insulin molecule. These three mutants showed only 0.2 – 5 % binding to the insulin receptor as compared to normal insulin and confer an increased risk for NIDDM development.

In front of the human insulin gene, a region of variable number of tandem repeats (VNTR) was found adjacent to the well studied insulin promoter/enhancer sequences.[21,22] This VNTR displays three classes of alleles in the population and, therefore, can be used as a marker for the insulin gene region in population association studies of diabetes. In spite of some initial positive reports on association between NIDDM and allele 3 of the insulin 5' VNTR locus the majority of studies have not demonstrated any association with NIDDM.[23] Two independent studies suggest that allele 3 is associated with low insulin response to glucose during sustained glucose loading.[24,25] On the other hand, in newly diagnosed NIDDM patients, no differences were detected in the ability to secrete insulin during oral glucose loading in genotype groups defined by the insulin 5' VNTR.[26] However, several studies demonstrate that allele 1 is associated and linked with IDDM.[27,28]

The chromosomal region encompassing the tyrosine hydroxylase gene, the insulin gene, and the insulin-like growth factor 2 gene was characterized by typing of six different polymorphic markers within the region (Fig. 1). The RFLP patterns were followed by segregation in nuclear families, and 26 haplotype patterns were identified and their distribution compared between healthy individuals and NIDDM patients.[29] No statistically significant differences were found, but five haplotypes were overrepresented in the patient group; and two haplotypes were more abundant in the group of healthy individuals. In conclusion, even if some studies have suggested that genetic variants in the insulin gene itself, or another gene in its vicinity, might have some influence on insulin response to glucose,[30,31] no definite involvement of this chromosomal region has been demonstrated in common NIDDM.

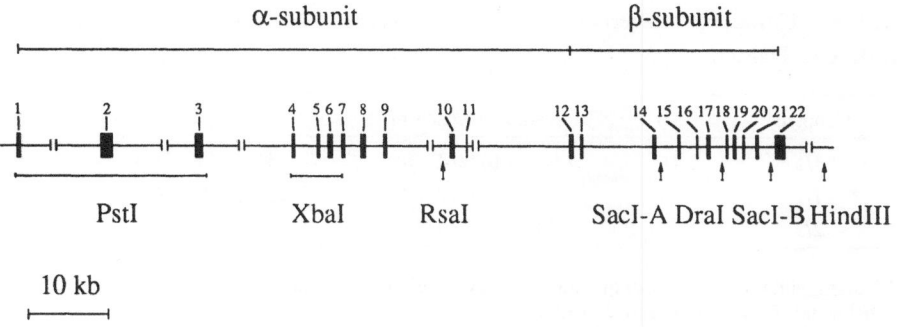

Figure 2. The human insulin receptor gene. Exons encoding the a α- and β-subunits are indicated with bars. The locations and names of the used polymorphic markers are shown below.

Insulin Receptor Gene

Severe insulin resistance due to mutations of the insulin receptor gene has been demonstrated in combination with such rare diseases as Leprechaunism, Rabson-Mendenhall syndrome, and congenital lipodystrophy. To date, more than 35 different mutations have been characterized with severe insulin resistance as phenotype.[32] However, these insulin receptor mutations cannot explain the high frequency of insulin resistance found in the majority of NIDDM patients or obese subjects. In addition, a number of silent mutations and polymorphic positions have been identified but little is known whether they confer any physiological effects.[33] Furthermore, in a number of linkage studies between diabetes as phenotype and the insulin receptor gene no evidence for linkage has been obtained.[34-36] Polymorphic markers (RFLPs) from the insulin receptor gene have also been used for population association studies comparing healthy controls with NIDDM patients. Positive associations, indicating linkage disequilibrium between genetic markers of the insulin receptor and a genetic variation of some functional importance, have been found in some of the investigated populations, but these have been difficult to reproduce in other populations (Fig. 2).[26,37-41] During our investigation of insulin receptor RFLP genotypes in groups of NIDDM patients with different treatments, we noted that patients on diet treatment were overrepresented in the 2/2-genotype group as defined by RsaI (Table 2). The fasting glucose levels were considerably lower in diet treated patients with the 2/2-genotype in comparison with the two other genotypes (Table 3). Given the inherent problems with this kind of exploratory statistics, the observation should be repeated before any firm conclusions can be drawn.

Table 2. Type of treatment in relation to insulin receptor RsaI genotype.

RsaI	Number of individuals		
Genotype	Diet	Oral	Insulin
1/1	14 (0.56)	7 (0.28)	4 (0.16)
1/2	21 (0.54)	8 (0.20)	10 (0.26)
2/2	8 (0.80)	1 (0.10)	1(0.10)

Frequencies are indicated within parenthesis.

Table 3. Clinical characteristics and RsaI genotypes of NIDDM patients with different RsaI genotypes.

Genotype	No	Fasting glucose (mM)		Fasting IRI * (pM)	
1/1	14	7.90	(6.60–9.50)	134	(86–215)
1/2	21	6.90	(6.00–8.70)	122	(86–162)
2/2	8	5.70	(4.63–6.45)	95	(67–125)

Medians with lower and upper quartiles are shown within parenthesis
* IRI stands for immunoreactive insulin

Insulin receptor RNA from the human and the rat is spliced in two alternative forms, including or excluding exon 11 of the human gene.[42-46] The alternatively spliced RNA molecules give rise to two isoforms of the extracellular α-subunit of the mature receptor protein. The two insulin receptor isoforms have been ascribed some differences in activity as measured *in vitro*. The functional importance of these differences in relation to insulin action *in vivo* is not clear. However, we have demonstrated a significant difference in the percentage of Ex 11- RNA between insulin resistant NIDDM patients and healthy controls.[47] It is possible that the synthesis of two insulin receptor forms is a means for fine tuning of receptor function similar to that described for other receptors, e. g., the thyroid hormone receptor.[48] The Ex 11- variant of the insulin receptor has been reported to bind insulin with 1.5–3 times higher affinity,[44,49-51] and in contrast, the Ex 11+ variant has been shown to have 2-fold higher tyrosine autophosphorylation and kinase activity.[50] These differences may suggest that the somewhat better binding characteristics of Ex 11- is a mechanism to reduce the amount of insulin that is available for binding to the insulin receptor variant with higher phosphorylating activity. Further studies of the regulation of the alternatively spliced mRNA species and the function of the two receptor isoforms will be needed to interpret their impact on *in vivo* insulin action.

Glucokinase

Glucokinase was early suggested as the prime "glucose sensor" in the pancreatic ß-cell.[52-55] This proposal was based on the finding that hexoses must be metabolized to elicit insulin secretion, and on the fact that glucokinase has a K_m for glucose within the physiological range. The high K_m will make the phosphorylation entirely dependent on the extracellular glucose concentration, and thus, glucokinase would be a possible pacemaker for glycolysis in the liver and pancreatic ß-cells.

Tight linkage between the glucokinase gene and early onset NIDDM was demonstrated in French families selected for the presence of fasting hyperglycemia in at least two family members.[56] Subsequently, a nonsense mutation in exon 7 of the glucokinase gene in one large pedigree was identified,[57] followed by several reports identifying linkage and glucokinase mutations in families with early-onset diabetes.[58-62] However, linkage analysis of a large number of pedigrees displaying dominant segregation of NIDDM have not been able to demonstrate linkage between the glucokinase gene and NIDDM.[58,63] Thus far, the most successful criteria for selecting families with derangements of the glucokinase gene seems to be early onset (<25 yr.) of mild fasting hyperglycemia and impaired glucose tolerance during oral glucose tolerance tests.[64] In our own experience, the most important clinical findings in a family segregating early-onset NIDDM with a mutation in the glucokinase gene, as compared with age matched offsprings of NIDDM patients in general, were fasting hyperglycemia already during puberty and relatively well controlled diabetes still decades after onset. In another study, concerning patients with early-onset diabetes from four families cosegregating the disease and the glucokinase gene, an attenuated insulin secretion was recorded during continuous glucose infusion.[65] Nevertheless, mutations in the glucokinase gene apparently explain only a minor fraction of all familial NIDDM with the possible exception of American Blacks and Mauritian Creols where associations have been detected between NIDDM and polymorphic markers for the glucokinase gene.[64,66]

AGE-RELATED DEGENERATIVE DISEASES – MITOCHONDRIAL DNA MUTATIONS

Over the last few years, a field of medicine which may be termed "age-related degenerative diseases" has emerged. The common denominators in this group of diseases are derangements of the respiratory chain accompanied by mutations of the mitochondrial DNA (mtDNA). The dysfunctions progress with age, and strike the mitochondria in tissues most reliant on continuous aerobic energy production, such as CNS, cardiac and skeletal muscle, pancreatic islets, kidney, and liver.[67,68] Obviously, the pathophysiology of NIDDM involves a number of these tissues. The pancreatic islets are involved in insulin release, and muscle and liver are the principal sites of insulin sensitivity. We may speculate that, with age, derangement of oxidative phosphorylation due to exogenous and endogenous influences will increase the mutational burden of the mitochondrial genome, which will accelerate the lowering of mitochondrial functions in these tissues and, in turn, lead to hampering of insulin secretion and insulin sensitivity. We may also speculate that the age-related processes may lead to mitochondrial alterations earlier in genetically susceptible individuals bearing inherited shortcomings in the defense and reparation of mitochondrial mutations, alone, or in combination with genetic dysfunctions in key steps of glucose homeostasis. So far, three mutations in mitochondrial DNA – one deletion of 10.4 kilobase- pairs, one partial duplication, and one mutation affecting the transfer RNA^{Leu} gene[69-71] – have been identified in families segregating maternally inherited diabetes as a part of more complex syndromes.

CONCLUSIONS

Studies of the pathogenesis of NIDDM have passed through different stages: from descriptive epidemiology via physiological studies of disease predictors, to studies of the molecular factors determining these characters. Application of molecular genetics to NIDDM has made possible the identification of mutations in candidate genes encoding vital functions in glucose homeostasis. However, these mutations are only relevant to a minor fraction of the NIDDM population. The area of candidate genes has to be widened by studies of additional genes in well selected groups of patients and controls. Our knowledge of this area might be greatly enhanced by unbiased genetic linkage studies in families segregating NIDDM as a monogenetic trait, as well as in the application of animal models with inherited NIDDM-like syndromes.

The well known significance of age on the risk to develop NIDDM can be understood in terms of factors governing the aging process in general. Thus, we may visualize NIDDM as a consequence of inherited limitations of crucial functions in glucose homeostasis in combination with the gradual decline of cellular functions with age. The latter may to a certain extent be ascribed to the incremental accumulation of mutations in the mitochondrial genome.

ACKNOWLEDGEMENTS

This work was supported by the Swedish Medical Research Council, the Knut and Alice Wallenberg Foundation, the Berth von Kantzow Foundation, the Nordic Insulin Foundation, the King Gustaf V and Queen Victoria Foundation, the Magnus Bergvall Foundation, and the Emil and Wera Cornell Foundation. A similar presentation was delivered at the VI World Congress on Diabetes in the Tropics and Developing Countries, 1993, Bombay, India.

REFERENCES

1. Pyke DA, Nelson PG, Diabetes mellitus in identical twins, in: "The Genetics of Diabetes Mellitus," Cruetzfeldt W, Köbberling J, Neel JV, ed., Springer-Verlag, Berlin (1976).
2. Barnett AH, Eff C, Leslie RDG, Pyke DA, Diabetes in identical twins, *Diabetologia.* 20:87-93 (1981).
3. Barnett AH, Eff C, Leslie RDG, Pyke DA, Diabetes in identical twins: a study of 200 Pairs, *Diabetologia.* 17:333-343 (1979).
4. Köbberling J, Tillil H, Empirical risk figures for first degree relatives of non-insulin dependent diabetics, in: "The Genetics of Diabetes Mellitus," Köbberling J, Tattersall R, ed., Academic Press, London (1982).
5. Zimmet P, Type 2 (non-insulin-dependent) diabetes - an epidemiological overview, *Diabetologia.* 22:399-411 (1982).
6. Zimmet P, King H, Taylor R, The high prevalence of diabetes mellitus, impaired glucose tolerance and diabetic retinopaty in Naru- the 1982 survey, *Diabetes Research.* 1:13-18 (1984).
7. Bennett PH, Epidemiology of diabetes mellitus, in: "Diabetes Mellitus: Theory and Practice," Rifkin H, Porte DJ, ed., Elsevier Science Publishing Co., Inc., New York (1990).
8. Cerasi E, Luft R, "What is inherited-what is added", hypothesis for the pathogenesis of diabetes mellitus, *Diabetes.* 16:615-627 (1967).
9. Iselius L, Lindsten J, Morton NE, Efendic S, Cerasi E, Haegermark A, Luft R, Genetic regulation of the kinetics of glucose-induced insulin release in man, *Clin Genetics.* 28:8-15 (1985).
10. Luft R, Cerasi E, Hamberger CA, Studies on the pathogenesis of diabetes in acromegaly, *Acta Endocinol (Copenh).* 56:593-607 (1967).
11. Luft R, Cerasi E, Andersson B, Obesity as an additional factor in the pathogenesis of diabetes, *Acta Endocinol (Copenh).* 59:344-352 (1968).
12. Fajans SS, Scope and heterogeneous nature of MODY, *Diab Care.* 13:49-64 (1990).
13. Bell GI, Xiang K, Newman MV, Wu S, Wright LG, Fajans SS, Spielman RS, Cox NJ, Gene for non-insulin-dependent diabetes mellitus (maturity-onset diabetes of the young subtype) is linked to DNA polymorphism on human chromosome 20q, *Proc Natl Acad Sci USA.* 88:1484-1488 (1991).
14. Shafrir E, Renold AE, ed. "Frontiers in Diabetes Research. Lessons from Animal Diabetes. II," J Libbey, London (1988).
15. Shafrir E, Diabetes in animals, in: "Diabetes Mellitus. Theory and Practice," Rifkin H, Porte D, ed., Elsevier, New York (1990).

16. Lander ES, Botstein D, Mapping medelian factors underlying qantitative traits using RFLP linkage maps, *Genetics*. 121:185-199 (1989).

17. Hilbert P, Lindpaintner K, Beckmann JS, Serikawa T, Soubrier F, Dubay C, Cartwright P, De Gouyon B, Julier C, Takahasi S, Vincent M, Ganten D, Georges M, Lathrop GM, Chromosomal mapping of two genetic loci associated with blood-pressure regulation in hereditary hypertensive rats, *Nature*. 353:521-529 (1991).

18. Jacob HJ, Pettersson A, Wilson D, Mao Y, Lernmark Å, Lander ES, Genetic dissection of autoimmune type 1 diabetes in the BB rat, *Nature Genetics*. 2:56-60 (1992).

19. Haneda M, Polonsky KS, Bergenstal RM, al e, Familial hyperinsulinemia due to a structurally abnormal insulin. Definition of an emerging new clinical syndrome, *N Engl J Med*. 310:1288-1294 (1984).

20. Steiner DF, Tager HS, Chan SJ, Nanjo K, Sanke T, Rubenstein AH, Lessons learned from molecular biology of insulin-gene mutations, *Diab Care*. 13:600-609 (1990).

21. Bell GI, Selby MJ, Rutter WJ, The highly polymorphic region near the human insulin gene is composed of simple tandemly repeating sequences., *Nature*. 295:31-35 (1982).

22. Walker MD, Edlund T, Boulet AM, Rutter WJ, Cell-specific expression controlled by the 5´-flanking region of insulin and chymotrypsin genes, *Nature*. 306:557-561 (1983).

23. Bell GI, Horita S, Karam JH, A polymorphic locus near the human insulin gene is associated with insulin-dependent diabetes mellitus, *Diabetes*. 33:176-183 (1984).

24. Cocozza S, Riccardi G, Monticelli A, Capaldo B, Genovese S, Krogh V, Celentano E, Farinaro E, Varrone S, Avvedimento VE, Polymorphism at the 5' end flanking region of the insulin gene is associated with reduced insulin secretion in healthy individuals., *Eur J Clin Invest*. 18:582-586 (1988).

25. Luthman H, Wedell A, Norgren S, Hamsten A, Luft R, Efendic S, Lindsten J, Human insulin gene 5'VNTR polymorphism: association to blood glucose and insulin response during glucose loadings in non-diabetic subjects., *Manuscript.*.

26. Morgan R, Bishop A, Owens DR, Luzio SD, Peters JR, Rees A, Allelic variants at the insulin-receptor and insulin gene loci and susceptibility to NIDDM in Welsh population, *Diabetes*. 39:1479-1484 (1990).

27. Julier C, Hyer RN, Davies J, Merlin F, Soularue P, Briant L, Cathelineau G, Deschamps I, Rotter JI, Froguel P, Boitard C, Bell JI, Lathrop GM, Insulin-IGF2 region on chromosome 11p encodes a gene implicated in HLA-DR4-dependent diabetes susceptibility, *Nature*. 354:155-159 (1991).

28. Bain SC, Prins JB, Hearne CM, Rodrigues NR, Rowe BR, Pritchard LE, Ritchie RJ, Hall JRS, Undlien DE, Ronningen KS, Dunger DB, Barnett AH, Todd JA, Insulin gene region-encoded susceptibility to type-1 diabetes is not restricted to HLA-DR4-positive individuals, *Nature Genetics*. 2:212-215 (1992).

29. Sten-Linder M, Wedell A, Iselius L, Efendic S, Luft R, Luthman H, DNA polymorphisms in the tyrosine hydroxylase/insulin/insulin-like growth factor II chromosomal region in relation to glucose and insulin responses, *Diabetologia*. 36:25-32 (1993).

30. Olansky L, Janssen R, Welling C, Permutt MA, Variability of the insulin gene in American Blacks with NIDDM - analysis by single-strand conformational polymorphisms, *Diabetes*. 41:742-749 (1992).

31. Olansky L, Welling C, Giddings S, Adler S, Bourey R, Dowse G, Serjeantson S, Zimmet P, Permutt MA, A variant insulin promoter in non-insulin-dependent diabetes-mellitus, *J Clin Invest*. 89:1596-1602 (1992).

32. Taylor SI, Molecular mechanisms of insulin resistance - lessons from patients with mutations in the insulin-receptor gene, *Diabetes*. 41:1473-1490 (1992).

33. O'Rahilly S, Krook A, Morgan R, Rees A, Flier JS, Moller DE, Insulin receptor and insulin-responsive glucose transporter (GLUT-4) mutations and polymorphisms in a Welsh type-2 (non-insulin-dependent) diabetic population, *Diabetologia*. 35:486-489 (1992).

34. Elbein S, Ward W, Beard J, Permutt M, Molecular-genetic analysis and assessment of insulin action and pancreatic b-cell function, *Diabetes*. 37:377-382 (1988).

35. Cox NJ, Epstein PA, Spielman RS, Linkage studies on NIDDM and the insulin and insulin-receptor genes, *Diabetes.* 38:653-658 (1989).

36. Elbein SC, Sorensen LK, Taylor M, Linkage analysis of insulin-receptor gene in familial NIDDM, *Diabetes.* 41:648-656 (1992).

37. Takeda J, Seino Y, Yoshimasa Y, Fukumoto H, Koh G, Kuzuya H, Imura H, Seino S, Restriction fragment length polymorphism (RFLP) of the human insulin receptor gene in Japanese: its possible usefulness as a genetic marker, *Diabetologia.* 29:667-669 (1986).

38. McClain DA, Henry RR, Ullrich A, Olefsky JM, Restriction-fragment-length polymorphism in insulin receptor gene and insulin resistance in NIDDM, *Diabetes.* 37:1071-1075 (1988).

39. Raboudi SH, Mitchell BD, Stern MP, Eifler CW, Haffner SM, Hazuda HP, Frazier ML, Type II diabetes mellitus and polymorphism of insulin-receptor gene in Mexican Americans, *Diabetes.* 38:975-980 (1989).

40. Xiang K-S, Cox NJ, Sanz N, Huang P, Karam JH, Bell GI, Insulin-receptor and apolipoprotein genes contribute to development of NIDDM in Chinese Americans, *Diabetes.* 38:17-23 (1989).

41. Sten-Linder M, Vilhelmsdotter S, Wedell A, Stern I, Pollare T, Arner P, Efendic S, Luft R, Luthman H, Screening for insulin receptor gene DNA polymorphisms associated with glucose intolerance in a Scandinavian population, *Diabetologia.* 34:265-270 (1991).

42. Seino S, Bell GI, Alternative splicing of human insulin receptor messenger RNA, *Biochem Biophys Res Commun.* 159:312-316 (1989).

43. Moller DE, Yokota A, Caro JF, Flier JS, Tissue-specific expression of two alternatively spliced insulin receptor mRNAs in man, *Mol Endocrinol.* 3:1263-1269 (1989).

44. Mosthaf L, Grako K, Dull TJ, Coussens L, Ullrich A, McClain DA, Functionally distinct insulin receptors generated by tissue-specific alternative splicing, *EMBO J.* 9:2409-2413 (1990).

45. Goldstein BJ, Dudley AL, The rat insulin receptor: primary structure and conservation of tissue-specific alternative messenger RNA splicing, *Mol Endocrinol.* 4:235-244 (1990).

46. Goldstein BJ, Dudley AL, Heterogeneity of messenger RNA that encodes the rat insulin receptor is limited to the domain of exon 11, *Diabetes.* 41:1293-1300 (1992).

47. Norgren S, Zierath J, Galuska D, Wallberg-Henriksson H, Luthman H, Differences in the ratio of RNA encoding the A- and B-forms of the insulin receptor between control subjects and NIDDM patients, *Diabetes.* (1993).

48. Brent GA, Moore DD, Larsen PR, Thyroid hormone receptor regulation of gene expression, *Annu Rev Physiol.* 53:17-35 (1991).

49. McClain DA, Different ligand affinities of the 2 human insulin receptor splice variants are reflected in parallel changes in sensitivity for insulin action, *Mol Endocrinol.* 5:734-739 (1991).

50. Kellerer M, Lammers R, Ermel B, Tippmer S, Vogt B, Obermaier-Kusser B, Ullrich A, Häring HU, Distinct α-subunit structures of human insulin receptor-A and receptor-B variants determine differences in tyrosine kinase activities, *Biochemistry.* 31:4588-4596 (1992).

51. Yamaguchi Y, Flier JS, Yokota A, Benecke H, Backer JM, Moller DE, Functional properties of two naturally occurring isoforms of the human insulin receptor in chinese hamster ovary cells, *Endocrinol.* 129:2058-2066 (1991).

52. Randle PJ, Rate of release of insulin in vitro, in: "Ciba Foundation Colloquia Endocrinol," ed., (1964).

53. Matschinsky FM, Ellerman JE, Metabolism of glucose in the islets of Langerhans, *J Biol Chem.* 243:2730-2736 (1968).

54. Meglasson MD, Matschinsky FM, New perspectives on pancreatic islet glucokinase, *Am J Physiol.* 246:E1-13 (1984).

55. Matschinsky FM, Glucokinse as glucose sensor and metabolic signal generator in pancreatic beta-cells and hepatocytes, *Diabetes.* 39:647-652 (1990).

56. Froguel P, Vaxillaire M, Sun F, Velho G, Zouali H, Butel MO, Lesage S, Vionnet N, Clement K, Fougerousse F, Tanizawa Y, Weissenbach J, Beckmann JS, Lathrop GM, Passa P, Permutt MA, Cohen D, Close linkage of glucokinase locus on chromosome 7p to early-onset non-insulin-dependent diabetes mellitus, *Nature*. 356:162-164 (1992).

57. Vionnet N, Stoffel M, Takeda J, Yasuda K, Bell GI, Zouali H, Lesage S, Velho G, Iris F, Passa P, Froguel P, Cohen D, Nonsense mutation in the glucokinase gene causes early-onset non-insulin-dependent diabetes-mellitus, *Nature*. 356:721-722 (1992).

58. Hattersley AT, Turner RC, Permutt MA, Patel P, Tanizawa Y, Chiu KC, Orahilly S, Watkins PJ, Wainscoat JS, Linkage of type-2 diabetes to the glucokinase gene, *Lancet*. 339:1307-1310 (1992).

59. Stoffel M, Froguel P, Takeda J, Zouali H, Vionnet N, Nishi S, Weber IT, Harrison RW, Pilkis SJ, Lesage S, Vaxillaire M, Velho G, Sun F, Iris F, Passa P, Cohen D, Bell GI, Human glucokinase gene - isolation, characterization, and identification of 2 missense mutations linked to early-onset non-insulin-dependent (type-2) diabetes-mellitus, *Proc Natl Acad Sci USA*. 89:7698-7702 (1992).

60. Stoffel M, Patel P, Lo YMD, Hattersley AT, Lucassen AM, Page R, Bell JI, Bell GI, Turner RC, Wainscoat JS, Missense glucokinase mutation in maturity-onset diabetes of the young and mutation screening in late-onset diabetes, *Nature Genetics*. 2:153-156 (1992).

61. Katagiri H, Asano T, Ishihara H, Inukai K, Anai M, Miyazaki J, Tsukuda K, Kikuchi M, Yazaki Y, Oka Y, Nonsense mutation of glucokinase gene in late-onset non-insulin-dependent diabetes-mellitus, *Lancet*. 340:1316-1317 (1992).

62. Sakura H, Eto K, Kadowaki H, Simokawa K, Ueno H, Koda N, Fukushima Y, Akanuma Y, Yazaki Y, Kadowaki T, Structure of the human glucokinase gene and identification of a missense mutation in a Japanese patient with early-onset non-insulin-dependent diabetes-mellitus, *J Clin Endocriol Metabol*. 75:1571-1573 (1992).

63. Cook JTE, Hattersley AT, Christopher P, Bown E, Barrow B, Patel P, Shaw JAG, Cookson WOCM, Permutt MA, Turner RC, Linkage analysis of glucokinase gene with NIDDM in caucasian pedigrees, *Diabetes*. 41:1496-1500 (1992).

64. Permutt MA, Chiu KC, Tanizawa Y, Glucokinase and NIDDM - a candidate gene that paid off, *Diabetes*. 41:1367-1372 (1992).

65. Velho G, Froguel P, Clement K, Pueyo ME, Rakotoambinina B, Zouali H, Passa P, Cohen D, Robert JJ, Primary pancreatic beta-cell secretory defect caused by mutations in glucokinase gene in kindreds of maturity onset diabetes of the young, *Lancet*. 340:444-448 (1992).

66. Chiu KC, Province MA, Permutt MA, Glucokinase gene is genetic marker for NIDDM in American Blacks, *Diabetes*. 41:843-849 (1992).

67. Wallace DC, Mitochondrial genetics: a paradigm for ageing and degenerative diseases?, *Science*. 256:628-632 (1992).

68. Wallace DC, Diseases of the mitochondrial DNA, *Annu Rev Biochem*. 61:1175-1212 (1992).

69. Ballinger SW, Shoffner JM, Hedaya EV, Trounce I, Polak MA, Koontz DA, Wallace DC, Maternally transmitted diabetes and deafness associated with a 10.4 kb mitochondrial DNA deletion, *Nature Genetics*. 1:11-15 (1992).

70. Rötig A, Bessis JL, Romero N, Cormier V, Saudubray JM, Narcy P, Lenoir G, Rustin P, Munnich A, Maternally inherited duplication of the mitochondrial genome in a syndrome of proximal tubulopathy, diabetes-mellitus, and cerebellar ataxia, *American Journal of Human Genetics*. 50:364-370 (1992).

71. van den Ouweland JMW, Lemkes HHPJ, Ruitenbeck W, Sandkuijl LA, de Vijlder MF, Struyvenberg PAA, van de Kamp JJP, Maassen JA, Mutation in mitochondrial transfer RNA(Leu(UUR)) gene in a large pedigree with maternally transmitted type-II diabetes-Mellitus and deafness, *Nature Genetics*. 1:368-371 (1992).

AUTOREGULATION OF GLUCOSE TRANSPORT: EFFECTS OF GLUCOSE ON GLUCOSE TRANSPORTER EXPRESSION AND CELLULAR LOCATION IN MUSCLE

Shlomo Sasson, Yaqoub Ashhab, Danielle Melloul and Erol Cerasi

From the Departments of Endocrinology & Metabolism, Pharmacology and Immunology, Hebrew University Hadassah Medical Center, Jerusalem, Israel

Decreased peripheral utilization of glucose is an important pathogenic mechanism in diabetes. Although it is universally acknowledged that insulin resistance plays a major role in the reduction of glucose consumption by peripheral tissues, and many defects have been described in the function of insulin receptors (see chapter by Olefsky), indirect clinical and in vivo experimental observations suggest that hyperglycemia per se may participate in inducing and/or maintaining a reduced glucose uptake (summarized in chapter by De Fronzo). The idea occurred to us some years ago that a certain analogy may exist between the downregulation of hormone receptors by augmented hormone concentrations, and the reduction of glucose uptake by hyperglycemia. The extraordinary redundancy of compensatory events that operate in vivo make the testing of such a hypothesis near-impossible. We therefore chose to work in vitro, and because muscle is the main glucose consumer of the periphery, we focused on in vitro muscle preparations and myocyte lines.

Glucose Downregulates Muscle Glucose Utilization

To test whether glucose itself, in the absence of hormones or other metabolic effectuators, may modify the utilization of glucose by muscle, rat soleus muscles were isolated, washed, and preincubated for 3 h in the presence of varying glucose concentrations, washed again, and their rate of glucose utilization measured with 5-^3H glucose under standard incubation conditions (1). Figure 1 shows that the preexposure glucose concentration had a dramatic effect on the subsequent glucose utilization rate of the muscle: while glucose withdrawal increased sharply the rate, elevation of the glucose concentration

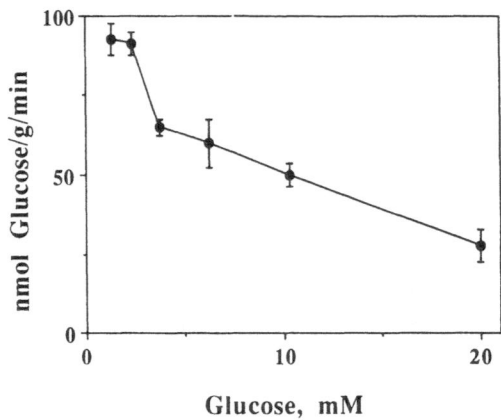

Fig. 1. Dose dependency of glucose effect on the rate of glucose utilization. Isolated rat soleus muscles were preexposed to the indicated glucose concentrations for 3 h, then washed for 5 min in glucose-free Krebs-Ringer bicarbonate buffer. Utilization assays contained 5.0 mM glucose; incubation time was 5 min. Values are mean ± SE (n = 5).

above 4-5 mM gradually reduced the utilization (approximately by 2% per mM increase of the glucose concentration). The soleus muscle's in vitro viability being limited, no sequential incubations could be performed to assess the reversibility of the modulation induced by glucose. Nevertheless, some information could be obtained by pair comparisons of muscles freshly isolated from rats, and those incubated for 2.5 h in vitro: the glucose utilization rate of muscles, measured immediately after excision from rats whose blood glucose level was 9.6± 1.1 mM, was 154±0.7 nM. g^{-1}. min^{-1}, while those of muscles incubated at 1.0 or 17.0 mM glucose were 20.3±1.3 and 12.3±1.5 nM . g^{-1}. min^{-1}, respectively (1).

Glucose Transport is Regulated by Glucose

Glucose utilization is the result of several steps in glucose metabolism. To assess whether the observed changes were mainly due to changes in the transport rate of glucose, soleus muscles were incubated for 3 h with 1.0 or 15.0 mM glucose, then washed, and the transport rate of hexose measured with the glucose analogues 2-deoxyglucose and 3-0-methylglucose. It could be shown that there was excellent parallelism between the modulation of glucose utilization, and the modification of hexose transport rates: the utilization was reduced from 17.1±1.7 to 9.5±1.5 nM. g^{-1}. min^{-1} in muscles incubated at 15 vs. 1 mM glucose; the corres-ponding transport rates were 4.9±0.25 to 2.6±0.07 nM . g^{-1}. min^{-1} for 2-deoxyglucose, and 2.5±0.13 to 1.4±0.13 nM . g^{-1}. min^{-1} for 3-0-methylglucose, indicating that in percentual terms the glucose preincubation affected the hexose transport and utilization rates similarly (around 45% reduction) (1). Together with other data, these findings suggest that, in our experimental conditions, the rate-limiting step in the modulation of muscle glucose utilization by glucose is the transmembrane transport of the hexose.

Whole rat muscles are not convenient systems to study in detail hexose transport kinetics. We therefore used extensively a highly differentiated rat skeletal muscle cell line, the L8 line (2). Also in this system, it was demonstrated that the preexposure glucose concentration modulated the 2-deoxy-glucose uptake rate; furthermore, the rate-limiting step was indeed the transport of glucose. Thus, myocytes modulated by 17 mM glucose for 20 h and subjected to the uptake assay for 10 min with 10 mM 2-deoxyglucose contained 0.75±0.13 mM free hexose, while their concentration of 2-deoxyglucose-6-phosphate was 2.52±0.19 mM; in contrast, the corresponding values for myocytes preincubated at 1 mM glucose were 0.67±0.05 and 10.63±0.32 mM, respectively (3). These findings, together with the demonstration in L8 cells as well as in soleus muscle that also the uptake of the non-phosphorylated glucose analogue 3-0-methylglucose is regulated in a similar manner (1,3), clearly indicate that the transport step, and not hexokinase activity, is modulated by the glucose concentration of the extracellular space.

Mechanism of Glucose-Induced Downregulation of Glucose Transport

The modulatory effect of glucose is not due to a general modification of plasma membrane transport characteristics. As an example, while exposure to high glucose downregulated the myocyte 2-deoxyglucose uptake from 1.26±0.09 to 0.70±0.11 nM/10^6 cells . min, the uptake of the amino acid analogue AIB remained grossly unaltered, the values being 0.83±0.03 vs. 0.75±0.01 nM/10^6 cells min, respectively.

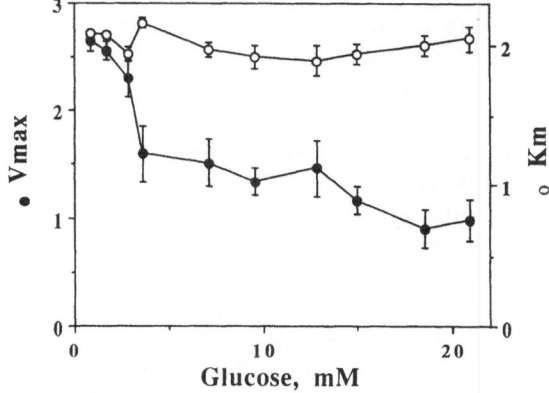

Fig. 2. Effect of preincubation glucose concentration on the Vmax (nmol dGlc/10^6 cells/min) and Km (mM) of 2-deoxyglucose uptake in L8 myocytes. The points represent mean ± SE of 2-15 experiments.

The glucose transport rate can be modulated either by changes in the affinity of the transport system for the hexose, or by changes in the maximal capacity of transport, which reflects the number of transport units. Figure 2 summarizes the results of a large number of experiments in which the kinetic characteristics of hexose transport were defined in myocytes preconditioned with varying glucose levels (3). It is clearly seen that the extracellular glucose concentration had no effect on the affinity of the transport system (Km) for 2-deoxyglucose; in contrast, the Vmax of transport was modulated in a biphasic manner, with a sharp increase on reduction of glucose below 3-4 mM, and a gradual decrease at higher glucose concentrations. Note the similarity between the glucose-dependency of the Vmax of transport in myocytes (Fig. 2) on the

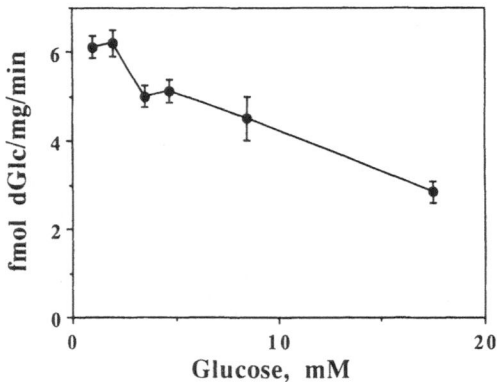

Fig. 3. Effect of preincubation glucose concentration on the uptake of 2-deoxyglucose by soleus muscle. Muscles were preincubated for 3 h at 37°C in DMEM containing 1.2-17.0 mM glucose. After washing, the muscles were taken for the standard uptake assay (0.5 mM 2-deoxyglucose and 33.3 μCi of [3H] 2-deoxyglucose). Mean ± SE (n = 5).

one hand, and that of the glucose utilization rate (Fig. 1) as well as glucose uptake rate (Fig. 3) in the soleus muscle on the other hand. These results strongly suggest that the physiological regulatory mechanism described here, i.e. autoregulation of glucose transport, expresses itself at the level of plasma membrane transporter numbers.

Since the pioneering work of Cushman (4) and Kono (5) it has been well established that the glucose transporter protein exists in a dynamic state in all cells studied to-date, the equilibrium between plasma membrane-located and intracellular transporters being influenced by many factors. Most importantly, insulin induces the translocation of intracellular transporters to the plasma membrane in insulin-sensitive cells (6). The possible existence of such a dynamic situation in muscle glucose autoregulation is suggested by the reversibility of the glucose effect on myocyte hexose transport (Fig. 4). This hypothesis was tested by assessing the number of glucose transporters in the plasma membranes and intracellular membranes (microsomal fraction) of myocytes whose glucose uptake was up- or downregulated by conditioning the cells in the presence of low or high glucose concentrations, respectively, using the cytochalasin B binding assay to identify the total population of glucose transporters (7). Table 1 summarizes the results, but also points to the complexity of the issue. The table shows, as expected, that the hexose uptake of myocytes kept for 25 h at a high glucose concentration was about 50% that of cells kept at low glucose. The specific, glucose-displaceable cytochalasin B binding capacity of the plasma membrane fraction of myocytes conditioned at 2 mM glucose was about 3-fold higher than the binding in the microsomal fraction, while in cells conditioned at 20 mM, binding sites were evenly distributed between the two membrane fractions (column A in Table 1). Also, when the conditioning medium was changed to allow reversal of the autoregulation (columns c and d in Table 1), together with modulation of the glucose uptake rates, cytochalasin B binding was redistributed towards the plasma membranes in myocytes switched to low glucose, and towards the microsomal membranes in cells transferred to high glucose (Table 1 column A). These results would suggest that glucose exerts an

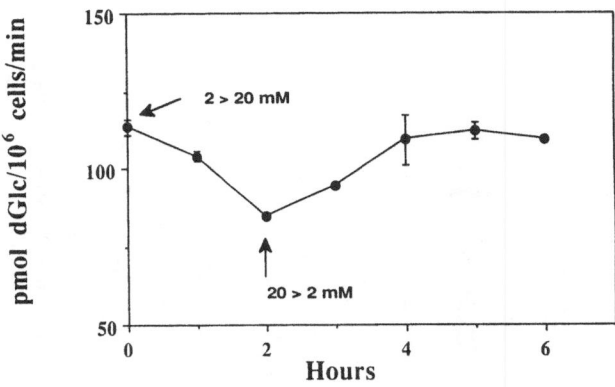

Fig. 4. Time-scale of the glucose effect on muscle glucose uptake. L8 myocytes were incubated overnigh at 2 mM glucose. At 0 time the medium was changed to 20 mM glucose. After a further 2-h incubation, the glucose was again changed to 2 mM. At the end of each incubation period, batches of culture dishes were rinsed, and the [^{3}H] 2-deoxyglucose uptake rate was determined. Mean ± SEM of three experiments (where not shown, the SE was smaller than the size of the symbol).

action opposite to that described for insulin (6), namely reverse translocation when its concentration is high, and recruitment of the transporters to the plasma membrane when glucose is withdrawn (see also chapter by Klip).

The above results were obtained when cytochalasin B binding was expressed in the conventional manner, i.e. per mg membrane protein. However, the methodology for purification of membranes, especially in muscle cells, is far from optimal. We therefore took into consideration the real yields and degrees of purification obtained for each fraction, and recalculated the cytochalasin B binding in terms of total quantities per myocyte, which better reflects the in situ glucose transport capacity of the cell (7). These results are presented in Table 1 column B. The findings in the plasma membrane fraction are qualitatively similar to those of binding expressed per mg protein (column A), albeit with a greater amplitude in the degree of downregulation (56% vs. 37%, and 68% vs. 40% for columns a - b and c - d, respectively). However, the results in the microsomal fractions were surprising. Indeed, whereas the ratio between cytochalasin B binding sites of total plasma membranes and those of microsomal membranes in upregulated myocytes was similar to that calculated per mg of membrane protein, in myocytes exposed to high glucose total binding sites in microsomal membranes were much more abundant. Thus, compared with myocytes exposed to 1-2 mM glucose, down-regulation of transport by 17-20 mM glucose augmented the total microsomal pool of cytochalasin B binding sites 5-6 fold.

Cytochalasin B binds to all glucose transporters. To characterize the molecular species of transporters involved in glucose autoregulation, specific antibodies were utilized. L8 myocytes express the GLUT-4 isoform of transporters at a very low level, therefore we could not evaluate their contribution to this regulatory mechanism. The cellular location of GLUT-1, the main transporter in this cell line, was evaluated by Western blot analysis of the same

TABLE 1. Effect of glucose preconditioning on cellular distribution of cytochalasin B binding sites in L8 myocytes

Preconditioning glucose concentration	2-deoxyglucose uptake (nM/10^6 cells min)	Cytochalasin B binding (pM)			
		A: per mg membrane protein		B: per 10^6 cells	
		PM	MM	PM	MM
a) 2 mM	0.21±0.03	9.8±1.0	3.6±0.5	46.5±4.7	25.6±3.6
b) 20 mM	0.11±0.02	6.2±0.2	6.9±0.4	20.6±0.8	152±9.7
c) 20 mM → 1 mM	0.17±0.03	9.7±2.0	4.1±0.3	54.1±11.3	26.3±2.1
d) 2 mM → 17 mM	0.13±0.03	5.8±1.4	7.3±0.4	17.5±4.2	128.6±6.3

The uptake of [^{3}H]2-deoxyglucose was measured in L8 myocytes incubated as follows: (a) 2 mM glucose for 25 h, (b) 20 mM glucose for 25 h, (c) 20 mM glucose for 20 h followed by 1 mM glucose for 5 h, and (d) 2 mM glucose for 20 h followed by 17 mM-glucose for 5 h. Specific [^{3}H] cytochalasin B binding was measured in the plasma membrane-enriched fractions (PM) and in the microsomal membrane-enriched fractions (MM). Results are expressed in a conventional manner (per mg of membrane protein) under A. Under B, the total glucose-displaceable cytochalasin B binding for each subcellular compartment was calculated from the fold purifications of the respective marker enzyme activities and their recoveries in each membrane fraction. Mean ± SEM of three different experiments.

membranes used for cytochalasin B binding (Fig. 5). The findings in plasma membrane fractions were quantitatively similar to those obtained with cytochalasin B binding, densitometric evaluation of the blots showing that downregulation of transport resulted in 41-51% decrease in the signal, whereas upregulation increased it approximately 2-fold (Fig. 5 a and c). In contrast with the cytochalasin B data, however, the microsomal membranes reacted poorly with the C-terminal GLUT-1 antibodies, distinct signals being obtained only after prolonged exposure of the films (Fig. 5 b). Nevertheless, this signal was also modulated by the glucose conditioning of the myocytes, 5 to 10-fold stronger signals being registered in myocytes whose glucose transport was downregulated by high glucose (Fig. 5 c). Western blots of the total myocyte homogenates did not show significant differences between cells condi-tioned at high or low glucose concentrations (7).

What conclusions can be drawn from these studies? All our data is consistent with the idea that glucose regulates the number of glucose transporters at the plasma membrane compartment of the L8 myocyte stoichiometrically with the regulation of the transport rate. In contrast, Walker et al. (8) found that in L6 myotubes complete glucose withdrawal (in the presence of xylose) resulted in 630% stimulation of glucose uptake in the face of 75% increase in plasma membrane cytochalasin B binding. They therefore concluded that glucose withdrawal affects mainly the intrinsic activity of the

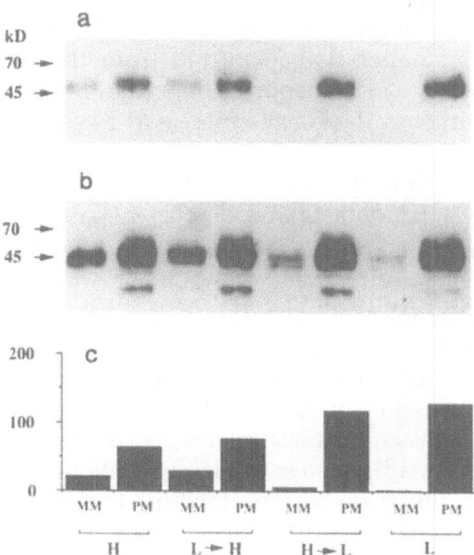

Fig. 5. Western blots of plasma membrane- and microsomal membrane-enriched fractions of L8 myocytes obtained with anti- (GLUT-1) antibodies. L8 myocytes were incubated for 25 h in 2 mM (L) or 20 mM (H) glucose. Other batches were incubated for 20 h in 2 mM glucose, then transferred to medium containing 17 mM glucose for 5 h (L H), and other cultures were incubated for 20 h in 20 mM glucose and switched to 1 mM glucose for 5 h (H L). At the end of the incubation periods, the cells were fractionated to obtain plasma membrane- (PM) and microsomal membrane-enriched fractions (MM), which were subjected to PAGE and Western blotted with polyclonal antibodies aganist the C-terminus of GLUT-1. (a) Short exposure and (b) long exposure of the gels. (c) Laser densitometric reading of the blots.

transporters. The divergence of the techniques utilized (including the use of serum-free medium and prolonged complete glucose starvation by Walker et al.) makes the comparison of the respective data difficult. The full parallelism that we observed between the changes in the rate of hexose transport and in the number of glucose transporters in the plasma membranes, determined either by glucose-displaceable cytochalasin B binding or Western blotting of the GLUT-1 protein under the physiological regulatory conditions of our study, indicates that no significant change in the intrinsic activity of GLUT-1 occurs during autoregulation.

Does reverse translocation (or internalization) of transporters occur during downregulation of hexose transport of muscle cells exposed to high glucose concentrations? This is unlikely, since total cytochalasin B binding sites were not constant (Table 1). Thus, in myocytes incubated for 25 h at 20 mM glucose, cyto-chalasin B binding in the calculated total microsomal pool was 126.4 ± 10.53 pM/10^6 cells higher than in myocytes cultured at 2 mM glucose, whereas the decrease in binding in the total plasma membrane pool was only 25.9 ± 6.42 pM/10^6 cells. These results suggest that in myocytes whose hexose transport was downregulated by high concentrations of extracellular glucose, a decreased rate of transporter translocation from internal pools to the plasma membrane causes intracellular accumulation of transporters and a decrease in their number at the plasma membrane. During upregulation of transport at low glucose, the rate of transporter movement to the plasma membrane is augmented and leads to an appreciable depletion of the intracellular pool. Since the number of cytochalasin B binding sites added to the plasma membrane is smaller than expected from the depletion of the intracellular pool, a higher rate of transporter degradation in the plasma membrane could be partially compensating for the increased rate of translocation.

In accordance with earlier observations, Haney et al. (9) demonstrated recently that in 3T3-L1 fibroblasts and HepG2 hepatoma cells overexpressing human GLUT-1, the transporters are mainly localized at the plasma membrane. The Western blots of L8 myocyte fractions presented here support their observation. It is therefore of great interest that when transporter distribution is assessed by the cytochalasin B binding technique, a different picture emerges. One possibility is that antibodies directed at the C-terminal region of GLUT-1 (as in our study and in that of Haney et al.) underestimate the transporters when they are associated with intracellular vesicles. Such a possibility has been suggested for GLUT-4 in adipocytes: by using antibodies against the C- and N-terminal peptides of the transporter, it could be demonstrated that the C-terminus of GLUT-4 is masked in the intracellular compartment, and becomes available only after translocation to the plasma membrane (10). Our findings suggest that a similar situation may exist for GLUT-1.

It may be questioned to what extent the cellular traficking of GLUT-1 in a muscle cell line relates to the physiological regulation of glucose transport in skeletal whole muscle, which expresses almost exclusively the insulin-regulatable transporter GLUT-4. The striking similarity in the characteristics of autoregulation that exists between the soleus muscle and the L8 cell line (see previous paragraphs in this chapter) definitely suggests that mechanisms similar to those described for the L8 myocyte could be operating in the whole muscle. We have not been able to set up in vitro soleus incubation systems that could allow the purification of plasma membrane fractions for Western blotting, due to the poor in vitro viability of the muscle, and the large amounts of tissue needed for fractionation. However, indirect evidence does suggest that the hypothesis presented here may be valid for skeletal muscle in vivo: indeed, the group of Häring (11) has recently shown that in muscle specimens of hyperglycemic Type 2 diabetics plasma membrane cytochalasin B binding and GLUT-4 density (by Western blotting) are markedly reduced compared to those in non-diabetics. Taking together the information discussed above, we suggest that autoregulatory control of glucose transport is a physiological regulatory system that operates through action on the cellular location of the glucose transporter protein.

Chronic Effects of Glucose on the Muscle Transport System

The molecular biology of glucose transporters has seen a dramatic development over the past few years. A large number of studies have shown that the expression of GLUT-4 in adipocytes is markedly influenced by metabolic factors and hormones; the diabetic state is one typical situation of reduced transporter expression in these cells (for a recent review, see ref. 12). In contrast, the expression of GLUT-4 in skeletal muscle seems to be quite stable,

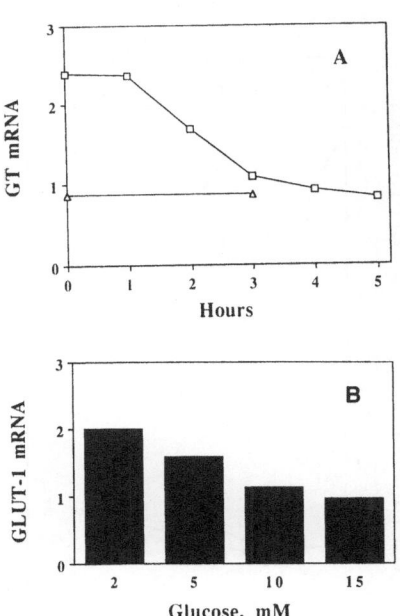

Fig. 6. Downregulation of GLUT-1 mRNA levels in L8 myocytes by glucose.
A- Myocytes were cultured in DMEM medium with 2 mM glucose for 20 h. At this time, the medium was changed to 20 mM glucose, and the mRNAs of GLUT-1 (squares), GLUT-4 (triangles) and ß2-microglobulin measured at the times shown. ß2-microglobulin levels (not shown) remained constant, and were used to normalize the GLUT-1 and GLUT-4 mRNA signals in the dot blots. **B-** Myocytes were cultured for 24 h in DMEM containing the shown concentrations of glucose. GLUT-1 mRNA levels were normalized against the mRNA of ß2-microglobulin, which remained constant (not shown).

most authors finding normal levels of GLUT-4 mRNA and protein in muscles from diabetic animals and patients (12).

In L8 cells, where mainly GLUT-1 is expressed, glucose does influence the expression of this transporter. Figure 6 shows that the mRNA levels of GLUT-1 were about twice as high in myocytes maintained at low glucose levels compared to those in high glucose (13). When the glucose concentration was augmented, GLUT-1 mRNA levels were reduced time- (Fig. 6A) and dose-dependently (Fig. 6B). It is remarkable that the time-course of glucose-induced downregulation of GLUT-1 mRNA is almost superimposable on the time-course of downregulation of hexose transport in these myocytes. However, these two events are not causally related, since the amount of GLUT-1 protein in the myocytes is not reduced within this time-frame (7). Furthermore, as shown in Figure 7, reduction of the glucose concentration upregulated the myocyte GLUT-1 mRNA levels only after a lag period of 8-10 h, while hexose transport was maximally upregulated already at 3 h. Thus, while glucose definitely exerts a chronic effect at the level of GLUT-1 expression, its acute effects on hexose transport are apparently mediated through the post-translational action on cellular location of GLUT-1 (see previous sections of this chapter). In L8 myocytes, the expression of GLUT-4 was not modified by the extracellular glucose concentration (Fig. 7).

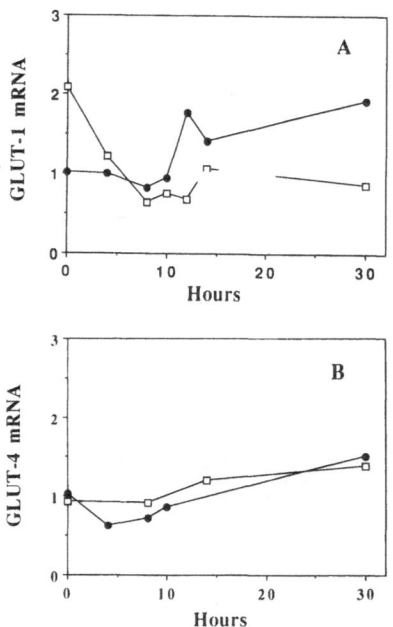

Fig. 7. Chronic effect of glucose concentration on glucose transporter mRNA levels in L8 myocytes. Myocytes were cultured for 20 h in DMEM containing 2 (squares) or 20 mM (circles) glucose. At time zero, the media were changed to contain the opposite glucose concentration (i.e., squares changed from 20 to 2 mM, circles from 2 to 20 mM glucose). At the given times, the mRNA levels of GLUT-1 **(A)**, GLUT-4 **(B)** and ß2-microglobulin (not shown) were determined by slot blot technique. Results are normalized to ß2-microglobulin mRNA values, which were not influenced by the medium glucose concentration.

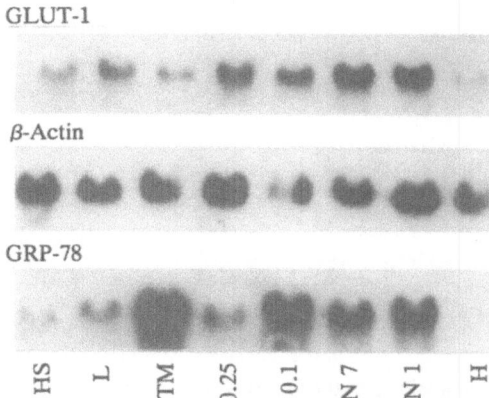

GLUT-1

β-Actin

GRP-78

HS L TM 2-ME 0.25 2-ME 0.1 ION 7 ION 1 H

Fig. 8. Effect of various stress stimuli on the levels of GLUT-1, GRP-78, and ß-actin mRNAs in L8 cells. The cells were incubated for 10 h in DMEM containing 25 mM glucose alone (H) or with the addition of either 1 and 7 μM calcium ionophore A23187 (ION), 0.1% and 0.25% 2-mercaptoethanol (2-ME) or 2 μg of tunicamycin per ml (TM). Additional treatments were glucose deprivation for 24 h (L) and heat shock for 2 h at 42°C (HS). Representative Northern blots are shown.

Why is GLUT-1 ubiquitously expressed, and why is it glucose-regulated? The response of GLUT-1 mRNA levels to glucose withdrawal (Fig. 7) is strongly reminiscent of the regulation of GRP-78, the glucose-regulated stress protein which is a resident of the endoplasmic reticulum (ER) and functions as a molecular chaperone (14). GRP-78 is induced not only by prolonged glucose starvation, but also by agents that deplete the calcium stores of the ER, or disrupt protein glycosylation or sulfhydryl bond formation (14). We wondered whether these stressful stimuli could induce the expression also of GLUT-1. Figure 8 shows that this is indeed the case (15): all the manipulations that induce cellular stress, including glucose starvation, augmented the mRNA levels of GRP-78 and GLUT-1 in parallel, while ß-actin (Fig. 8) and GLUT-4 mRNAs (not shown) were not modified. Another cellular stressor, heat shock, which induces the closely related HSP family of stress proteins but lacks effect on GRP (14), had no effect on GLUT-1 mRNA (Fig. 8). The GRP-like induction of GLUT-1 could be demonstrated in many cell types, and could be blocked by actinomycin D, suggesting that it occurs at the transcriptional level (15). The specificity of the GLUT-1 response was demon-strated in a physiological preparation, the rat soleus muscle: the stress of excision and ex vivo incubation of the muscle caused a striking increase in the GRP-78 and in the initially low GLUT-1 mRNA levels, while the predominant species of transporter mRNA, GLUT-4, was not modified (15).

What role does GLUT-1 play in cellular stress? Although the type of severe stress induced in our experiments is known to inhibit the translation of most mRNAs (16), the production of the highly-glycosylated (55 kDa) species of GLUT-1 was markedly stimulated (Fig. 9). The GLUT-1 proteins were apparently correctly inserted into the plasma membrane, since the glucose uptake of these myocytes was stimulated 2-3 fold (15).

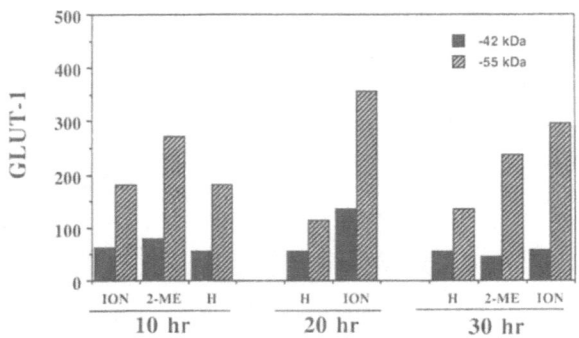

Fig. 9. Western blot analysis of the GLUT-1 protein in L$_8$ myocytes. The myocytes were incubated for 10 h in DMEM containing 10% FCS and 25 mM glucose (H) with or without 1 μM calcium ionophore (ION) or 0.1% 2-mercaptoethanol (2-ME). Following stiimulation, cells were washed and further incubated for 14 h or 20 h in fresh DMEM with glucose and FCS alone. Quantitation of the GLUT-1 protein was performed by scanning densitometry and is expressed in arbitrary units.

GRPs are induced by the presence of malfolded proteins and are bound to prefolded and malfolded proteins that accumulate in the ER (14). It is therefore accepted that GRPs act as chaperones in the assembly of exocytotic proteins, both during normal synthesis and under conditions that promote protein denatu-ration and aggregation. Although regulated by the concentration of glucose in the culture medium, no GRP has been shown to be involved in glucose transport through the plasma membrane. Our demonstration that mature GLUT-1 molecules are correctly synthesized under these conditions and lead to the sustained (at least 20 h) stimulation of hexose uptake (15) indicates that GLUT-1 plays an important role in recovery from cellular stress. Indeed, the increased glucose uptake may provide ATP for GRPs, whose chaperone function is activated through their intrinsic ATPase activity (14). In principle, the stimulation of any isoform of tissue specific glucose transporter could fullfil the needs for increased glucose uptake; however, evolutionary economic pressures may have forced many cell types to rely upon only few genes under conditions endangering cell survival. Our findings suggest that GLUT-1 is one of these genes.

Transcriptional Regulation of GLUT-1 -- Some Considerations

The GLUT-1 protein is structurally homologous to the other transporter proteins, including GLUT-4 (17), but shows no structural homology to GRPs. The latter are related to the family of the HSPs; yet, despite the structural similarity, GRPs and HSPs are not regulated in a coordinated manner. Similarly, GLUT-1 and GLUT-4 fail to respond to identical stimulations, whereas similar stimuli seem to coinduce the structurally unrelated GLUT-1 and GRP-78 proteins. However, GLUT-1 is not the only structurally dissimilar protein that responds to cellular stress. For example, it was found that trypsin gene expression is stimulated in an exocrine pancreatic cell line by glucose withdrawal or by calcium ionophores, while the same treatments reduced the mRNA levels of other cell-specific proteins like chymotrypsin and amylase (18). The induced expression of the GRP-78 and trypsin genes was found at the level of transcription and controlled by DNA sequences located in the 5' flanking region of the genes.

Such observations necessarily imply that there may exist common regulatory elements in the 5' flanking regions of these genes. Indeed, we observed ~60% sequence identity in a ~500 nt 5' flanking region of the GLUT-1 and trypsin genes, and a consensus sequence, **5' GGAGGAGGCGCTT 3'**, was found in GRP-78 (-164 to -177), GLUT-1 (-609 to -621) and trypsin (-406 to -419) 5' flanking regions. However, the biological significance of this type of sequence matching needs to be tested.

Another approach would be to select regulatory regions shown experimentally to convey stressor responsiveness to GRP-78, and search for candidate motifs common to chaperones and GLUT-1. Recent work by Lee and colleagues (19) has shown the central role played by multiple CCAAT motifs located in the promoter of GRP-78 within 160 nt from the transcription initiation site for the stress-inducibility of the gene. Taking into consideration the importance ascribed to the CCAAT box in metabolic control (20), this observation could be relevant also for GLUT-1. Indeed, the GLUT-1 promoter contains two CCAAT boxes, at -49 and -193. In addition to GRP-78, two other major proteins of the ER lumen, GRP-94 and protein disulfide isomerase (PDI), are regulated by cellular stressors (14). GRP-78 and GRP-94 are related proteins which are regulated in concert (14); also PDI is induced by similar stimuli, and its promoter region has recently been described (21). All these genes, including GLUT-1, have multiple copies of the CCAAT motif, and are GC-rich in the neighbouring sequences.

The above description relates to preliminary work towards understanding the unique regulation of GLUT-1 expression. It has to be stressed, however, that we have focused on relatively limited 5' flanking regions, and that functionally significant sequences may exist in distant 5' as well 3' regions in mammalian genes. As an example, recently Murakami et al. (22) identified an enhancer element within the second intron of the GLUT-1 gene. Bearing in mind these limitations, work is in progress with the construction of chimeric genes carrying some of the sequences described above for transfection into L8 myocytes and other cell lines. Future work may hopefully lead to means applicable in vivo for the upregulation of GLUT-1 expression in tissues of physiological significance, such as skeletal muscle.

Acknowledgments

We are grateful to Ms. Rachel Oron for the long-standing devoted and expert technical help provided. Our sincere thanks are also due to Ms. Ina Perlov, The Hebrew University of Jerusalem, for the guidance given in the use of the GCG program. These studies were supported by grants from the Wolfson Foundation, the US-Israel Binational Science Foundation, and the Israel Academy of Sciences.

References

1. S. Sasson, D. Edelson, and E. Cerasi, In vitro autoregulation of glucose utilization in rat soleus muscle, *Diabetes*. 36:1041 (1987).

2. D. Yaffe, Cellular aspects of muscle differentiation in vitro, *Curr Top Dev Biol*. 4:37 (1969).

3. S. Sasson, and E. Cerasi, Substrate regulation of the glucose transport system in rat skeletal muscle: characterization and kinetic analysis in isolated soleus muscle and skeletal muscle cells in culture, *J Biol Chem*. 261:16827 (1986).

4. S.W. Cushman, and L.J. Wardzala, Potential mechanism of insulin action on glucose transport in the isolated rat adipose cell, *J Biol Chem*. 255:4758 (1980).

5. K. Suzuki, and T. Kono, Evidence that insulin causes translocation of glucose transport activity to the plasma membrane from an intracellular storage site, *Proc Natl Acad Sci. USA* 77:2542 (1980).

6. G.D. Holman, I.J. Kozka, A.E. Clark, C.J. Flower, J. Saltis, A.D. Habberfield, I.A. Simpson, and S.W. Cushman, Cell surface labeling of glucose transporter isoform GLUT 4 by bis-mannose photolabel: correlation with stimulation of glucose transport in rat adipose cells by insulin and phorbol ester, *J Biol Chem*. 265:18172 (1990).

7. R. Greco-Perotto, E. Wertheimer, B. Jeanrenaud, E. Cerasi, and S. Sasson, Glucose regulates its transport in L8 myocytes by modulating cellular trafficking of the transporter GLUT-1, *Biochem J*. 186(1): 157 (1992).

8. P.S. Walker, T. Ramlal, V. Sarabia, U.M. Koivisto, P.J. Bilan, J.E. Pessin, and A. Klip, Glucose transport activity in L_6 muscle cells is regulated by the coordinate control of subcellular glucose transporter distribution, biosynthesis, and mRNA transcription, *J Biol Chem*. 265:1516 (1990).

9. P.M. Haney, J.W. Slot, R.C. Piper, D.E. James, and M. Mueckler, Intracellular targeting of the insulin-regulatable glucose transporter (GLUT-4) is isoform specific and independent of cell type, *J Cell Biol* 114:689 (1991).

10. R.S. Smith, M.J. Charron, N. Shah, H.F. Lodish, and L. Jarrett, Immunoelectron microscopic demonstration of insulin-stimulated translocation of glucose transporters to the plasma membrane of isolated rat adipocytes and masking of the carboxyl-terminal epitope of intracellular Glut 4, *Proc Natl Acad Sci USA* 88:6893 (1991).

11. B. Vogt, C. Mühlbacher, J. Carrescosa, B. Obermaier-Kusser, E. Seffer, J. Mushack, D. Pongratz, and H.U. Häring, Subcellular distribution of GLUT 4 in the skeletal muscle of lean Type 2 (non-insulin-dependent) diabetic patients in the basal state, *Diabetologia*. 35:456 (1992).

12. W.T. Garvey, Glucose transport and NIDDM, *Diabetes Care*. 15:396 (1992).

13. E. Wertheimer, Autoregulation of glucose uptake in skeletal muscle cells. Ph.D. Thesis, The Hebrew University of Jerusalem (1991).

14. A.S. Lee, Mammalian stress response: induction of the glucose-regulated protein family, *Curr Opinion Cell Biol*. 4:267 (1992).

15. E. Wertheimer, S. Sasson, E. Cerasi, and Y. Ben-Neriah, The ubiquitous glucose transporter GLUT-1 belongs to the glucose regulated protein family of stress-induced proteins, *Proc Natl Acad Sci USA*. 88:2525 (1991).

16. D.G. Macejak, and P. Sarnow, Internal initiation of translation mediated by the 5' leader to a cellular mRNA, *Nature*. 353:90 (1991).

17. G.I. Bell, T. Kayano, J.B. Buse, C.F. Burant, J. Takeda, D. Lin, H. Fukumoto, and S. Seino, Molecular biology of mammalian glucose transporters, *Diabetes Care*. 13: 198 (1990).

18. C. Stratowa, and W.J. Rutter, Selective regulation of trypsin gene expression by calcium and by glucose starvation in a rat exocrine pancreas cell line, *Proc Natl Acad Sci USA*. 83:4292 (1986).

19. S.K. Wooden, L-J. Li, D. Navarro, I. Qadri, L. Pereira, and A.S. Lee, Trans-activation of the *grp 78* promoter by malfolded proteins, glycosylation block, and calcium ionophore is mediated through a proximal region containing a CCAAT motif which interacts with CTF/NF-I, *Mol Cel Biol*. 11:5612 (1991).

20. D. Cheneval, R.J. Christy, D. Geiman, P. Cornelius, and M.D. Lane, Cell-free transcription directed by the 422 adipose P2 gene promoter: Activation by the CCAAT/enhancer binding protein, *Proc Natl Acad Sci USA*. 88:8465 (1991).

21. K. Tasanen, J. Oikarinen, K.I. Kivirikko, and T. Pihlajaniemi, Promoter of the gene for the multifunctional protein disulfide isomerase polypeptide. Functional significance of the six CCAAT boxes and other promoter elements, *J Biol Chem*. 267:11513 (1992).

22. T. Murakami, T. Nishiyama, T. Shirotani, Y. Shinohara, M. Kan, K. Ishii, F. Kanai, S. Nakazuru, and Y. Ebina, Identification of two enhancer elements in the gene encoding the type 1 glucose transporter from the mouse which are responsive to serum, growth factor, and oncogenes, *J Biol Chem*. 267:9300 (1992).

INSULIN RESISTANCE AND THE PATHOGENESIS OF NON-INSULIN DEPENDENT DIABETES MELLITUS: CELLULAR AND MOLECULAR MECHANISMS

Jerrold M. Olefsky

University of California, San Diego
Department of Medicine
9500 Gilman Drive
La Jolla, CA 92093
and
Veterans Administration Medical Center
Medical Service (111G)
3350 La Jolla Village Drive
San Diego, CA 92161

INTRODUCTION

Non-insulin dependent diabetes mellitus (NIDDM) is a complex metabolic disorder of heterogeneous etiology (1-4). There is clearly a strong hereditary component to the disease, but the exact genetic abnormalities are likely to differ among different population groups (1,5). In addition, NIDDM is likely to be multigenic, in that more than one discrete gene defect needs to complement before the NIDDM phenotype manifests. Thus, from a genetic point of view, NIDDM is heterogeneous and polygenic making identification of "diabetes" genes particularly difficult. Numerous biochemical abnormalities have been identified in NIDDM and the relative contribution of different physiologic or cellular defects differs among different patient groups (6,7). Regardless of the exact pathophysiologic sequence in a particular patient, once full blown fasting hyperglycemia develops, a characteristic set of metabolic derangements can be identified in the great majority of NIDDM patients (Fig. 1). This consists of abnormalities at the level of the pancreatic islets, the liver, and peripheral insulin target tissues which, taken together, represent a final common metabolic pathway for the pathogenesis of hyperglycemia (1,8).

GENERAL OVERVIEW

Figure 1 summarizes these abnormalities. and represents NIDDM patient. To begin at the hepatic level, the role of

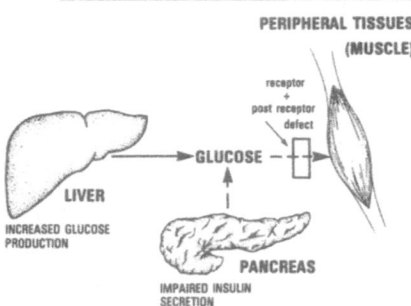

FIGURE 1. Summary of the metabolic abnormalities in NIDDM which contribute to the hyperglycemia. Increased hepatic glucose production, impaired insulin secretion, and insulin resistance due to receptor and postreceptor defects all combine to generate the hyperglycemic state. From Ref. 1.

the liver in the pathogenesis of NIDDM is overproduction of glucose. Increased basal hepatic glucose production is a characteristic feature of essentially all NIDDM patients with fasting hyperglycemia (9-12). The Figure depicts skeletal muscle as the prototypical peripheral insulin target tissue, since in the in vivo insulin stimulated state, 80-90% of all glucose uptake is into skeletal muscle. Target tissues are insulin resistant in NIDDM, and this has been well described in most (1-3,9-17), but not all (4), population groups. Lastly, abnormal islet cell function plays a central role in the development of hyperglycemia in NIDDM. Decreased beta cell function, and increased glucagon secretion are frequent concomitants of the NIDDM state (4,18). Taken together, abnormalities in these three organ systems account for the NIDDM syndrome. In the following sections, each of these abnormalities will be considered in further detail.

Increased Hepatic Glucose Output: Basal hepatic glucose production rates (HGO) can be measured in man by the primed continuous infusion of ^{3}H-glucose in the post-absorptive state. Figure 2 depicts results of earlier studies (9) on this subject demonstrating a marked increase in HGO in NIDDM patients with fasting hyperglycemia compared to controls. The magnitude of the increase in HGO was comparable whether or not the NIDDM subjects were obese, and importantly, no increase in HGO was observed in patients with simple impaired glucose tolerance. Over the years, this observation has been verified in many different laboratories (9-12) and appears to hold across a large variety of NIDDM subpopulations. When individual data are examined (Figure 3), a striking correlation is noted between the increase in HGO and the height of the fasting glucose level. This relationship strongly suggests that it is the increase in HGO

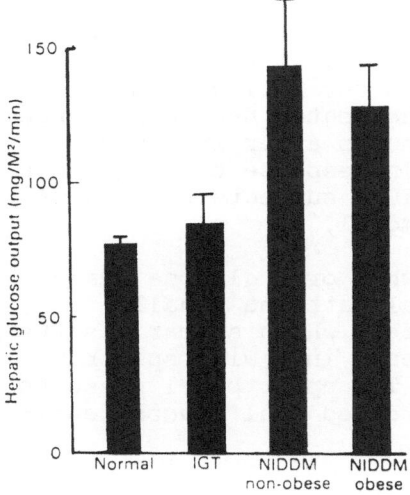

FIGURE 2. Rates of hepatic glucose production in the basal state (7 to 9 AM following an overnight fast) in normal subjects, subjects with impaired glucose tolerance, and obese or nonobese subjects with NIDDM. HGO is normal in patients with IGT but is markedly increased in NIDDM. From Ref. 9.

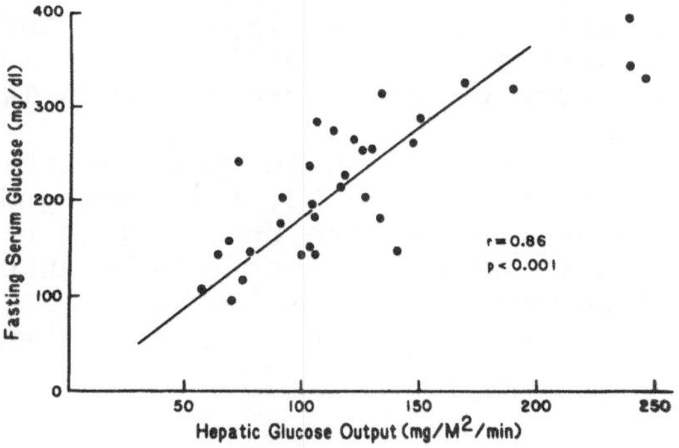

FIGURE 3. Relationship between individual hepatic glucose production rates and fasting plasma glucose level in NIDDM subjects. From Ref. 9.

(1,9). This conclusion is particularly compelling, when one considers that in the basal state approximately 80% of total glucose disposal is through non-insulin mediated tissues such as the CNS. Consequently, since only a minority of basal glucose uptake involves insulin mediated pathways, a decrease in insulin mediated glucose uptake (due to decreased insulin secretion, insulin resistance, or both) would have only a small effect on the plasma glucose level (1). Consequently, overproduction of glucose is the culprit, and the liver is the responsible organ. The mechanisms of increased HGO are complex, but likely involve some combination of: (1) hepatic insulin resistance, (2) increased glucagon secretion, (3) increased FFA levels, and (4) increased gluconeogenic precursor supply. Regardless of the precise mechanism, it is likely that the hepatic abnormalities are secondary, since they are fully reversible and can be corrected by several forms of anti-diabetic therapy (19-21).

Insulin Secretion: This subject is discussed in more detail elsewhere in this symposium and no attempt at a comprehensive review will be attempted. Rather, only a couple of general issues will be illustrated. Stimulated insulin levels can be low, normal, or high in NIDDM, depending on the type of stimulus employed, the degree of obesity, and the severity of the diabetes (see Table 1 in Ref. 22 for review). The most uniform deficits occur following intra-venous glucose, particularly when the acute insulin response (insulin release in the first 10 minutes after IV glucose) is measured. The first phase insulin response to IV glucose is generally completely absent in NIDDM subjects with fast-ing hyperglycemia greater than 140 mg/dl.

The situation is more variable when oral glucose chal-lenges are employed, but some general patterns usually emerge. These are depicted in Figure 4 which summarizes the results of oral glucose tolerance tests in a wide spectrum of normal and NIDDM subjects (23). The upper panel presents the glucose levels following standardized oral glucose chal-

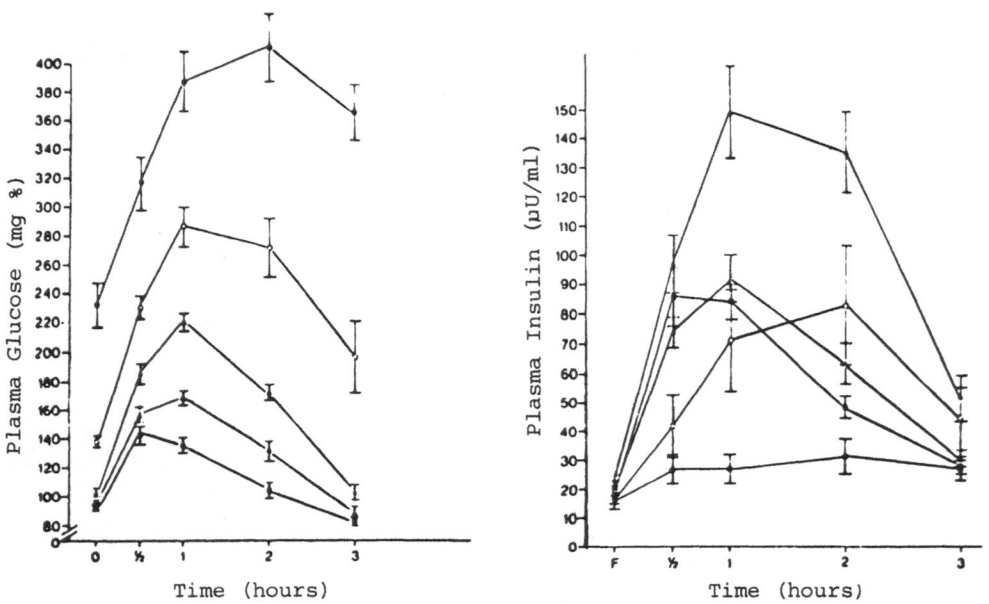

FIGURE 4. (A) Mean (+ SE) plasma glucose response to oral glucose in the five patient groups (●) = normal; (▲) = borderline tolerance; (■) = impaired glucose tolerance; (0) = fasting hyperglycemia (110 to 150 mg/dl); (●) = fasting hyperglycemia (> 150 mg/dl). (B) Mean (+SEM) plasma insulin response to oral glucose in the five patient groups. Symbols the same as in A. From Ref. 23.

lenges in subjects ranging from very normal, to mild dia-betes, to severe NIDDM. The lower panel presents the cor-responding insulin profiles and, as can be seen, insulin levels are higher than normal when a very mild state of glucose intolerance exists, but progressively fall as the degree of hyperglycemia worsens and in the group of NIDDM patients with substantial fasting hyperglycemia, insulin secretion is markedly reduced compared to controls. Although these are crosssectional studies, if one thinks of them logitudinally, the notion could be advanced that early on in the natural history of the development of NIDDM, hyperinsulinemia exists, whereas in later stages of this disease β cell function is markedly subnormal. In NIDDM, the defect in β cell function is relatively (but not completely) specific for glucose stimuli. When mixed meals are used as the insulinogenic stimulus, the absolute insulin levels are higher in the NIDDM subjects compared to the values during OGTTs (24), although the general trend of hyperinsulinemia accompanying mild glucose intolerance and hypoinsulinemia in severe NIDDM still obtains.

Peripheral Insulin Resistance: Insulin resistance can be studied in vivo by means of the euglycemic glucose clamp technique (25). Using this technique, dose response curves for insulin stimulation of in vivo glucose disposal can be constructed, and Figure 5 summarizes such dose response studies in a group of control, IGT, and NIDDM subjects (9).

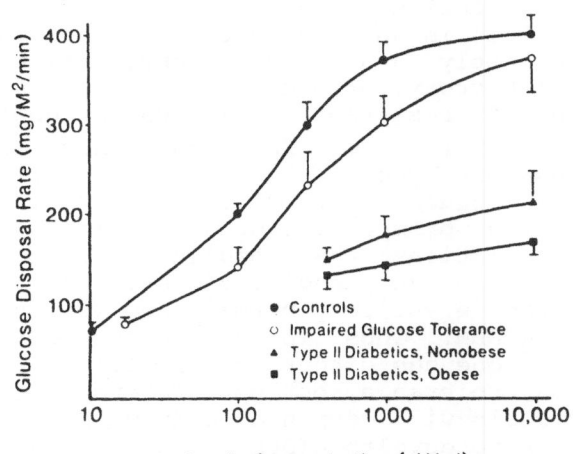

Insulin Concentration (μU/ml)

FIGURE 5. Mean dose-response curves for control subjects (closed circles), subjects with impaired glucose tolerance (open circles), and nonobese (closed triangles) and obese (closed squares) Type II diabetics. Data are the mean (+SEM) glucose disposal rates from multiple euglycemic glucose clamp studies performed at the indicated steady state plasma insulin levels. From Ref. 9.

Glucose disposal rates were markedly reduced in the NIDDM subjects (mean fasting glucose levels 224 mg/dl), compared to controls at all insulin levels and, although the insulin resistance was somewhat more severe in the obese NIDDM subjects, the non-obese subjects were markedly insulin resistant. In the IGT group, the insulin resistance can be clearly demonstrated, although it is not nearly as severe as in the NIDDM subjects, and consists of a rightward shift in the dose response curve with no significant change in the maximal response to insulin (9).

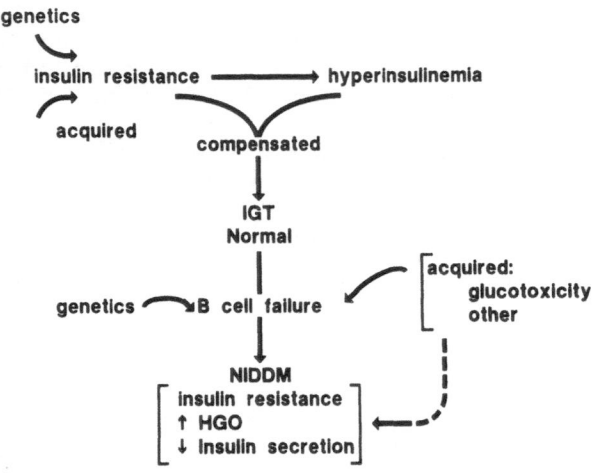

FIGURE 6. Proposed etiology for the development of NIDDM.

Etiology of NIDDM: The pathophysiologic findings depicted in Figure 1 represent a single point in time, after the full NIDDM syndrome has developed. However, such an analysis does not tell us about the progressive evolution of this disease. This is an area which has received considerable attention lately, and Figure 6 represents a schematic for the natural history, or progression, towards NIDDM. One starts with insulin resistance, which can be genetic, or acquired, or both. If β cell function is normal, this will lead to compensatory hyperinsulinemia which will maintain relatively normal glucose metabolism. The major evidence for this comes from prospective studies which have shown, in a variety of populations, that insulin resistance and hyperinsulinemia exist in the pre-diabetic state at a time when glucose tolerance is entirely normal, and well before the onset of frank NIDDM. Thus, in the compensated insulin resistant state, one has either normal glucose tolerance or impaired glucose tolerance, but not diabetes. A subpopulation of individuals with compensated insulin resistance eventually go on to develop NIDDM. The magnitude of this subpopulation depends on the particular ethnic groups studied and the means of assessment. During the transition from the compensated state to frank NIDDM, at least three pathophysiologic changes can be observed. First, is a marked fall in β cell function and insulin secretion. Whether this is due to preprogrammed genetic abnormalities in β cell function, or acquired defects (such as glucotoxicity), or both, remains to be elucidated. Nevertheless, a marked decrease in insulin secretion accompanies this transition and is clearly a major contributor to the development of NIDDM. A second metabolic change is at the level of the liver. Patients with IGT have normal basal rates of HGO, whereas patients with fasting hyperglycemia have increased HGO. Thus, the capacity of the liver to overproduce glucose is an important contributory factor (albeit secondary) to the pathogenesis of NIDDM. Finally, many, but not all studies, have indicated that patients with NIDDM are more insulin resistant that those with IGT. Whether this increment in insulin resistance is secondary to glucotoxicity or other acquired factors remains to be determined.

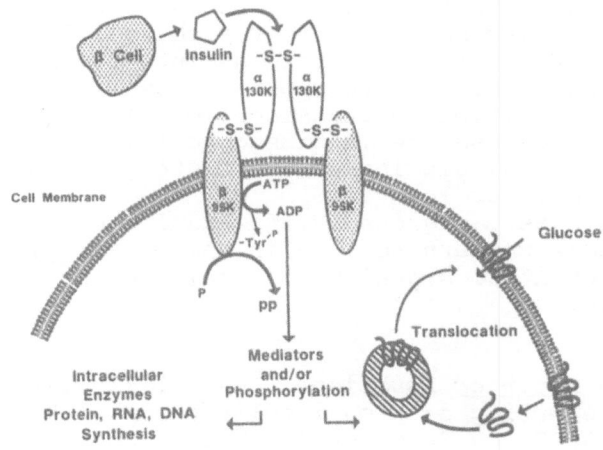

FIGURE 7. Model of cellular insulin action.

Cellular Mechanisms of Insulin Resistance: Before considering the mechanisms of insulin resistance in NIDDM, it is useful to briefly review some of our current knowledge about insulin action (Figure 7). The first step in insulin's cellular action involves binding to its cell surface receptor. The insulin receptor is a heterotetrameric protein composed of two identical alpha and two identical beta subunits (26-29). The alpha subunits are entirely extracellular and they are responsible for insulin binding. The beta subunits are transmembrane proteins containing extracellular, transmembrane, and cytoplasmic domains. All four subunits are covalently linked together by disulfide bonds. After insulin binding to the alpha subunits, a signal is transmitted through the membrane, to the cytoplasmic beta subunit domain. This domain of the beta subunit contains an intrinsic tyrosine kinase activity, and after insulin binding, the beta subunit undergoes autophosphorylation at characteristic tyrosine residues (26-30). Once the receptor is autophosphorylated, its intrinsic tyrosine kinase catalytic activity is markedly enhanced and it now can phosphorylate endogenous protein substrates (26-32). Putative endogenous substrates are then phosphorylated on tyrosine residues initiating a phosphorylation cascade and/or subsequent steps of insulin action. It is possible that other signaling events occur, independent of tyrosine phosphorylation, which mediate or modulate some of insulin's actions.

With respect to NIDDM, one of the most important effects of insulin is to stimulate glucose disposal. Under most physiologic circumstances, glucose transport is rate limiting for overall glucose disposal (33-35) and therefore, the ability of insulin to increase cellular glucose transport has received a great deal of attention. Based on the work of several laboratories, it is well established that there are at least 5 species of glucose transporters with tissue specific distribution (36,37). One of these species, termed the "insulin sensitive glucose transporter", or Glut 4, is highly expressed only in skeletal muscle, adipose tissue, and cardiac muscle, and is the unique transporter responsible for most of insulin's effects to stimulate overall glucose uptake (36,37). In the unstimulated state, most of the Glut4 proteins are located in an intracellular vesicular

pool. Upon insulin stimulation, recruitment or translocation of these glucose transporter-rich vesicles occurs,
causing insertion of Glut4 proteins into the plasma membrane
where they now begin to transport glucose into the cell
(38,39). In addition to translocation, insulin may also
activate or increase the intrinsic activity of Glut4. Once
the insulin signal dissipates the Glut4 proteins return to
their intracellular location.

With this as background, lets examine some of our current knowledge concerning insulin receptors, glucose transport, and potential signaling molecules which may couple the
two in NIDDM.

Insulin Receptor Function: Several studies have been
completed reporting data on insulin receptor kinase activity
in NIDDM (40-45).   In general, this is done by isolating
receptors from human tissues and then measuring receptor
kinase activity in a cell free environment. Results from
such a study are presented in Figure 8 (40). The upper
panel represents data on autophosphorylation and the lower
panel depicts the ability of receptors to phosphorylate an
exogenous substrate Glu:Tyr4:1. As can be seen, receptors
isolated from NIDDM patients have severely compromised auto
phosphorylation/kinase activity. Interestingly, receptors

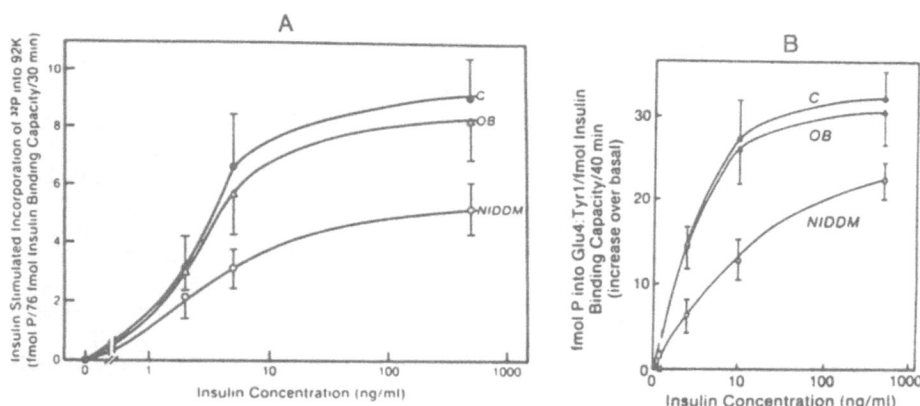

FIGURE 8.   (A)   Insulin dose response of autophosphorylation. Aliquots of receptor preparations were preincubated
with zero, 1.5, or 500 ng/ml insulin after which autokinase
reactions were conducted for 30 minutes at 4°C. The results
are graphed as the mean (+SEM) increase over basal from 10
control (C, solid circle), 13 obese (OB, open triangle), and
13 NIDDM (open circle) subjects.   (B) Insulin-stimulated
phosphorylation of Glu4:Tyr1 by receptors from control,
obese, and NIDDM subjects. Insulin receptors were preincubated with zero, 1.5, 10, or 500 ng/ml unlabeled insulin in
the absence or presence of 0.5 ng/ml ^{125}I-insulin. Insulin
binding and Glu4:Tyr1 phosphorylation were determined.
From Ref. 40.

isolated from insulin resistant, obese, non-diabetic subjects have perfectly normal kinase activity. Similar findings have been observed using receptors isolated from liver
(41), skeletal muscle (42,43), and erythrocytes (44) in
NIDDM subjects. Thus, this receptor kinase defect appears
to be generalized to all insulin target tissues and relatively specific for the hyperglycemic insulin resistant
state seen in NIDDM. Further studies have demonstrated that
this kinase defect is associated with an increased proportion of receptors which are kinase inactive (45,46).

Since NIDDM has a strong genetic component, and since
receptor function is deranged in this disease, it is important to ask whether genetic variations in the insulin receptor are a primary etiologic factor accounting for NIDDM. In
other words, is the insulin receptor a diabetes gene? Several lines of evidence argue strongly that the answer is no
(47-50). Firstly, when obese NIDDM patients are induced to
lose weight, the hyperglycemia is markedly improved, and
insulin receptor kinase function reverts to normal (47).
Secondly, when fibroblasts are propagated in tissue culture,
so that they are several generations removed from the in
vivo milieu, insulin receptor kinase activity in fibroblasts
derived from NIDDM subjects is normal (48). Thirdly, and
perhaps most compellingly, are the results of direct studies
analyzing the insulin receptor gene in NIDDM.

Figure 9 depicts the gene structure of the insulin
receptor, as first identified by Bell and colleagues (51).
In analyzing this gene in NIDDM, we used a PCR strategy to
amplify exons 16 through 22. This approach was chosen
because exons 16 through 22 encompass the entire cytoplasmic extension of the β subunit, containing the kinase domain, as well as other signalling properties of the insulin
receptor. Since, no abnormalities of binding affinity or
structure of the ectodomain have been reported in NIDDM, it
seemed reasonable to assume that any sequence variations
would reside within these exons. The amplified DNA fragments were then sequenced and these studies were performed
in 7 "garden variety" NIDDM patients (49). These studies
revealed no coding abnormalities in the sequence of the
insulin receptor gene in any of the patients (49). Other
studies using molecular scanning approaches to screen
larger numbers of patients for mutations have revealed that
the degree of genetic variation in the insulin receptor is
exceedingly small in NIDDM, on the order of 1-2% (50).
Whether the few mutations that have been identified in

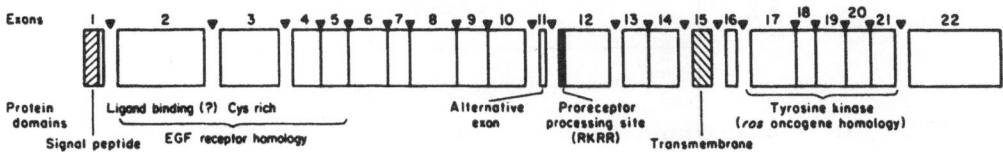

FIGURE 9. Insulin-receptor gene intron/exon structure, depicting 22 exons, their intronic boundaries, and corresponding encoded regions of the insulin receptor (adapted from
Ref. 51).

insulin receptor genes from NIDDM patients are functionally significant, or whether they exceed the prevalence of these muta tions in the normal population, is not clear, but one can readily conclude from the aggregate of these studies (49,50) that the overwhelming majority (98-99%) of NIDDM subjects do not carry mutations in the insulin receptor gene. Thus, the coding region of the insulin receptor does not appear to be a diabetes gene.

Glucose Transport System: Let us now consider another major element in abnormal cellular action in NIDDM, namely the glucose transport system. Figure 10 shows results of glucose transport studies in adipocytes isolated from normal, IGT, and NIDDM patients (52). The results are strikingly similar to the in vivo glucose clamp studies (Fig. 4), and show a large decrease in insulin stimulated glucose transport at all insulin concentrations in the NIDDM groups (52). Thus, a defect in insulin stimulated glucose transport exists in isolated cells from NIDDM subjects and this abnormality parallels the in vivo findings of decreased insulin stimulated glucose disposal. However, it should be pointed out that adipose tissue is responsible for only a small component of in vivo glucose disposal, and that most of the glucose uptake during the glucose clamp studies involves skeletal muscle. Obviously then, the results in adipocytes are only relevant to in vivo physiology insofar as adipocyte glucose transport represents a paradigm for skeletal muscle. Fortunately, the available data indicate that this is the case. For example, Caro and his colleagues (53)have conducted studies of isolated human rectus muscle strips, and Figure 11 summarizes some of their results. Clearly, there is a major decrease in skeletal muscle glu-

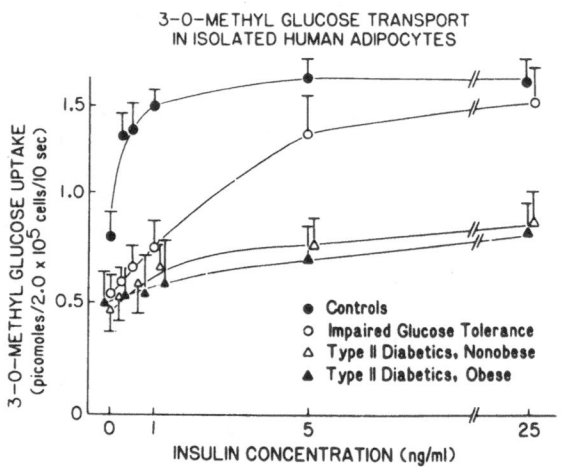

3-0-METHYL GLUCOSE TRANSPORT
IN ISOLATED HUMAN ADIPOCYTES

- Controls
- Impaired Glucose Tolerance
- Type II Diabetics, Nonobese
- Type II Diabetics, Obese

FIGURE 10. **Dose-response curve for insulin's ability to stimulate glucose transport (3-0-methylglucose uptake) in isolated adipocytes prepared from normals, patients with impaired glucose tolerance, and obese or nonobese subjects with NIDDM. The functional form of these dose-response curves is quite comparable to the shape of the dose-response curves for in vivo insulin-stimulated glucose disposal. From Ref. 1.**

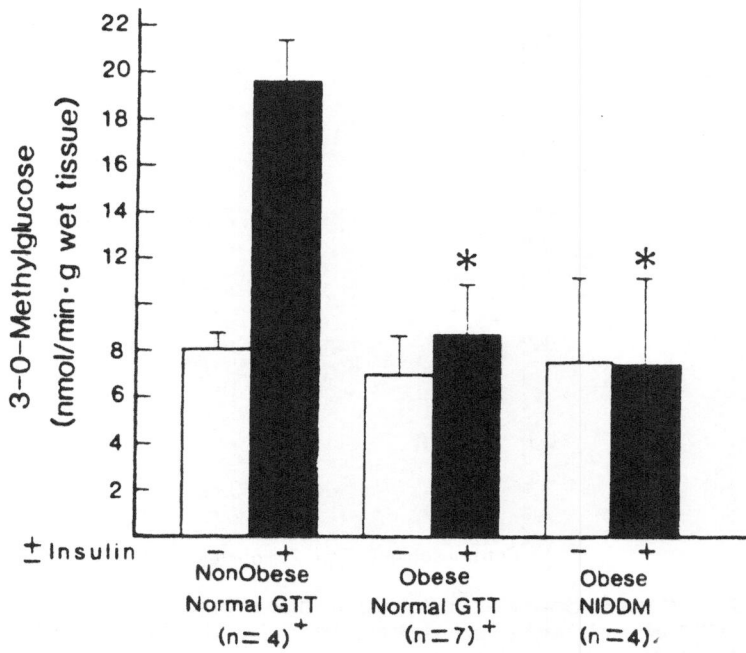

FIGURE 11. 3-0-methylglucose transport in muscle fiber strips from nonobese subjects, morbidly obese subjects with normal glucose tolerance, and morbidly obese subjects with NIDDM. *Significantly different (p<0.05) from nonobese group (+ insulin). From Ref. 53.

cose transport in NIDDM, compared to control, and these results are similar to what is seen in isolated adipocytes.

What is the mechanism of this decrease in insulin stimulated glucose transport in NIDDM? At this point in the discussion, at least 3 possibilities exist. There could be a decrease in the ability of insulin to signal recruitment, or translocation, of Glut4 to the cell surface. Secondly, recruitment could be normal, but there could be a marked decrease in the intrinsic activity of Glut4, and, finally, there could be a deficiency of Glut4 proteins in this disease. With regards to the latter point, a number of laboratories have examined the level of Glut4 mRNA as well as Glut4 protein in NIDDM skeletal muscle, and the general consensus of these data is that no deficiency exists (54-57). A representative example of these studies is depicted in Figure 12. Here we see that the amount of total skeletal muscle Glut4 is comparable among the study groups, demonstrating that an absolute deficiency of Glut4 is not the cause of decreased skeletal muscle transport in NIDDM. It should be emphasized that these studies assess total cellular Glut 4 and do not tell us about Glut4 localization within the cell, nor recruitment to the plasma membrane.

Based on these studies, our attention must focus on possibilities one and two; i.e., a translocation (signalling) defect, or an intrinsic activity defect in Glut4. When one considers the possibility of a defect in Glut4 intrinsic activity, one must think about a structural defect in the protein itself. Such an abnormality could be due to genetic variation in the Glut4 sequence and since NIDDM has a strong hereditary component, one can speculate on the possibility

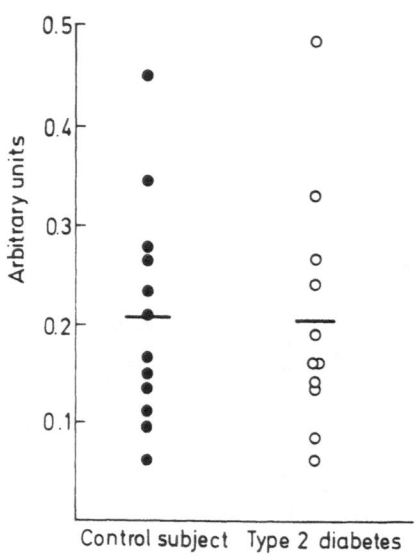

FIGURE 12. Glut4 content of human skeletal muscle from control and NIDDM subjects as determined by Western blotting with anti Glut4 Abs. From Ref. 57.

of Glut4 as a candidate "diabetes" gene. To approach this problem, we sought to identify the primary sequence of the Glut4 gene in a series of typical insulin resistant NIDDM subjects. These studies were made possible through the work of Bell and his colleagues, who succeeded in elucidating the intron/exon structure of the human Glut4 gene (58). As adapted from the work of Bell and colleagues the Glut4 gene structure is depicted in Figure 13. Based on the available intron-exon sequences, we utilized a PCR strategy which would amplify all 11 exons of this gene in 5 fragments (A-C). The appropriate sized fragments were readily amplified in all patients and directly sequenced according to previously published methods (49). In 5 of the 6 patients no changes in the amino acid sequence were observed. Two patients were heterozygous and one homozygous for a silent polymorphism at nucleotide position 535 and this polymorphism is also commonly seen in non-diabetic subjects (59). Interestingly, one of the 7 patients was heterozygous for an A for T substitution at nucleotide position 1292 (Fig. 14). This led to an isoleucine for valine substitution at codon 383 in the 5th extracellular loop of the Glut4 protein, according to its proposed topological organization within the plasma membrane (Figure 15). The exact biologic significance of this substitution is unclear, but larger scale molecular scanning studies have verified that this mutation exists in only a small percentage of NIDDM subjects (1-2%), and this mutation also exists in low frequency in non-diabetic subjects (59). Whether the frequency of this mutation is higher in NIDDM

INTRON-EXON STRUCTURE OF

THE HUMAN GLUT4 GENE

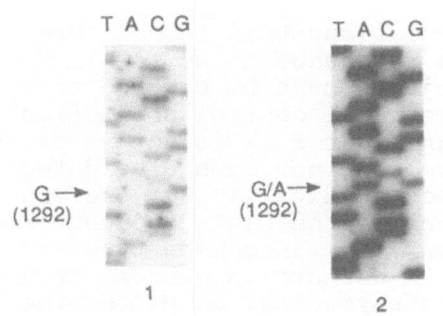

FIGURE 13.   Intron/Exon structure of the Glut4 gene as derived from the work of Bell and colleagues (Ref 58).

FIGURE 14.   Partial nucleotide sequences of sense strand of GLUT-4 exon 9 from one of the six NIDDM subjects, compared to control. There are two bands (G and A) at nucleotide position 1292 (lane 2) in the NIDDDM patient instead of just G as in the normal subject (Lane 1). From Ref. 49.

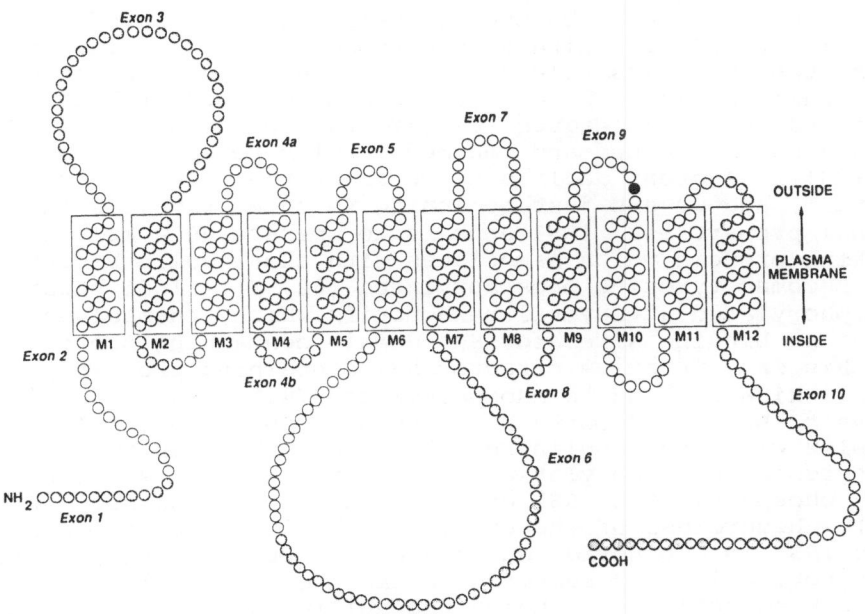

FIGURE 15.   Modified model for the exon-intron organization of the insulin-sensitive human glucose transporter (GlUT-4) gene. Adapted from the model originally proposed for the ancestral facilitative glucose transporter gene by Bell et al. (60). The amino acids encoded by each of the 11 exons in the GLUT-4 gene are indicated by circles. 12 putative membrane-spanning domains are numbered M1-M12 and are shown as rectangles. A black circle indicates the position of the amino acid in GLUT-4 (383 according to numbering by Fukumoto et al. (61)), where the NIDDM subject was heterozygous for a conservative amino acid replacement.

populations than in controls is unclear at this time; since the frequency is so low in NIDDM, a very large number of subjects must be studied to see if the frequency is even lower in a control population. Furthermore, the functional significance -- if any -- of this mutation at the protein level has not been elucidated. Nevertheless, based on these studies (49,59) one can conclude that genetic variations in the Glut4 sequence are exceedingly uncommon (less than 1-2%) in common type NIDDM subjects. Consequently, while defects in glucose transport are important to the pathophysiology of NIDDM, the coding sequence of Glut4 does not appear to be a diabetes gene locus.

Transmembrane Signalling: Let us next examine potential signalling events which link the insulin receptor to glucose transport stimulation. A variety of post-receptor signalling systems and mediators, have been studied with respect to insulin action (20-29), but one of the most intensively investigated signalling mechanisms involves tyrosine phosphorylation. Following insulin binding and receptor autophosphorylation a number of endogenous protein substrates undergo tyrosine phosphorylation in different cell systems (26-32). The most well studied of these is pp185, an endogenous protein which becomes tyrosine phosphorylated after insulin stimulation with a dose response and time course consistent with its role as an important signalling molecule in insulin action. Recently, through the work of White and Kahn and their colleagues (62,63), this protein has been cloned and sequenced and renamed insulin receptor substrate 1 (IRS1). A great deal is known about the structure of this protein, and we are just beginning to understand its functional properties. IRS1 appears to operate as a multisite docking protein which associates with the insulin receptor and becomes tyrosine phosphorylated (62,63). IRS1 becomes phosphorylated at several YMXM motifs which, themselves, serve as binding sites for SH2 domain containing proteins. In this way, putative effector SH2 containing proteins, such as PI3 kinase, bind to phosphorylated IRS1. This could serve to bring SH2 containing proteins into a molecular complex with the insulin receptor where they can then undergo tyrosine phosphorylation, or alternatively, association with phosphorylated IRS1 could be the activating step itself (64). Regardless of the precise mechanisms, it is clear that IRS1 is a bonifide endogenous substrate of the insulin receptor, which may subserve an important signalling role.

Consequently, we studied the ability of insulin to induce phosphorylation of IRS1 in adipocytes from control, obese, and NIDDM subjects (65). The basic experimental method is depicted in Figure 16. Cells are incubated with or without insulin, solubilized, followed by Western blotting with an anti-phosphotyrosine antibody (65). As can be seen, insulin stimulation leads to marked tyrosine phosphorylation of the 95K beta subunit as well as pp185/IRS1 in both rat and human adipocytes (65).

The dose response curve for these bioeffects in human adipocytes is seen in Figure 17. The curves for autophos-

Rat   Human
Adipocyte   Adipocyte

Insulin — +   — +

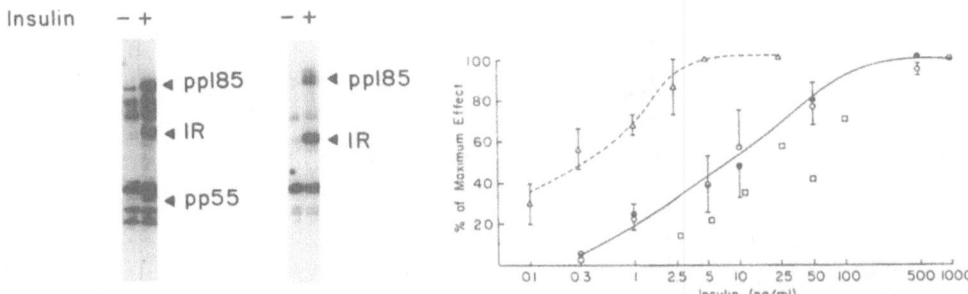

◄ ppl85   ◄ ppl85

◄ IR   ◄ IR

◄ pp55

FIGURE 16. <u>Insulin-induced</u>
tyrosine phosphorylation of
endogenous proteins in rat
(left) and human (right)
adipocytes. Isolated adi-
pocytes from rats and human
subjects were incubated
with (+) or without (−) 100
ng/ml insulin for 10 min at
37°C. In both species,
insulin-receptor (IR) β-
subunit, running at 95,000
Mr, and a 185,000 Mr pro-
tein, designated pp185,
were tyrosine phosphorylat-
ed in response to insulin
treatment. From Ref. 65.

FIGURE 17. Dose response of
insulin-stimulated 3-0-methyl
glucose transport, substrate
phosphorylation, and insulin-
receptor binding at 37°C. 3-
0-methylglucose transport; Δ,
insulin-receptor phosphoryla-
tion; ●, pp185 phosphoryla-
tion; 0, insulin binding;□.
From Ref. 65.

phorylation and IRS1 phosphorylation are nearly superimposable, consistent with the idea that IRS1 is an immediate
substrate of the phosphorylated insulin receptor beta subunit. Interestingly, when these curves are compared to the
dose response for insulin stimulated glucose transport in
these same cells, one can see that the glucose transport
curve is left shifted compared to autophosphorylation and
IRS1 phosphorylation (65). This illustrates the concept of
spare kinase activity; in other words, 50% of glucose
transport stimulation is reached at an insulin concentration which activates only a small proportion (< 10%) of the
available kinase activity, whereas maximal transport stimulation is observed when ˜30% of the kinase is stimulated.
     Data in the various study groups are seen in Figure
18. The upper panel shows the antiphosphotyrosine Western
blotting results. In this analysis, we sought to examine
the coupling between phosphorylated β-subunit and phosphorylated IRS1. Consequently, both bands were quantitated in
cell extracts from all individuals and expressed as a ratio
of autophosphorylated 95K beta subunit to phosphorylated
IRS1. As is evident, no statistically significant difference in this coupling ratio was observed in the 3 study

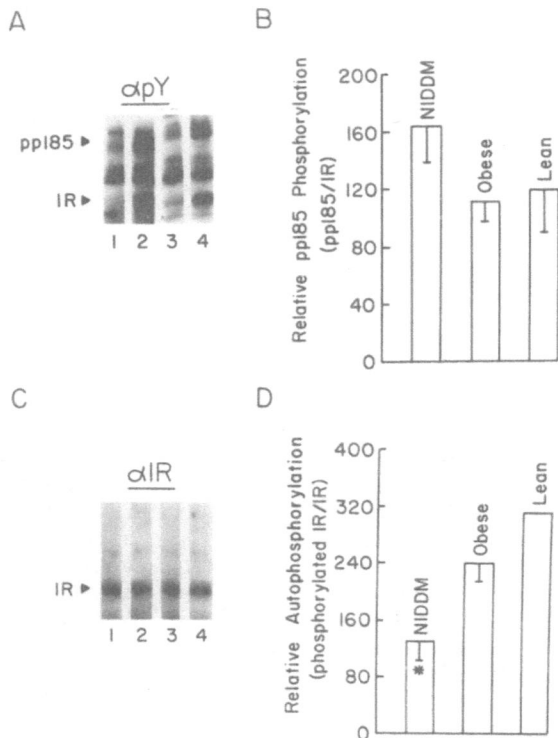

FIGURE 18. Insulin-stimulated phosphorylation of insulin receptor (IR) and pp185 in adipocytes from non-insulin-dependent diabetic (NIDDM), obese, and lean subjects. (A) Representative autoradiograph. Adipocytes from NIDDM (lanes 1 and 2) and obese (lanes 3 and 4) subjects incubated with (lanes 2 and 4) or without (lanes 1 and 3) insulin were solubilized and immunoblotted with anti-phosphotyrosine antibody. (B) To estimate coupling efficiency between insulin-receptor activation and pp185 phosphorylation, autoradiographs of antiphosphotyrosine immunoblots from cells treated with 10 ng/ml insulin for 5 min were subjected to scanning densitometry, and the ratio of phosphorylated pp185 to phosphorylated insulin-receptor β-subunit was expressed as relative densitometry units. There is no statistically significant difference between cells from NIDDM (n=8), obese (n=8), or lean (n=3) subjects in efficiency of pp185 phosphorylation. (C) Representative autoradiograph. Cells from NIDDM (lanes 1 and 2), obese (lane 3), and lean (lane 4) subjects incubated with (lanes 2,3, and 4) or without (lane 1) insulin were solubilized and immunoblotted with anti-insulin-receptor antibody. (D) To estimate efficiency of insulin-receptor autophosphorylation, ratio of phosphorylated β-subunit, from antiphosphotyrosine immunoblots, to total β-subunit in gel lanes, from anti-insulin-receptor immunoblots, is expressed as relative densitometry units. Insulin-stimulated receptor autophosphorylation is significantly lower (p<0.02) in cells from NIDDM subjects vs. obese or lean nondiabetic subjects. From Ref. 65

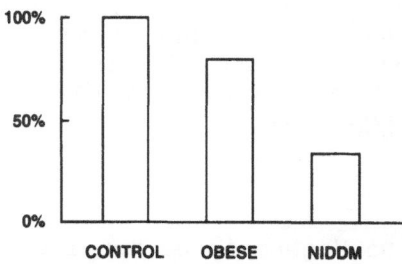

pp185/IRS-1 CONTENT IN ADIPOCYTES

FIGURE 19. Total estimated amount of insulin stimulated (10 ng/ml) phosphorylated pp185/IRS1, per cell, in adipocytes from control, obese, and NIDDM subjects. Data are derived from Figure 18 along with estimates of insulin binding capacity and represent the mean ± SEM.

groups, leading to the conclusion that the ability of auto-phosphorylated beta subunits to mediate IRS1 phosphorylation is normal in NIDDM. On the other hand, as is seen in the lower panel, the absolute amount of phosphorylated β-subunit is less in NIDDM. This analysis was accomplished by Western blotting with an anti-insulin receptor antibody, to ascertain beta subunit mass, and then comparing these results to the amount of phosphorylated beta subunit (65). It is clear that decreased autophosphorylation exists in adipocytes from NIDDM subjects, consistent with the receptor kinase defect shown earlier (Fig. 8). Since coupling between autophosphorylated beta subunits and IRS1 is normal in NIDDM, whereas the amount of autophosphorylated beta subunit is decreased, it is evident that the total amount of phosphorylated IRS1 is decreased in the NIDDM cells. When estimated on a per cell basis (Figure 19), the striking decrease in phosphorylated IRS1 content following insulin stimulation in cells from NIDDM subjects is quite apparent.

The importance of this defect in insulin stimulated IRS1 phosphorylation to the pathogenesis of cellular insulin resistance in NIDDM remains to be determined. However, it is apparent that the defect is rather substantial, and, insofar as IRS1 proves to be a key downstream signalling molecule of the insulin receptor, this abnormality could represent an important aspect of insulin resistance.

Since the insulin receptor and Glut4 genes do not appear to be major contributors to the genetic component of insulin resistance in NIDDM, it seems appropriate to turn our attention to the signalling events which link these two components of the insulin action system. As such, IRS1, or other signalling molecules become potential "candidate genes" in NIDDM. The degree of insulin resistance is not equal for all aspects of insulin action in NIDDM and insulin stimulated glucose transport is paticularly affected. Consequently, those elements of the insulin action cascade which signal to the glucose transport system should receive particular attention as potential "diabetes" genes contributing to the hereditary aspects of insulin resistance in this disorder.

ACKNOWLEDGEMENT

This study was supported in part by a research grant from the National Institutes of Health NIDDK NIH DK 33651, GCRC RR-00827 from the Veterans Administration Medical Center, Medical Research Service, and by the Sankyo Company Diabetes Research Foundation.

REFERENCES

1. J.M. Olefsky. Pathogenesis of non-insulin depednent diabetes (Type II), in: DeGroot: Endocrinology Second Edition. L.J. DeGroot. G.M. Besser. G.F. Cahill, J.C. Marshall, D.H. Nelson, W.D. Odell, J.T. Potts, Jr., A.H. Rubenstein, and E. Steinberger, W.B. Saunders Co., Philadelphia (1989).

2. R.A. DeFronzo, The triumvirate: β-cell, muscle, liver: A collusion responsible for NIDDM, Diabetes 37:667 (1988).

3. G.M. Reaven, Role of insulin resistance in human disease, Diabetes 37:1495 (1988).

4. S. Efendic, R. Luft, and A. Wajngot, Aspects of the pathogenesis of Type 2 diabetes, Endoc. Rev. 5:395 (1984).

5. J.L. Rotter, C.M. Vadheim, and D.L. Rimoin, Genetics of diabetes mellitus. in: Diabetes Mellitus: Theory and Practice, H. Rifkin, D. Prote, Jr., Elsevier, New York (1990).

6. J.M. Olefsky. Insulin action and insulin resistance in non-insulin dependent diabetes mellitus, in: Recent Advances in Insulin Action and its Disorders, Y. Shigeta, M. Kobayashi, and J.M. Olefsky), Excerpta Medica, Amsterdam (1991).

7. J.F. Caro, L.G. Dohm, W.J. Pries, and M.K. Sinha, Cellular alterations in liver, skeletal muscle, and adipose tissue responsible for insulin resistance in obesity and Type II diabetes, Diab/Metab Rev. 5:665 (1989).

8. J.M. Olefsky. Diabetes mellitus, in: Cecil Textbook of Medicine, 18th Edition, Volume 2, Chapter 231, J.B. Wyngaarden, L.H. Smith, Jr., J.C. Bennett, and F. Plum, W.B. Saunders, Philadelphia, PA (1991).

9. O.G. Kolterman, R.S. Gray, J. Griffin, P. Burstein, J. Insel, J.A. Scarlett, and J.M. Olefsky, Receptor and post-receptor defects contribute to the insulin resistance in non-insulin dependent diabetes mellitus. J. Clin. Invest. 68:957 (1981).

10. R.R. Revers, R. Fink, J. Griffin, J.M. Olefsky, and O.G. Kolterman, Influence of hyperglycemia on insulin's in vivo effects of Type II diabetes. J. Clin. Invest. 73:664 (1984).

11. C. Bogardus, S. Lillioja, B.V. Howard, G. Reaven, and D. Mott, Relationships between insulin secretion, insulin action, and fasting plasma glucose concentration in non-diabetic and non-insulin-dependent diabetic subjects. J. Clin. Invest. 74:1238 (1984).

12. S. Dinneen, J. Gerich, and R. Rizza, Carbohydrate metabolism in non-insulin-dependent diabetes mellitus, N. Engl. J. Med. 327:707 (1992).

13. J.F. Caro, Clinical review 26: Insulin resistance in obese and nonobese man, J. Clin. Endocrin. Rev. 73:691 (1991).

14. O. Hother-Nielsen, O. Schmitz, P.H. Andersen, H. Beck-Nielsen, and O. Pedersen, Metformin improves peripheral but not hepatic insulin action in obese patients with type I diabetes, Acta. Endocrinol. (Copenh.), 120:257 (1989).

15. R.C. Bonadonna, and R.A. DeFronzo, Glucose metabolism in obesity and Type 2 diabetes, Diab. & Metab. 17:112 (1991).

16. G.M. Reaven, Resistance to insulin-stimulated glucose uptake and hyperinsulinemia: Role in non-insulin-dependent diabetes. high blood pressure. dyslipidemia and coronary heart disease. Diab. & Metab (Paris) 17:78 (1991).

17. C. Bogardus, S. Lillioja, B.V. Howard, G. Reaven and D. Mott, Relationships between insulin secretion, insulin action, and fasting plasma glucose concentration in non-diabetic and non-insulin-dependent diabetic subjects, J. Clin. Invest. 74:1238 (1984).

18. W.K. Ward, J.C. Beard, J.B. Halter, M.A. Pfeifer, and D. Porte, Jr., Pathophysiology of insulin secretion in non-insulin-dependent diabetes mellitus, Diabetes Care 7:491 (1984).

19. W.T. Garvey, J.M. Olefsky, J. Griffin, R. Hammon, and O.G. Kolterman, The effects of insuin treatment on insulin secretion and action in Type II diabetes mellitus. Diabetes 34:222 (1985).

20. R.R. Henry, P. Wallace and J.M. Olefsky, The effects of weight loss on the mechanisms of hyperglycemia in obese noninsulin-dependent diabetes mellitus, Diabetes 35:990 (1986).

21. O.G. Kolterman, R.S. Gray, G. Shapiro, J.A. Scarlett, J. Griffin, and J.M. Olefsky, The acute an chronic effects of sulfonylurea therapy in Type II diabetics, Diabetes 33:346 (1984).

22. R.A. DeFronzo, and E. Ferrannini, The pathogenesis of non-insulin-dependent diabetes, Medicine 61:125 (1982).

23. G.M. Reaven, and J.M. Olefsky, Relationship between heterogeneity of insulin responses and insulin resistance in normal subjects, Diabetologia 13:201 (1977).

24. R.R. Henry, L. Scheaffer and J.M. Olefsky, Glycemic effects of short-term intensive dietary restriction and isocaloric refeeding in non-insulin dependent diabetes mellitus, J. Clin. Endocrinol. Metab. 61:917 (1985).

25. R.A. DeFronzo, J.D. Tobin, and R. Andres, Glucose clamp technique: A method for quantifying insulin secretion and resistance, Am. J. Physiol. 237:E214 (1979).

26. O.M. Rosen, After insuln binds, Science 237:1452 (1987).

27. C.R. Kahn and M.F. White, The insulin receptor and the molecular mechanism of insulin action, J. Clin. Invest. 82:1151 (1988).

28. J.M. Olefsky, The insulin receptor: A multi-functional protein, Diabetes 39:1009 (1990).

29. H.U. Haring, The insulin receptor: Signalling mechan-

ism and contribution to the pathogenesis of insulin resistance, Diabetologia 34:848 (1991).

30. M. Kasuga, F.A. Karlsson, and C.R. Kahn, Insulin stimulates the phosphorylation of the 95,000 Dalton subunit of its own receptor, Science 215:185 (1982).

31. H.E. Tornqvist, M.W. Pierce, and A.R. Frackelton, Identification of insulin receptor tyrosine residues autophosphorylated in vitro, J. Biol. Chem. 262:10212 (1987).

32. M.F. White, S.E. Shoelson, H. Keutmann, and C.R. Kahn, A cascade of tyrosine autophosphorylation in the subunit activates the insulin receptor, J. Biol. Chem. 263:2969 (1988).

33. R.I. Fink, P. Wallace, G. Brechtel, and J.M. Olefsky, Evidence that glucose transport is rate-limiting for in vivo glucose uptake, Metabolism 41:897 (1992).

34. J-P. Idstrom, M.J. Rennie, T. Schersten, and A-C. Bylund-Fellenius, Membrane transport in relation to net uptake of glucose in the perperfused rat hindlimb, Biochem. J. 233:131 (1986).

35. K. Kubo, and J.E. Foley, Rate-limiting steps for insulin-mediated glucose uptake into perfused rat hindlimb, Am. J. Physiol. 250:E100 (1986).

36. J.E. Pessin, and G.I. Bell, Mammalian facilitative glucose transporter family: Structure and molecular regulation, Ann. Rev. Physiol. 84:911 (1992).

37. M. Muckler, Family of glucose transporter genes, Diabetes 39:6 (1990).

38. S.W. Cushman, and L.J. Wardzala, Potential mechanism of insulin action on glucose transport in the isolated rat adipose cell, J. Cell Biochem. 255:4758 (1980).

39. T. Kono, K. Suzuki, L.E. Dansey, F.W. Robinson, and T.L. Blewis, Energy-dependent and protein synthesis-independent recycling of the insulin-sensitive glucose transport mechanism in fat cells, J. Biol. Chem. 256:6400 (1981).

40. G.R. Freidenberg, R.R. Henry, H.H. Klein, D.R. Reichart, and J.M. Olefsky, Decreased kinase activity of insulin receptors from adipocytes of non-insulin dependent diabetic (NIDDM) subjects, J. Clin. Invest. 79:240 (1987).

41. J.F. Caro, O. Ittoop, W.J. Pories, D. Meelheim, E.G. Flickinger, F. Thomas, M. Jenquin, J.F. Silverman, P.G. Khazanie, and M.K. Sinha, Studies on the mechanism of insulin resistance in the liver from humans with non-insulin-dependent diabetes, J. Clin. Invest. 78:249 (1986).

42. J.F. Caro, M.K. Sinha, S.J. Raju, O. Ittoop, W.J. Pries, E.G. Flickinger, D. Meelheim, and G.L. Dohm, Insulin receptor kinase in human skeletal muscle from obese subjects with and without non-insulin-dependent diabetes, J. Clin. Invest. 79:1330, (1987).

43. P. Arner, T. Pollare, H. Lithell, and J.N. Livingston, Defective insulin receptor tyrosine kinase in human skeletal muscle in obesity and Type 2 (non-insulin-dependent) diabetes mellitus, Diabetologia 30:437 (1987).

44. R.J. Comi, G. Grunberger, and P. Gorden, The relation-

ship of insulin binding and insulin-stimulated
tyrosine kinase activity is altered in Type II
diabetes, J. Clin. Invest. 79:453 (1987).

45. H. Maegawa, Y. Shigeta, K. Egawa, and M. Kobayashi,
    Impaired autophosphorylation of insulin receptors
    from abdominal skeletal uscles in nonobese subjects
    with NIDDM, Diabetes 40:815 (1991).

46. D.J. Brillon, G.R. Freidenberg, R.R. Henry, and J.M.
    Olefsky, Mechanism of defective insuln receptor
    kinase activity in NIDDM: Evidence for two recep-
    tor populations, Diabetes 38:397 (1989).

47. G.R. Freidenberg, D. Reichart, J.M. Olefsky, and R.R.
    Henry, Reversibility of defective adipocyte insulin
    receptor kinase activity in non-insulin dependent
    diabetes mellitus: Effect of weight loss, J. Clin.
    Invest. 82:1398 (1988).

48. G.R. Freidenberg, D. Reichart, and J.M. Olefsky,
    Insulin receptor kinase activity is not reduced in
    fibroblsts from subjects with non-insulin depen-
    dent diabetes mellitus (NIDDM), Clin. Res. 38:119A
    (1990).

49. J. Kusari, U.S. Verma, J.B. Buse, R.R. Henry, and J.M.
    Olefsky, Analysis of the gene squences of the in
    sulin receptor and the insuln sensitive glucose
    transporter (Glut-4) in patients with common type
    non-insulin dependent diabetes mellitus, J. Clin.
    Invest. 88:1323 (1991).

50. S. O'Rahilly, W.H. Choi, P. Patel, R.C. Turner, J.S.
    Flier, and D.E. Moller, Detection of mutations in
    insulin-receptor gene in NIDDM patients by analysis
    of single-stranded conformation polymorphisms,
    Diabetes 40:777 (1991).

51. S. Seino, M. Seino, S. Nishi, and G.I. Bell, Structure
    of the human insulin receptor gene and characteri-
    zation of its promoter, Proc. Natl. Acad. Sci.
    U.S.A. 86:114 (1989).

52. T.P. Ciaraldi, O.G. Kolterman, J.A. Scarlett, M. Kao,
    and J.M. Olefsky, Role of th glucose transport
    system in the post-receptor defect of non-insulin
    dependent diabetes mellitus, Diabetes 31:1016
    (1982).

53. G.L. Dohm, E.B. Tapscott, W.J. Pories, D.J. Dabbs, E.G.
    Flickinger, D. Meelheim, T. Fushiki, S.M. Atkinson,
    C.W. Elton, and J.F. Caro, An in vitro human muscle
    preparation suitable for metabolic studies: De-
    creased insulin stimulation of glucose transport in
    muscle from morbidly obese and diabetic subjects,
    J. Clin. Invest. 82:486 (1988).

54. O. Pedersen, J.F. Bak, P.H. Andersen, S. Lund, D.E.
    Moller, J.S. Flier, and B.B. Kahn, Evidence against
    altered expression of GLUT1 or GLUT4 in skeletal
    muscle of patients with obesity or NIDDM, Diabetes
    39:865 (1990).

55. J. Eriksson, L. Koranyi, R. Bourey, C. Schalin-Jantti,
    E. Widen, M. Mueckler, A.M. Permutt, and L.C.
    Groop, Insulin resistance in type 2 (non-insulin-
    dependent) diabetic patients and their relatives is
    not associated with a defect in the expression of
    the insulin-responsive glucose transporte (GLUT-4)
    gene in human skeletal muscle, Diabetologia 35:143
    (1992).

56. W.T. Garvey, L. Maianu, J.A. Hancock, A.M. Golichowski, and A. Baron, Gene expression of GLUT4 in skeletal muscle from insulin-resistant patients with obesity, IGT, GDM, and NIDDM, Diabetes 41:465 (1992).

57. A. Handberg, A. Vaag, P. Damsbo, H. Beck-Nielson, and J. Vinten, Expression of insulin regulatable glucose transporters in skeletal muscle from type 2 (non-insulin dependent) diabetic patients, Diabetologia 33:625 (1990).

58. J.B. Buse, K. Yasuda, T.P. Lay, T.S. Seo, A.L. Olson, J.E. Pessin. J.H. Karam. S. Seino. and G.I. Bell. Human GLUT4/muscle-fat glucose transporter gene: Characterization and genetic variation, Diabetes (In press).

59. W-H. Choi, S. O'Rahilly, J.B. Buse, A. Rees, R. Morgan, J.S. Flier, and D.E. Moller, Molecular scanning of insulin-responsive glucose transporter (GLUT4) gene in NIDDM subjects, Diabetes 40:1712 (1991).

60. G.I. Bell, T. Kayano, J.B. Buse, C.F. Burant, J. Takeda, D. Lin, H. Fujumoto, and S. Seino, Molecular biology of mammalian glucose transporters, Diabetes Care 13:198 (1990).

61. Fukumoto, H., T. Kayano, J.B. Buse, Y. Edwards, P.F. Pilch, G.I. Bell and S. Seino, Cloning and characterization of the major insulin-redponsive glucose transporter expressed in human skeletal muscle and other insulin-responsive tissues, J. Biol. Chem. 264:7776 (1989).

62. X. Kian, P. Rothenberg, C.R. Kahn, J.M. Backer, E. Araki, P.A. Wilden, D.A. Cahill, B.J. Goldstein, and M.F. White, Structure of the insulin receptor substrate IRS-1 defines a unique signal transduction protein, Nature 35:73 (1991).

63. S.E. Shoelson, S. Chatterjee, M. Chaudhuri, and M.F. White, YMXM motifs of IRS-1 define substrate specificities of the insulin receptor kinase, Proc. Natl. Acad. Sci. 89:2027 (1992).

64. J.M. Backer, M.G. Myers, Jr., X-J. Sun, J. Schlessinger, and M.F. White, The insulin receptor substrate IRS-1 associates with and activates the phosphatidylinositol 3-kinase. Diabetes 41 (Suppl. 1):166 #582 (1992).

65. R.S. Thies, J.M. Molina, T.P. Ciaraldi, G.R. Freidenberg, and J.M. Olefsky, Insulin receptor autophosphorylation and endogenous substrate phosphorylation in human adipocytes from control, obese, and non-insulin dependent diabetic subjects, Diabetes 39:250 (1990).

INDIRECT EFFECTS OF INSULIN IN REGULATING GLUCOSE FLUXES

Z.Q. Shi, A. Giacca, S.J. Fisher, M. Lekas, D. Bilinski, M. Van Delangeryt,
H.L.A. Lickley*, and M. Vranic

Department of Physiology
Department of Surgery*
University of Toronto
Toronto, Ontario, Canada, M5S 1A8

INTRODUCTION

Insulin can regulate glucose metabolism directly via its effects on glucose transport and metabolic pathways. However, these effects cannot account for all the metabolic events that occur following insulin secretion or administration. A number of laboratories, including our own, have suggested that many of the known effects of insulin on glucose metabolism may be exerted partially via indirect pathways [1]. Cumulative evidence indicates that the indirect actions of insulin not only contribute to, but sometimes may play a predominant role in the overall effects of insulin on glucose fluxes.

The mechanisms of insulin's direct actions on glucose metabolism in insulin sensitive tissues have been studied in some detail. The direct actions take place after insulin binds to the insulin-receptor (MW 152-154 kda) on the cell surface of the insulin-sensitive tissues. This leads to receptor autophosphorylation and activation of protein kinases which in turn trigger an intracellular biochemical cascade. Insulin can act on skeletal muscle to increase glucose transport by recruiting glucose transporters (mainly GLUT 4) from intracellular pool(s) to the plasma membrane [2-5] or to the transverse tubules [6]. This has been shown using both the cytochalasin B binding assay and/or the Western blot assay for glucose transporter proteins. An increase in the glucose transporter mRNA and cellular expression of the glucose transporter protein has also been demonstrated in muscle in response to long-term stimulation by insulin [7]. In isolated adipocytes, insulin can also stimulate glucose transport by recruiting glucose transporters [8]. Intracellularly, insulin stimulates oxidative metabolism of glucose by activating pyruvate dehydrogenase (PDH) phosphatase [9].

The indirect actions of insulin can be defined as the secondary metabolic effects on one tissue or metabolite brought about by insulin's effects on the other tissue(s) and metabolite(s) [1]. Figure 1 presents some of our recent findings and hypotheses on the indirect roles of insulin in glucose utilization during exercise and in regulating postabsorptive glucose production and cycling

New Concepts in the Pathogenesis of NIDDM, Edited by
C. G. Östenson *et al.*, Plenum Press, New York, 1993

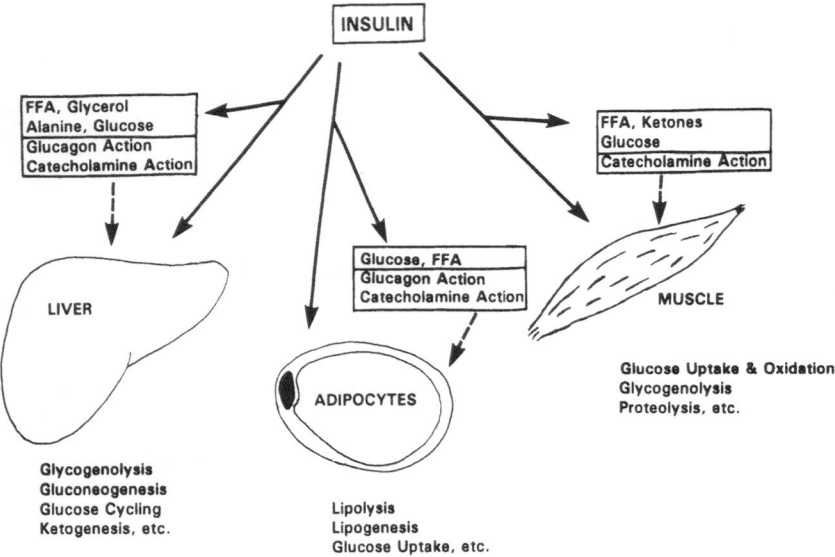

Figure 1. Schematic diagram of the direct (⟶) and indirect (----→) effects of insulin. The indirect (secondary) effects of insulin on the tissues and metabolites are produced following insulin's direct (immediate) actions on the other tissues and metabolites, and are influenced by insulin's interactions with counterregulatory hormones.

in the liver in dogs. The effects of insulin on glucose production and uptake are likely achieved by coordinated direct and indirect actions on various effector tissues. The mechanisms of the indirect effects have not yet been explored in sufficient detail. This is because the direct and indirect actions normally take place concomitantly in vivo, and the two are therefore difficult to distinguish from one another. In order to isolate the indirect effects of insulin, it is necessary to create models of insulin-deficiency in which a specific, putative aspect of insulin's indirect effects can be reproduced by a proper replacement regimen. For example, antilipolysis is a direct effect of insulin on adipose tissue, but changes in free fatty acid (FFA) levels exert an indirect effect on carbohydrate metabolism in the liver and muscle. The effects of antilipolysis and inhibition of ketogenesis on carbohydrate metabolism have been shown to represent some of the indirect actions of insulin and can be reproduced by β-adrenergic blockade and methylpalmoxirate even in the absence of insulin [10-12]. On the other hand, restriction of substrate supply to the liver may also be an important indirect effect of insulin, acting to reduce hepatic glucose production [13,14]. Under certain circumstances, the indirect effects of insulin on glucose metabolism may also take the form of a "permissive role" [15]. This can occur when a metabolic event is brought about by hormones other than insulin but whose occurence still requires the presence of a given amount of insulin. Whatever the primary stimulator, the presence of insulin is indispensable in permitting the subsequent biochemical events to occur.

INDIRECT EFFECTS OF INSULIN DURING EXERCISE

Restraint of the Fatty Acid-Glucose Cycle

It has been known for some time that both the flux and the metabolism of a given fuel (e.g., glucose) can be interfered with by other substrates (e.g., fatty acids and ketones). In diabetes, the

152

interference of FFA in glucose metabolism is two-fold. On one hand, increased mobilization and oxidation of FFA inhibits glucose uptake by the skeletal muscle (the FFA-glucose cycle). On the other hand, increased plasma FFA levels may lead to increased hepatic glucose production. Both these changes lead to hyperglycemia. McGarry et al proposed a bihormonal model in which the decrease in the insulin/glucagon ratio stimulates lipolysis and ketogenesis [16]. This model can be extended to include catecholamines, in addition to glucagon, especially in situations where fuel metabolism is perturbed by stressful challenges, such as moderate to strenuous exercise [10-12]. A precursor-product relationship has been demonstrated between plasma FFA and circulating ketone bodies [17]. Thus, a fall in plasma insulin or rise in glucagon and the catecholamines results in mobilization of FFA from adipose tissue and this in turn increases production of ketone bodies from the liver. When plasma ketone bodies reach 4 to 6 mM concentrations, insulin release from β-cells is stimulated [16]. This blunts lipolytic activity in fat cells and plasma FFA levels are then fixed at about 0.7 to 1.0 mM which allows only moderate ketone production but does not result in ketoacidosis. In diabetes, however, unrestrained lipolysis and ketogenesis can progress to a very high extent due to deficiency in insulin secretion and/or action. When FFA flux to the liver is increased, the level of malonyl-CoA, a derivative of the glycolytic metabolite citrate, is decreased. This metabolite is positioned as the first committed intermediate in long-chain fatty acid synthesis. Thus, increased FFA availability decreases malonyl-CoA, diminishes the synthesis of FFA and enhances FFA oxidation [16]. Exercise as a form of stress can exaggerate these metabolic derangements in diabetes. We have explored the interrelationship between FFA and glucose during exercise and our findings are elaborated in the following discussion.

During exercise under physiological conditions, the increase in muscle glucose uptake and clearance is matched by an increase in glucose production. In the meantime, the increased demand for energy is also met with by increased mobilization and oxidation of FFA. Proteolysis may also be stimulated with increasing intensity and duration of exercise. Insulin secretion falls moderately to allow augmentation of lipolysis and hepatic glucose production. However, plasma insulin is still kept at an effective level so as to maintain a hormonal balance between insulin and glucagon, and a balance of fuel consumption between glucose and FFA. In diabetes, insulin deficiency or resistance results in the loss of these balances. In the overnight insulin-deprived, depancreatized dogs, the total increase in glucose uptake during exercise is only 50% that of the normal. This residual increment in glucose uptake is due to muscle contraction, a mechanism that is independent of insulin. It is unquestionable that the impairment in glucose uptake in exercising diabetic dogs relates to lack of insulin action since such an impairment in the insulin-deprived dog can be corrected by insulin infusion at a basal rate (245 or 200 $\mu U.kg^{-1}.min^{-1}$) when plasma glucose was normalized [18,19]. The question is what actions of insulin are deficient or defective in this situation? In the in vitro experiments, muscle contraction per se can increase glucose uptake without insulin [20-22]. Clearly, the hormonal (sympathetic activity) and metabolic (lipolysis and glycogenolysis) perturbations induced by exercise as a moderate form of stress are absent in vitro. The discrepancy between the in vitro and in vivo studies implies the importance of some indirect effects of insulin in restraining the actions of the counterregulatory hormones, which are not required in vitro, but are necessary in vivo for providing a favorable hormonal and metabolic milieu. These effects are considered "indirect" since they do not directly act on the glucose transport or metabolic systems. The increase in lipolysis and plasma FFA has been demonstrated by ourselves and others to inhibit muscle glucose uptake [10,12,23]. In a classical study some thirty years ago, Randle et al proposed the concept of the "fatty acid-glucose cycle" (FFA-G cycle) in which the two components of the cycle as two main sources of fuel, reciprocally influence the fluxes of each other [24,25]. They demonstrated that an increased supply of FFA causes an increment in citrate concentration and increased activity in the tri-carboxylic acid cycle which in turn inhibits phosphofructokinase and leads to reduced glycolysis and glucose uptake. On the other hand, increased glucose flux can reduce the release of FFA into circulation by increasing the formation of glycerolphosphate hence facilitating re-esterification in adipose tissue and muscle. The inhibitory effect of increased FFA on glucose oxidation has since been supported by many other in situ and in vivo studies [26-29]. However, the operation of the FFA-G cycle depends on the hormonal and metabolic background. For example, during a constant glucose infusion at a rate high enough to ensure that the body relies entirely on glucose as an energy substrate (i.e. R.Q.=1), the administration of FFA, by intralipid and heparin infusion, had no effect on glucose oxidation but decreased glycogen oxidation [30]. However, the experiments were performed in normal man with supranormal glucose infusion (8 mg.kg^{-1}.min^{-1}) which resulted in hyperglycemia and increased insulin concentration. These conditions may restrict the inhibitory effect of FFA and favor an increase in carbohydrate oxidation.

Our studies in diabetic dogs revealed the inhibitory influence of the FFA-G cycle on glucose utilization during moderate exercise. We have reported a significant inverse correlation between levels of plasma FFA and the rates of glucose metabolic clearance (MCR) in the normal, alloxan-diabetic and depancreatized dogs [10,11]. Low FFA levels and high MCR in normal dogs were contrasted with high FFA levels and low MCR in the diabetic dogs with various degrees of insulin deficiency. We hypothesized that an indirect effect of insulin is required to augment glucose utilization through suppression of the FFA-G cycle during exercise. Our hypothesis was first studied by using a β-adrenergic blocker, propranolol, to inhibit lipolysis in both alloxan-diabetic and depancreatized dogs during exercise. In the alloxan-diabetic dogs (partially insulin deficient), the exercise-induced increase in MCR was normalized by propranolol [11]. This was accompanied by a substantial reduction in plasma FFA concentrations. Lactate concentration was also decreased, reflecting an inhibition of muscle glycogenolysis. In contrast, in the depancreatized dogs (totally insulin deficient), propranolol did not increase MCR during exercise [10]. Although lipolysis was inhibited to the same extent as in the alloxan-diabetic dogs, plasma FFA levels during exercise were still much higher in the depancreatized dogs.

A number of clues to this puzzle include: 1) The initial FFA levels were considerably higher in the depancreatized than the alloxan-diabetic dog, 2) There was residual insulin present in the alloxan-diabetic dogs but none in the depancreatized dog; and 3) Plasma glucose levels were much higher in the depancreatized dogs. The relationship between these findings and the uncorrected impairment in glucose clearance in the depancratized dog was then examined in the following experiments.

Since a limited reduction of lipolysis by propranolol did not improve glucose uptake in the depancreatized dog, we wanted to determine whether a direct inhibition of FFA oxidation may be more effective in suppressing the FFA-G cycle and improving glucose production and utilization. We therefore employed methylpalmoxirate (MP, inhibitor of carnitine palmitoyl transferase I), with or without β-blockade in the 24 h insulin-deprived, depancreatized dogs. MP suppresses the transport of long-chain FFA, which account for more than 90% of total FFA, across the inner mitochondrial membrane by irreversible binding to the active site of carnitine palmitoyl transferase

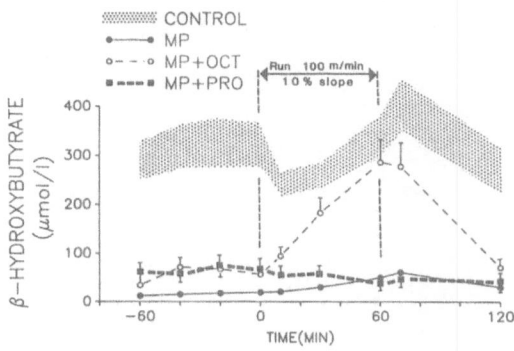

Figure 2. Plasma concentrations of β-hydroxybutyrate in the depancreatized dogs treated with methylpalmoxirate (MP, n=6), methylpaloxirate plus octanoate (MP+OCT, n=5) and methyl-palmoxirate plus propranolol (MP+PRO, n=6) at rest, during and after 60 min treadmill exercise (100 m/min, 10% slope). Data are shown as mean+SE. Shaded areas: mean+SE in untreated controls (n=6). (Modified from reference 12).

I [31-33]. Reduced entry of long-chain FFA into the mitochondria resulted in a diminution of FFA oxidation. MP's effects are manifest mainly in the liver and, to a much lesser extent, in muscle [31,32]. In our depancreatized dogs, MP (20 mg.kg^{-1}.day^{-1} orally for 2 days) suppressed hepatic FFA oxidation as evidenced by the nearly abolished ketosis (Figure 2) [12]. At the same time, plasma glucose concentrations were decreased both at rest (18%) and during exercise (15%, Figure 3). This was associated with a marked suppression of glucose production (39%). Our results are in line with observations by others that MP inhibits gluconeogenesis in both in vitro [34,35] and in vivo experiments [34,36] by diminishing FFA oxidation in the liver. In diabetes, gluconeogenesis is increased. In NIDDM patients and depancreatized dogs, gluconeogenesis accounts for about 50% of total glucose production [37,38]. Therefore, the MP-induced inhibition of glucose production could be primarily explained by suppression of gluconeogenesis, although some suppression of glycogenolysis may also be implicated. In another group of depancreatized dogs intravenous infusion of octanoate was given to MP-treated animals. Octanoate is a medium-chain (8-carbon) fatty acid which readily crosses the mitochondrial membrane without the need of carnitine palmitoyl transferase I for transport. The oxidation of octanoate can therefore offset the effects of suppression of long-chain FFA oxidation by MP. The doses of octanoate were chosen to restore β-hydroxybutyrate by the end of exercise to levels comparable to those in the control group (Figure 2). Correspondingly, glucose production was also restored to the control level (Figure 3) [12]. This implies that oxidation of FFA is necessary for glucose production in the liver. However, as with propranolol alone, MP alone did not improve the exercise-induced rise in glucose uptake and MCR (Figure 3), indicating that inhibition of FFA oxidation with MP is less effective in the muscle than in the liver.

The role of the FFA-G cycle was further examined by suppression of both lipolysis and FFA oxidation using combined MP and propranolol administration in order to more effectively suppress the FFA-G cycle. This resulted in a 38% decrease in FFA levels [12]. The exercise-induced rise in lactate was abolished, which could indicate an inhibition of muscle glycogenolysis. Glucose production did not further decrease from that seen with MP alone. The striking finding was that the exercise-induced increment in MCR was markedly augmented (Figure 3). Furthermore, this marked

155

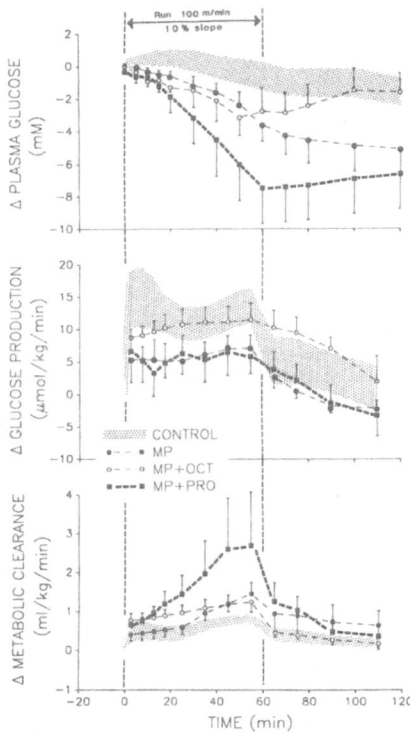

Figure 3. Plasma glucose concentrations and rates of glucose production and metabolic clearance (MCR) in the depancreatized dogs treated with methylpalmoxirate (MP, n=6), methypalmoxirate plus octanoate (MP+OCT, n=5) and methylpalmoxirate plus propranolol (MP+PRO, n=6) at rest, during and after 60 min treadmill exercise (100 m/min, 10% slope). Data are shown as mean±SE. Shaded area: mean±SE in untreated controls (n=6). (Modifed from reference 12).

findings that when same dose of propranolol was used to suppress lipolysis, the exercise-induced suppression of the FFA-G cycle normalized the exercise-induced increment in glucose uptake ($\Delta 29\pm3\%$). However, even with such combined suppression of lipolysis, FFA oxidation and muscle glycogenolysis, the maximal rate of MCR during exercise in the depancreatized dogs only reached 30% of normal controls (Figure 3) [12]. Thus, in the total absence of insulin, glucose uptake by the exercising skeletal muscle could be normalized by the suppression of the FFA-G cycle. However, this normalization of the exercise-induced increase in glucose uptake was due partly to the mass effects of uncorrected hyperglycemia and did not indicate a true normalization in glucose turnover since MCR was still largely uncorrected. MCR is an important and sensitive parameter which reflects the fractional extraction of circulating glucose in the animal as a whole [39]. Normalization of MCR can only be achieved when both glucose uptake and hyperglycemia are normalized. Therefore, suppression of the FFA-G cycle may only account for part of the role of insulin in regulation of glucose fluxes during exercise.

The Importance of Chronic Insulinization

We hypothesized that the residual insulin, which was chronically present in the alloxan-diabetic dogs (5.2±1.4 µU/ml) [11], may have played a pivotal role in normalizing the exercise-induced increase in both glucose uptake and MCR. Our hypothesis stems from the contrasting

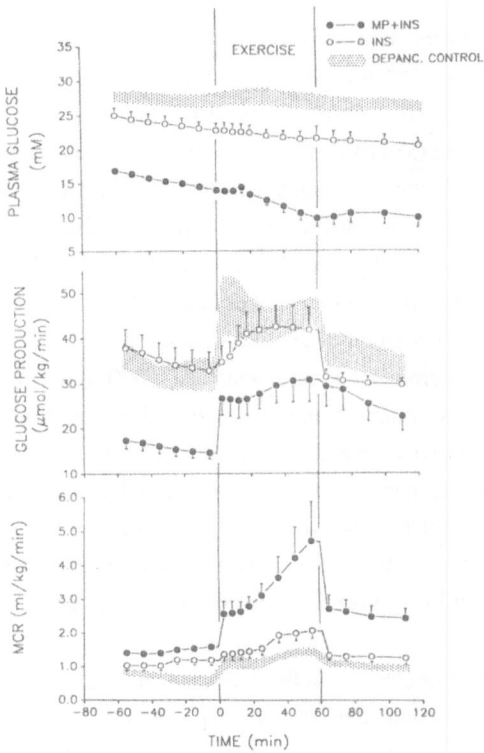

Figure 4. Plasma glucose concentrations and rates of glucose production and metabolic clearance (MCR) in the depancreatized dogs treated with subbasal insulin infusion (INS, n=6) and methylpalmoxirate plus subbasal insulin (MP+INS, n=6) at rest, during and after 60 min treadmill exercise (100 m/min, 10% slope). Data are shown as mean+SE. Shaded areas: mean+SE in untreated controls (n=6). (Modified from reference 42).

increments in glucose uptake and MCR were normalized in alloxan-diabetic dogs (partial insulin deficiency) [11], but not in depancreatized dogs (total insulin deficiency) [10]. In the latter, the exercise-induced increases in glucose uptake and MCR were not normalized even when lipolysis and fatty acid oxidation were both suppressed using a combination of MP+PRO [12,40]. Insulin and muscle contraction are known to act synergistically in increasing glucose uptake and oxidation during exercise. Hyperinsulinemia at various levels can reduce the exercise-induced FFA mobilization in normal humans [41]. However, little is known concerning the mechanisms of action of a minimal amount of insulin, such as was seen in the alloxan-diabetic dogs, in regulating glucose uptake during exercise in the diabetic state.

We examined the glucoregulatory effects of a 1/4 basal amount of insulin infusion at rest and during exercise in the 24h insulin-deprived, depancreatized dogs, with or without concomitant MP treatment to inhibit the FFA-G cycle [42]. The results are illustrated in Figures 4 and 5. With the subbasal intraportal insulin infusion (50 µU.kg⁻¹.min⁻¹) for 2 h, plasma insulin levels were restored to the levels seen in alloxan-diabetic dogs [42]. Resting plasma glucose was reduced by 13%, which was due to an enhancement in MCR, while glucose production was unaffected (Figure 4). The subbasal insulin had an evident antilipolytic effect as reflected by a 30% reduction in both FFA and glycerol, and an antiketotic effect as reflected by a 50% reduction in β-hydroxybutyrate in the basal state (Figure 5). With MP plus subbasal insulin, ketogenesis was almost completely suppressed throughout the entire experiment (Figure 5), although FFA levels were not decreased due to the FFA-sparing effect of MP. The addition of MP to the insulin infusion reduced resting plasma glucose as much as 44%, due not only to a marked increase in MCR ($128\pm4\%$), but also to a decrease in glucose production ($51\pm10\%$). During exercise, subbasal insulin alone did not reduce the exercise-induced increase in glucose production, although it blunted the initial sharp rise (Figure 4). FFA and glycerol levels were maintained lower during exercise with subbasal insulin than in the controls (Figure 5). The fact that the rise in FFA and glycerol during exercise was not suppressed was probably due to the small amount of insulin and/or short duration of insulin action. Subbasal insulin infusion and inhibition of the FFA-G cycle by MP did not decrease the exercise-induced increment in glucose production, but maintained a 30% lower rate of glucose production than the controls and achieved a marked improvement in the exercise-induced increase in MCR (Figure 4). The markedly reduced plasma glucose might also have facilitated the increment in MCR. This is in contrast to the effects of subbasal insulin alone or MP alone, neither of which augmented the exercise-induced rise in MCR. However, the improvement in MCR achieved by combined treatment of subbasal insulin and MP reached only one third normal values. MCR in the 1/4 basal insulin treated dogs was still lower than that observed in alloxan-diabetic dogs, despite the same insulin levels.

The above comparison suggests that the derangements in glucose and lipid metabolism due to chronic insulin deficiency could only be partially reversed by the acute insulin infusion in depancreatized dogs. The chronic effects of residual insulin, which is present in alloxan diabetic dogs but absent from depancreatized dogs, may be important for both direct and indirect actions on glucose transport and metabolism. Its indirect actions may be required in correcting hyperglycemia (as discussed later) and providing a favorable metabolic milieu upon which a suppression of lipid metabolism can effect the normalization of glucose turnover during exercise.

Improvement in Glucose Clearance with Normalization of Hyperglycemia

In both type I and II diabetes, the rate of glucose utilization may be maintained at normal or near normal values. However, this is achieved at the expense of an expanded glucose pool size characterized by hyperglycemia and a reduction in MCR [43]. In the postabsorptive state, diabetes is characterized by hyperglycemia which is due primarily to increased glucose production. In the presence of hyperglycemia the mass effect of increased glucose concentration in circulation compensates for the deficiency in MCR in working muscle. Thus the defect in tissue glucose extraction in diabetes and its correction is best indicated by the changes in MCR [39]. During exercise, an important variable involved in regulation of systemic MCR is the prevailing glucose level. Our data indicate that hyperglycemia contributed to the impairment in MCR in the diabetic dogs during exercise. Therefore, prevention and correction of hyperglycemia constitutes an important aspect of the indirect effects of insulin in normalization of MCR.

In our exercise experiments, normoglycemia and a high MCR in the normal dogs are contrasted with the hyperglycemia and diminished MCR in partial and total insulin-deficient dogs, although total glucose uptake can be near normal [10-12]. As shown in Figure 6, a highly significant inverse correlation exists between the exercise-induced rise in MCR and plasma glucose (range 6-25 mM) [42] when glycemia was decreased by several different means, both in the present study and in three earlier studies [10-12,23]. This appears consistent with previous findings in streptozotocin-diabetic

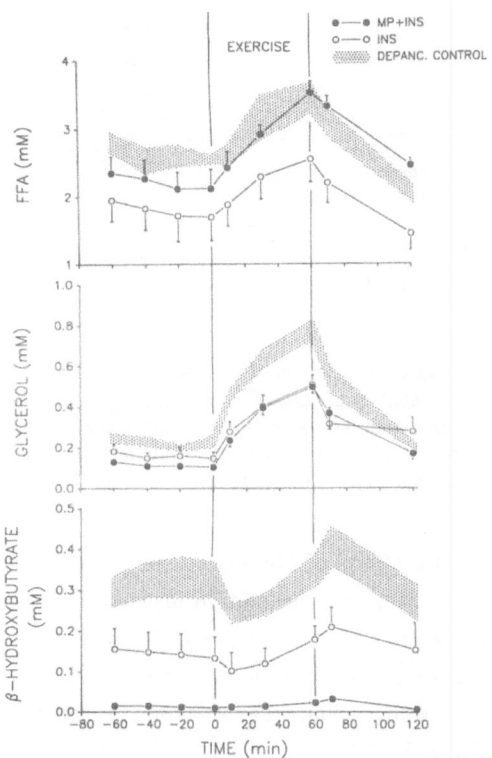

Figure 5. Plasma concentrations of free fatty acids (FFA), glycerol and β-hydroxybutyrate in the depancreatized dogs treated with subbasal insulin infusion (INS, n=6) and methylpalmoxirate plus subbasal insulin (MP+INS, n=6) at rest, during and after 60 min treadmill exercise (100 m/min, 10% slope). Data are shown as mean±SE. Shaded areas: mean±SE in untreated controls (n=6). (Reproduced with permission from reference 42).

rats at rest [4,44]. The mechanism of action of hyperglycemia on MCR may be different at rest and during exercise. In the diabetic rats, chronic hyperglycemia downregulated the glucose transport system and decreased the number of glucose transporters (GLUT 4) in the plasma membrane of skeletal muscle as evidenced by both glucose transporter protein and cytochalasin B binding assays. Correction of hyperglycemia using phlorizin, which induces glycosuria by inhibiting Na+-glucose co-transport in the renal tubules, completely restored the glucose transport system, reflected by normalized GLUT 4 number in the plasma membrane of the skeletal muscle [44]. This study suggested that with normalization of glycemia, GLUT4 transporters migrate from the intracellular compartment to the plasma membrane. In the partially pancreatized or alloxan-diabetic rats, correction of hyperglycemia with continuous administration of phlorizin improved glucose utilization and normalized tissue sensitivity to insulin [45]. Conversely, redistribution of glucose transporters from the plasma membrane to intracellular sites may be the mechanism by which hyperglycemia decreases resting MCR, despite normal or near normal glucose uptake rates due to the mass effect. During exercise, hyperglycemia might have a restraining effect on the exercise-induced increase in glucose transporters and MCR. The increase in MCR during exercise is stimulated by muscle contraction which requires increased energy provision from circulating glucose. It is conceivable that when the glucose requirement of the contracting muscle is satisfied by the greater mass effect of hyperglycemia, the signal that stimulates the rise in MCR during exercise can be turned off.

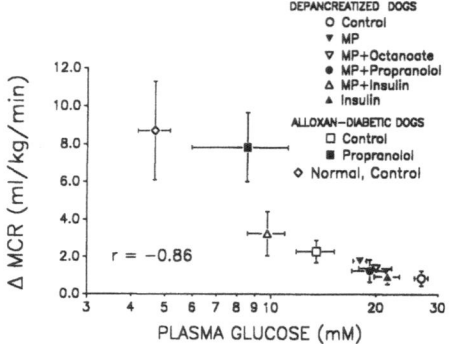

Figure 6. The inverse correlation (r=-0.86, p<0.01) between the circulating glucose levels and the exercise-induced increments in MCR at the end of 60 min moderate exercise in normal dogs, alloxan diabetic dogs (untreated or treated with propranolol), and depancreatized dogs (untreated or treated with MP, MP + octanoate, MP + propranolol, MP + subbasal insulin and subbasal insulin alone). Data were summarized from four studies. (Reference 10-12, 23, 42).

Therefore, the combined effects of MP and insulin on the improvement of MCR in our exercising, diabetic dogs can be partly accounted for by the glucose lowering effect of both agents. In conclusion, failure to fully correct hyperglycemia and increased levels of FFA are responsible, at least in part, for the failure of subbasal insulin infusion to normalize glucose uptake and MCR during exercise. However, adequate insulinization, at a dose substantially greater than the subbasal infusion given in the present study, may be essential not only for its indirect effects in inhibiting lipolysis and in lowering plasma glucose levels, but also for its direct effects in facilitating glucose uptake in muscle. In addition, it is probably important that insulin not only be acutely replaced but some insulin also be present chronically. The chronic effects of insulin on the regulation of muscle glucose uptake during exercise may in turn be direct or indirect.

INDIRECT EFFECTS OF INSULIN ON HEPATIC GLUCOSE CYCLING AND GLUCOSE PRODUCTION

Glucose Cycling is Inhibited by Insulin's Effects on Substrate Suppression

Increased rates of hepatic glucose cycling (glucose phosphorylation/dephosphorylation, GC) has been found to be indicative of glucose intolerance. Glucose phosphorylation and dephosphorylation reactions are catalyzed by glucokinase and glucose-6-phosphatase, respectively. The rate of GC is therefore a function of the enzyme that catalyzes the reaction opposite the net flux. In the postabsorptive state, GC is a function of glucokinase. Normal postabsorptive GC is reportedly 14% of hepatic glucose production in man [46] and 13% in dogs [47]. Insulin deprivation in the depancreatized dog (model of IDDM) resulted in a 7-fold higher rate of GC than in normal dogs [47,48], whereas glucose output (flux through glucose-6-phosphatase) and production were 3- and 2-fold greater, respectively. We have shown that GC also increased to a greater extent than glucose production in humans with glucose intolerance, and in mild, lean and obese type II diabetics [46,49]. In the postabsorptive state, when the net flux of glucose across the liver is toward glucose production, increased GC in diabetes may represent a compensatory mechanism limiting the extent of hyperglycemia [46]. Postprandially, when the net flux is toward liver glucose uptake, increased GC

160

may contribute to glucose intolerance. Increased GC in both type I and II diabetes appear to be related to hyperglycemia and defective insulin action. Counterregulatory hormones such as glucagon and prednisolone have been shown to increase GC [47]. However, the regulatory roles of insulin and glycemia with respect to GC are still unclear, since hyperglycemia alone does not increase GC while insulin treatment per se cannot decrease GC.

We hypothesize that improvement in GC may require the correction of both derangements in insulin deficiency and hyperglycemia. Our hypothesis was examined in a preliminary study in 24 h insulin deprived, anesthetized, depancreatized dogs [50]. We wished to determine whether insulin infusion with concurrent normalization of glycemia can decrease GC acutely and whether such decrease in GC, if achieved, is due to changes in activities of the enzymes involved. GC was determined as the difference between the rates of glucose appearance (Ra) measured with $[2\text{-}^3H]$ and $[6\text{-}^3H]$ glucose tracers. $[2\text{-}^3H]$ Glucose is detritiated in the phosphoglucose isomerase reaction, which is in equilibrium with the reactions involved in the GC. Freeze-clamped liver biopsies were taken at 30-60 min intervals for measurements of metabolites and enzyme activities. The decrease in plasma glucose was associated with a decrease in glucose production (from 22.4±2.6 to 9.8±0.2 $\mu mol.kg^{-1}.min^{-1}$) and an increase in MCR (from 1.0±0.2 to 2.4±0.3 $ml.kg^{-1}.min^{-1}$). Glucose cycling declined from 16.7±6.1 to 2.2±1.1 $\mu mol.kg^{-1}.min^{-1}$,p<0.01). In the meantime, hepatic hexose-6-phosphates decreased from 150±56 to 54±37 nmol/g, whereas total activities of glucose-6-phosphatase (23±11 vs 24±12 U/g) and glucokinase (1.2±0.8 vs 0.8±0.5 U/g) did not change significantly.

This preliminary study suggests: 1) Acute correction of both insulin deficiency and hyperglycemia significantly decreased glucose cycling in the depancreatized dog. Therefore, the elevated rates of glucose cycling in diabetes may be due to the combination of insulin deficiency (absolute or relative, with or without glucagon excess) and hyperglycemia; 2) Since the decrease in glucose cycling was accompanied by decreases in both plasma glucose and hepatic hexose-6-phosphates whereas glucose-6-phosphatase and glucokinase did not change, the reduction in glucose cycling with insulin-induced normoglycemia appears to be substrate-mediated. However, we cannot rule out the possibility that the activation states of the enzymes can be affected by the changes in insulin and glucose levels. Therefore, some of the actions of insulin on glucose cycling appear to be indirect, through lowering of glycemia (and consequently hepatocellular glucose) and glucose-6-phosphate. In addition, direct actions of insulin on glucokinase and/or glucose-6-phosphatase (content or actions at various substrate concentrations) may also be involved in the regulation of glucose cycling.

Inhibition of Hepatic Glucose Production by Peripheral Effects of Insulin

Our recent studies have shown that insulin can inhibit glucose production not only through a direct hepatic action, but also indirectly, possibly by reducing the provision of substrates and energy from peripheral tissues to the liver, thus diminishing gluconeogenesis, and/or by suppression of glucagon secretion.

A physiological gradient in insulin concentration exists between the portal circulation and the systemic circulation. Insulin secreted from the pancreatic β-cells is degraded in the liver by approximately 50%. The higher insulin concentration in the portal circulation is believed to be important in suppressing hepatic glucose production, postprandially or under other physiological circumstances, when needed. In diabetes treated with subcutaneous (peripheral) insulin injections this porto-peripheral gradient is lost. It is generally believed that without such a gradient, peripheral hyperinsulinemia is necessary to raise the level of portal insulin delivery in order to inhibit the increased rate of glucose production. Intraperitoneal insulin administrations have also been proposed to simulate a more physiological route of insulin delivery. However, it is questionable whether suppression of hepatic glucose production is dependent solely on portal insulin levels. Since gluconeogenesis is substantially dependent upon the availability of the precursors, such as alanine, glycerol and pyruvate, and the energy provided by oxidation of free fatty acids, we hypothesize that control of the peripheral precursors and free fatty acids may be an important aspect of insulin's actions on hepatic glucose production.

We have studied the differential effects of portal and peripheral insulin infusions in the following 3 groups of moderately hyperglycemic, depancreatized dogs: 1) Portal infusion group in which insulin was given in an initial bolus injection of 54 $pmol.kg^{-1}$ followed by a 5.4 $pmol.kg^{-1}.min^{-1}$ constant infusion (n=7); 2) Peripheral insulin infusion group which received an equimolar insulin infusion into the superior vena cava through a chronic cannulation in the jugular vein; 3) a 1/2 peripheral insulin group in which peripheral insulin was given at 1/2 dose of that in group 2 to provide a peripheral insulin concentration comparable to that seen in group 1. In this study, the total insulin load was matched between groups 1 and 2, but portal insulin levels were higher in group 1, while peripheral insulin levels were higher in group 2 (Figure 7). On the other hand, peripheral insulin levels were matched between groups 1 and 3, while portal insulin concentrations were more than 2-fold higher in group 1 than group 3. Glucose turnover was measured using HPLC-purified [6-^{3}H] and [2-^{3}H]-glucose. Glucose specific activity was maintained constant with the matched-step-tracer-infusion (MSTI) technique to prevent errors in estimating glucose production due to fluctuations in plasma specific activity. The design of this study thus allows us to identify whether the hepatic or the peripheral actions of insulin are of greater importance in suppressing hepatic glucose production. Our results demonstrated the predominant importance of peripheral insulin in suppressing hepatic glucose production in diabetes (Figure 7). When peripheral insulin levels were matched, not only the increment in glucose uptake, but also the suppression of glucose production were the same between groups 1 and 3 ($55\pm7\%$ vs $63\pm4\%$). At the same infusion rate, insulin delivered peripherally was significantly more potent in suppressing hepatic glucose production than with insulin delivered portally ($73\pm7\%$ vs $55\pm7\%$, $p<0.001$). Our results are in accordance with another study in which the differential effects of portal and peripheral insulin infusions were examined in normal dogs with euglycemic glucose clamps [14]. In that study, endogenous insulin and glucagon secretion were suppressed by somatostatin and replaced by exogenous infusions. When peripheral insulin levels were matched, glucose production was suppressed to the same extent by portal and peripheral insulin infusions, despite a much higher hepatic insulinization with the portal modality. Thus, the importance of peripheral effects of insulin on hepatic glucose production has been illustrated in both normal and diabetic states.

The mechanism for this presumably indirect effect of insulin in inhibiting glucose production has not been determined. However, with equimolar peripheral insulin infusion, which suppressed glucose production more than the portal and half dose peripheral infusions, we also observed a greater suppression of precursors (alanine and glycerol) and energy substrates (FFA) for gluconeogenesis and a greater, although not statistically significant suppression of peripheral glucagon concentrations.

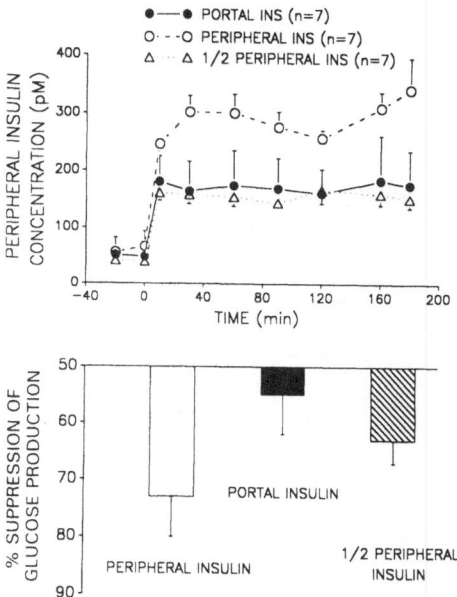

Figure 7. Peripheral insulin concentrations (upper panel) and percent suppression of hepatic glucose production (lower panel) achieved in the equimolar portal (n=7) and peripheral (n=7) insulin infusions, and half dose peripheral insulin (n=7) infusion experiments in the depancreatized dog. (Modified from reference 13).

THE PERMISSIVE EFFECT OF INSULIN IN MODULATING GLUCOSE UPTAKE IN STRESS

Fluxes and metabolism of carbohydrate, protein and fat are rapidly and markedly altered in response to stress [51-53]. Stress in general stimulates glycogenolysis, gluconeogenesis, proteolysis, lipolysis and ketogenesis. Glucose utilization may increase or decrease depending on the nature of stress. The overall glucoregulatory response is to ensure sufficient glucose provision to the brain, at the expense of increasing glucose release from glycogen stores and sparing glucose uptake from other tissues. The central nervous system plays an important role in coordinating the glucoregulatory responses to a variety of stresses, such as psychological stress [54,55], physical stress and trauma [56-58], hypoglycemia and exercise [59-61]. Both the hypothalamus-pituitary-adrenal cortex (HPA) axis and the sympathetic nervous system are stimulated during stress. Circulating levels of epinephrine, norepinephrine, cortisol and glucagon are increased to various extents depending on the cause and severity of the particular stress. Changes in these hormones alter the metabolic patterns of the effector tissues such as the liver, adipose tissue and skeletal muscle. Under normoinsulinemic conditions, the catabolic effects of these counterregulatory hormones can be restrained and modulated by direct and/or indirect effects of insulin. In diabetes, however, due to insulin deficiency and/or chronic insulin resistance, stress may worsen the existing metabolic derangements or trigger potential diabetic complications, such as hyperglycemia and ketoacidosis.

We have previously studied glucose metabolism in a stress model induced by intracerebroventricular (ICV) injection of carbachol, an acetylcholine analog, in the dog. ICV carbachol stimulated muscarinic cholinergic receptors in the hypothalamus and paraventricular nuclei which in turn activated the sympathetic nervous system and the HPA axis. In the ICV-carbachol

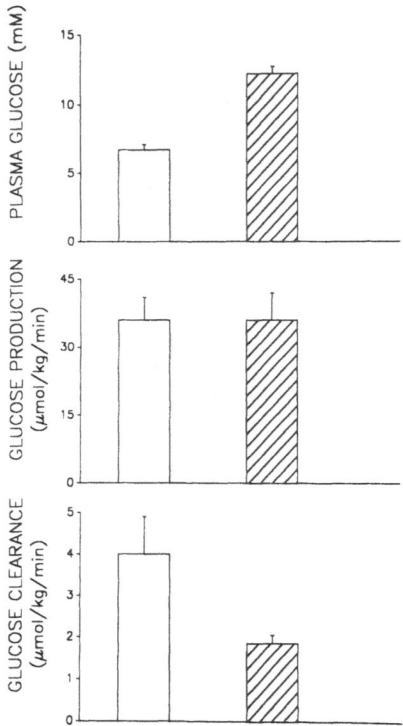

Figure 8. Plasma glucose concentrations, glucose production rates and glucose metabolic clearance rates in the normal and alloxan diabetic dogs treated with intra-cerebroventricular (ICV) injections of carbachol. (Modified from reference 15).

injected dog, there was a concomitant increase in glucose production, glucose utilization and MCR [15] (Figure 8). Meanwhile, epinephrine, norepinephrine and cortisol all rose markedly. Interestingly, the increase in glucose uptake and MCR occurred without a measurable increment in insulin levels. ICV administration of an octapeptide somatostatin (SS-8) prior to carbachol markedly suppressed the increments of epinephrine and cortisol, partially suppressed norepinephrine, and abolished the increase in glucose production and utilization [62]. In the same dogs after induction of diabetes by alloxan, ICV-carbachol injection resulted in a greater hyperglycemic response and a greater increase in norepinephrine, but neither epinephrine nor cortisol levels were different from those of normal dogs. This greatly augmented hyperglycemic response was due mainly to the fact that MCR did not increase in diabetic dogs (Fig 8). The increment in MCR during stress, as in exercise, presumably also requires a permissive effect of insulin [62]. These findings imply that the increased glucose utilization might be mediated by an unknown neural pathway, which is stimulated by ICV carbachol and inhibited by SS-8 [15,40,62]. This unknown neural pathway is hypothesized to be catecholaminergic, although catecholamines are usually thought to decrease rather than increase glucose clearance [63-65]. The rise and fall of catecholamine levels, especially epinephrine, paralleled the rise and fall of glucose utilization after ICV-carbachol with or without SS-8. The changes in cortisol probably cannot account for the rapid changes in glucose turnover due to the nature of its slow-action [15,62]. Our hypothesis, that an adrenergic neural pathway mediated the increased glucose uptake in stress, is supported by a number of recent studies. In the rat, stimulation of the ventromedial hypothalamus, which suppresses insulin secretion in the pancreas, increases glucose uptake in the heart and skeletal muscle and brown adipose tissue. This increase was prevented by adrenergic denervation of the tissues and potentiated by monoamino-oxidase inhibitors [66]. It has also been demonstrated that norepinephrine can stimulate glucose transport in isolated rat adipocytes via β-adrenergic pathways [67]. In the in vitro studies, it has been shown that in the absence of insulin and FFA, the catecholamines are able to stimulate glucose transport in the muscle via a β-adrenergic mechanism [68,69]. However, in the presence of FFA or insulin, the catecholamines were found to inhibit insulin [70] or non-insulin-stimulated [71] glucose transport. It thus appears that the putative neuro-adrenergic stimulation of glucose uptake is multifaceted and is greatly influenced by other hormone(s) and metabolite(s). Our model of stress induced by ICV carbachol can be of great use in identifying the neural or neuroendocrine factor(s) responsible for increased glucose uptake.

The permissive action of insulin on glucose uptake in this model of stress resembles the permissive action of insulin on the exercise-induced glucose uptake. Similarly, it might consist of both direct and indirect effects.

SUMMARY

Metabolism of fuels is driven by the energy demand of the organism and its regulation is influenced by many hormonal and metabolic factors. Insulin is of utmost importance in regulating glucose metabolism by promoting glucose uptake in the insulin-sensitive tissues for energy consumption and/or storage. The effects of insulin on glucose metabolism can be both direct and indirect. Ample evidence has indicated that insulin directly stimulates glucose transport systems in the target tissues. However, the changes in glucose fluxes can also be brought out by indirect effects of insulin which are produced secondary to the insulin-induced changes in other hormones and metabolites. In this chapter, we discussed a number of examples of insulin's indirect effects on glucose metabolism. We demonstrated that insulin can indirectly promote muscle glucose uptake during exercise by restraining the release and oxidation of fatty acids and decrease of hyperglycemia. We have presented some evidence for an indirect regulation of glucose cycling by insulin. We have also demonstrated the importance of the peripheral levels of insulin for insulin-induced inhibition of hepatic glucose production. This presumably indirect effect of peripheral insulin might consist of 1) suppression of the release of energy substrates and gluconeogenic precursors; and 2) suppression of glucagon secretion. In a carbachol-induced stress model, insulin is not required for a putatively neural regulation of an increase in systemic glucose uptake but a "permissive" effect of insulin is essential. These studies underscore the importance of the interactions between insulin and other hormones and metabolites as opposed to insulin's direct actions per se.

REFERENCES

1. M. Vranic, Glucose turnover: A key to understanding the pathogenesis of diabetes. Indirect effects of insulin. The Banting Lecture, *Diabetes* 41:1188 (1992).
2. A. Klip, T. Ramlal, D.A. Young and J.O. Holloszy, Insulin-induced translocation of glucose transporters in rat hindlimb muscles, *FEBS Lett* 224:224 (1987).
3. L.J. Warzala and B. Jeanrenaud, Potential mechanism of insulin action on glucose transport in isolated

diaphragm. Apparent translocation of intracellular transport units to the plasma membrane, *J Biol Chem* 256:7090 (1981).

4. A. Klip, A. Marette, D. Dimitrakoudis, T. Ramlal, A. Giacca, Z. Shi and M. Vranic, Effect of diabetes on glucoregulation. From glucose transporters to glucose metabolism in vivo, *Diabetes Care* 15(11):1747 (1992).

5. A. Klip, T. Ramlal, P.J. Bilan, G.D. Cartee, E.A. Gulve and J.O. Holloszy, Recruitment of GLUT-4 glucose transporter by insulin in diabetic rat skeletal muscle, *Biochem Biophys Res Comm* 172:728 (1990).

6. A. Marette, E. Burdett, A. Douen, M. Vranic and A. Klip, Insulin induces the translocation of GLUT4 glucose transporters from a unique intracellular organelle to transverse tubules in rat skeletal muscle, *Diabetes* 41(12):1562 (1992).

7. P.S. Walker, T. Ramlal, J.A. Donovan, T.P. Doering, A. Sandra, A. Klip and J.E. Pessin, Insulin and glucose-dependent regulation of the glucose trnaport system in the rat L6 skeletal muscle line, *J Biol Chem* 264:6587 (1989).

8. S.W. Cushman and L.J. Warzala, Potential mechanism of insulin action on glucose transport in the isolated rat adipose cell. Apparent translocation of intracellular transport systems to the plasma membrane, *J Biol Chem* 255:4748 (1980).

9. A.P. Thomas and R.M. Denton, Use of toluene-permeabilized mitochondria to study the regulation of adipose tissue pyruvate dehydrogenase in situ, *Biochem J* 238:93 (1986).

10. O. Bjorkman, P. Miles, D. Wasserman, L. Lickley and M. Vranic, Regulation of glucose turnover in pancreatectomized, totally insulin deficient dogs: effects of beta-adrenergic blockade, *J Clin Invest* 81:1759 (1988).

11. D.H. Wasserman, H.L.A. Lickley and M. Vranic, Role of beta-adrenergic mechanisms during exercise in poorly-controlled insulin deficient diabetes, *J Appl Physiol* 59:1282 (1985).

12. K. Yamatani, Z. Shi, A. Giacca, R. Gupta, S. Fisher, L. Lickley and M. Vranic, Role of FFA-glucose cycle in glucoregulation during exercise in total absence of insulin, *Am J Physiol* 263:E646 (1992).

13. A. Giacca, S. Fisher, Z.Q. Shi, R. Gupta, L. Lickley and M. Vranic, Importance of peripheral insulin levels for insulin-induced suppression of glucose production in depancreatized dogs, *J Clin Invest* 90:1769 (1992).

14. M. Ader and R.N. Bergman, Peripheral effects of insulin dominate suppression of fasting hepatic glucose production, *Am J Physiol* 245:E1020 (1990).

15. P. Miles, K. Yamatani, L. Lickley and M. Vranic, Mechanism of glucoregulatory responses to stress and their deficiency in diabetes, *Proc Natl Acad Sci U. S. A.* 88(4):1296 (1991).

16. J.D. McGarry and D.W. Foster, *Ketogenesis*, in:Diabetes mellitus: Theory and practice, H. Rifkin and D. Porte, eds.,Elsevier Science Publishing Co., Inc., New York (1990).pp.292.

17. L.V. Basso and R.J. Havel, Hepatic metabolism of free fatty acids in normal and diabetic dog, *J Clin Invest* 49:537 (1970).

18. M. Vranic, R. Kawamori, S. Pek, N. Kovacevic and G. Wrenshall, The essentiality of insulin and the role of glucagon in regulating glucose utilization and production during strenuous exercise in dogs, *J Clin Invest* 57:245 (1976).

19. D.H. Wasserman, J.L. Bupp, J.L. Johnson, D. Bracy and D.B. Lacy, Glucoregulation during rest and exercise in depancreatized dogs: role of the acute presence of insulin, *Am J Physiol* 262:E574 (1992).

20. T. Ploug, H. Galbo, J. Vinten, M. Jorgensen and E. Richter, Kinetics of glucose transport in rat muscle: effects of insulin and contractions, *Am J Physiol* 253:E12 (1987).

21. H. Wallberg-Henriksson, S.H. Constable, D.A. Young and J.O. Holloszy, Glucose transport into rat skeletal muscle: interaction between exercise and insulin, *J Appl Physiol* 65:909 (1988).

22. R. Nesher, I.E. Karl and K.M. Kipnis, Dissociation of the effect(s) of insulin and contraction on glucose transport in rat epitrochlearis muscle, *Am J Physiol* 249:C226 (1985).

23. D.H. Wasserman, H.L.A. Lickley and M. Vranic, Important role of glucagon during exercise and diabetes, *J Appl Physiol* 59:1272 (1985).

24. P.J. Randle, E.A. Newsholme and P.B. Garland, Regulation of glucose uptake by muscle. 8. Effects of fatty acids and ketone bodies and pyruvate, and of alloxan-diabetes and starvation on the uptake and metabolic fate of glucose in rat heart and diaphragm muscle, *Biochem J* 93:652 (1964).

25. P.J. Randle, P.B. Garland, C.N. Hales and E.A. Newsholme, The glucose-fatty acid cycle: its role in insulin sensitivity and the metabolic disturbances of diabetes mellitus, *Lancet* 1:785 (1963).

26. M.J. Rennie and J.O. Holloszy, Inhibition of glucose uptake and glycogenolysis by availability of oleate in well-oxygenated perfused skeletal muscle, *Biochem J* 168:161 (1977).

27. D. Thiebaud, R.A. DeFronzo, E. Jacot, A. Golay, K. Acheson, E. Maeder, E. Jecquier and J.P. Felber, Effect of long chain triglyceride infusion on glucose metabolism in man, *Metab* 31:1128 (1982).

28. P.J. Randle, *Molecular mechanisms regulating fuel selection in muscle*, in:Biochemistry of exercise, J.R. Poortsman and G. Niset, eds.,Baltimore University Park Press, Baltimore (1983).pp.13.

29. P.J. Randle, *Interaction of fatty acid and glucose metabolism*, in:Diabetes 1988, R. Larkins, P. Zimmet and D. Chisholm, eds.,Elsevier Science Publishers B.V. (Biomedical Division), (1989).pp.163.

30. B.M. Wolfe, S. Klein, E.J. Peters, B.F. Schmidt and R.R. Wolfe, Effect of elevated free fatty acids on glucose oxidation in normal humans, *Metab* 37(4):323 (1988).

31. G.F. Tutwiler, R. Mohrbecker and W. Ho, Methyl 2-tetradecylglycidate, an orally effective hyperglycemic agent that inhibits long chain fatty acid oxidation selectively, *Diabetes* 28:242 (1979).

32. J.C. Young, J.E. Bryan and S.H. Constable, Effects of the oral hypoglycemic agent methyl palmoxirate on exercise capacity of rats, *Can J Physiol Pharmacol* 62:815 (1984).

33. J.W. Bailey, M.D. Jensen and J.M. Miles, Effects of intravenous methyl palmoxirate on the turnover and oxidation of fatty acids in conscious dogs, *Metab* 40(4):428 (1991).

34. R.W. Tuman, G.F. Tutwiler, J.M. Joseph and N.H. Wallace, Hypoglycemic and hypoketonemic effects of single and repeated oral doses of methyl palmoxirate (methyl 2-tetradecylglycidate) in streptozotoic/alloxan-induced diabetic dogs, *Br J Pharmacol* 94:130 (1991).

35. G.F. Tutwiler and H.J. Brentzel, Relation of oxidation of long-chain fatty acids to gluconeogenesis in the perfused liver of the guinea pig: effect of 2-tetradecylglycidic acid (McN-3802), *Eur J Biochem* 124:465 (1982).

36. L. Mandarino, E. Tsalikian, S. Bartold, H. Marsh, A. Carney, E. Buerklin, G. Tutwiler, M. Haymond, B. Handwerger and R. Rizza, Mechanism of hyperglycemia and response to treatment with an inhibitor of fatty acid oxidation in a patient with insulin resistance due to antiinsulin receptor antibodies, *J Clin Endocrinol Metab* 59:658 (1984).

37. A. Consoli, N. Nurjhan, F. Capani and J. Gerich, Predominant role of gluconeogenesis in increased hepatic glucose production in NIDDM, *Diabetes* 38:550 (1989).

38. R.W. Stevenson, P.E. Williams and A.D. Cherrington, Role of glucagon suppression on gluconeogenesis during insulin treatment of the conscious diabetic dog, *Diabetologia* 30:782 (1987).

39. J. Radziuk and H.L.A. Lickley, The metabolic clearance of glucose: measurment and meaning, *Diabetologia* 28:315 (1985).

40. Z.Q. Shi, A. Giacca, K. Yamatani, P.D.G. Miles, S. Fisher, M. Van Delangeryt, L. Lickely and M. Vranic, *Glucose uptake in exercise and stress: The indirect role of insulin*, in:Proceedings of the Diabetes Mellitus and Exercise Symposium, J. Devlin, E.S. Horton and M. Vranic, eds.,Smith-Gordon and Company Ltd, London (1992).pp.101.

41. D.J. Wasserman, R.J. Geer, D.E. Rice, D. Bracy, P.J. Flakoll, L.L. Brown, J.O. Hill and N.N. Abumrad, Interaction of exercise and insulin action in humans, *Am J Physiol* 260:E37 (1991).

42. Z.Q. Shi, A. Giacca, K. Yamatani, S. Fisher, L. Lickley and M. Vranic, Effects of subbasal insulin infusion on resting and exercise-induced glucose uptake in depancreatized dogs, *Am J Physiol* 264:E334 (1993).

43. R.A. DeFronzo, R.C. Bonadonna and E. Ferrannini, Pathogenesis of NIDDM: A balanced overview, *Diabetes Care* 15(3):318 (1992).

44. D. Dimitrakoudis, T. Ramlal, S. Rastogi, M. Vranic and A. Klip, Glycemia regulates the glucose transporter number in the plasma membrane of rat skeletal muscle, *Biochem J* 284(2):341 (1992).

45. L. Rossetti, D. Smith, G.I. Shuman, D. Papachristou and R.A. DeFronzo, Correction of hyperglycemia with phlorizin normalizes tissue sensitivity to insulin in diabetic rats, *J Clin Invest* 79:1510 (1987).

46. S. Efendic, S. Karlander and M. Vranic, Mild type II diabetes markedly increases glucose cycling in the postabsorptive state and during glucose infusion irrespective of obesity, *J Clin Invest* 81:1953 (1988).

47. B. Issekutz, Studies on hepatic glucose cycles in normal and methylprednisolone-treated dogs, *Metab* 26:157 (1977).

48. H.L.A. Lickley, G.G. Ross and M. Vranic, Effects of selective insulin or glucagon deficiency on glucose turnover, *Am J Physiol* 236:E255 (1979).

49. S. Efendic, A. Wajngot and M. Vranic, Increased activity of the glucose cycle in the liver: Early characteristic of type 2 diabetes, *Proc Natl Acad Sci U. S. A.* 82:2965 (1985).

50. A. Giacca, Z.Q. Shi, S.J. Fisher, G. van de Werve and M. Vranic, Acute correction of both hyperglycemia and hypoinsulinemia decreases elevated glucose cycling in diabetes, *Diabetes* 41(suppl.):(1992).(Abstract)

51. H. Sasaki, s. Marubashi, Y. Yawata and et al, *Neuropeptides and glucose metabolism*, in:The brain as an endcrine organ, Endocrinology and Metabolism III, Progrss in Research and Clinical Practice, M.P. Cohen and P.P. Foa, eds.,Springer-Verlag, New York (1988).pp.150.

52. M.R. Brown and L.A. Fisher, Corticotropin-releasing factor: effects on the autonomic nervous system and visceral systems, *Fed Proc* 44:243 (1985).

53. M.R. Brown, L.A. Fisher, J. Spiess, J. Rivier, C. Rivier and W. Vale, Corticotropin-releasing factor: actions on the sympathetic nervous system and metabolism, *Endocrinol* 111:928 (1982).

54. D.A. Fisher and M.R. Brown, Somatostatin analog: Plasma catecholamine suppression mediated by the central nervous system, *Endocrinol* 107:714 (1980).

55. F.W. Kemmer, R. Bisping, H.J. Steingruber, H. Baar, F. Hartman, R. Schlagheche and M. Berger, Psychological stress and metabolic control in patients with type 1 diabetes mellitus, *N Engl J Med* 314:1078 (1986).

56. N.J. Christensen and O. Brandesborg, The relationship between plasma catecholamine concentration and pulse rate during exercise and standing, *Eur J Clin Invest* 3:299 (1973).

57. D.W. Wilmore, C.A. Lindsey, J.A. Moylan, G.R. Faloona, B.A. Pruitt and R.H. Unger, Hyperglucagonemia after burns, *Lancet* 1:73 (1974).

58. D.W. Wilmore, J.M. Long and A.D. Mason, Catecholamines: mediator of the hypermetabolic response to thermal injury, *Ann Surg* 180:653 (1974).

59. W.J. Kraemer, J.F. Patton, H.G. Knuttgen, L.J. Marchitalli, C. Cruthirds, A. Damokosh, E. Harman, P. Frykman and J.E. Dziados, Hypothalamic-pituitary-adrenal responses to short-duration high-intensity cycle exercise, *J Appl Physiol* 66(1):161 (1989).

60. A.J.W. Scheurink, A.B. Steffens, H. Bouritius, G.H. Dreteler, R. Bruntink, R. Remie and J. Zaagsma, Sympathoadrenal influence of glucose, FFA, and insulin levels in exercising rats, *Am J Physiol* 25:R161 (1989).

61. J. Vissing, G.A. Iwamoto, K.J. Rybicki, H. Galbo and J.H. Mitchell, Mobilization of glucoregulatory hormones and glucose by hypothalamic locomotor centers, *Am J Physiol* 20:E722 (1989).

62. P.D.G. Miles, K. Yamatani, H.L.A. Lickley and M. Vranic, The intracerebroventricular injection of a somatostatin analog (ODT8-SS) suppresses the stress response in normal and diabetic dogs, *Program of International Symposium on Somatostatin,Montreal,Canada* abstract #68:64 (1989).(Abstract)

63. D.E. Gray, H.L.A. Lickley and M. Vranic, Physiologic effects of epinephrine on glucose turnover and plasma free fatty acid concentrations mediated independently of glucagon, *Diabetes* 29:600 (1980).

64. R.A. Rizza, M. Haymond, P. Cryer and J. Gerich, Differential effects of epinephrine on glucose production and disposal in man, *Am J Physiol* 237:E356 (1979).

65. K.M.A. El-Tayeb, P.L. Brubaker, M. Vranic and H.L.A. Lickley, Beta-endorphin modulation of the glucoregulatory effects of repeated epinephrine infusion in normal dogs, *Diabetes* 34:1293 (1985).

66. Y. Minokoshi, M. Sudo and T. Shinazu, Neural control of glucose uptake in peripheral tissues, *Diabetes* 40(suppl. 1):153A (1991).(Abstract)

67. A. Marette and L. Bukowiecki, Stimulation of glucose transport by insulin and norepinephrine in isolated rat brown adipocytes, *Am J Physiol* 257:C714 (1989).

68. I. Bihler, P.C. Sawh and I.G. Sloan, Dual effect of adrenalin on sugar transport in rat diaphragm muscle, *Biochem Biophys Acta* 510:349 (1978).

69. I.G. Sloan, P.C. Sawh and I. Bihler, Influence of adrenalin on sugar transport in soleus, a red skeletal muscle, *Mol Cell Endorinol* 10:3 (1978).

70. J.L. Chiasson, H. Shikama, D.T.W. Chu and J.H. Exton, Inhibitory effect of epinephrine on insulin-stimulated glucose uptake by rat skeletal muscle, *J Clin Invest* 68:706 (1981).

71. D.A. Young, H. Wallberg-Henriksson, J. Cranshaw, M. Chen and J.O. Holloszy, Effects of catecholamine on glucose uptake and glycogenolysis in rat, *Am J Physiol* 248:C406 (1985).

THE INSULIN-ANTAGONISTIC EFFECT OF THE COUNTERREGULATORY HORMONES - CLINICAL AND MECHANISTIC ASPECTS

Ulf Smith, Stig Attvall, Jan Eriksson, Jesper Fowelin,
Peter Lönnroth and Christian Wesslau
Dept. of Internal Medicine, University of Göteborg
Göteborg, Sweden

INTRODUCTION

The counterregulatory hormones (catecholamines, glucagon, cortisol and growth hormone) can modulate the effects of insulin and, thus, influence the diabetic state under a variety of conditions. This can be beneficial to the diabetics when the hormones counteract the effect of an excessive insulin level, such as a hypoglycemia. However, under other situations they may be to a disadvantage since the glucose control deteriorates and the insulin requirements increase unpredictably.

The physiological effects of the counterregulatory hormones on hepatic glucose production are extensively dealt with by Cherrington and coworkers elsewhere in this issue and will not be further discussed. This short overview will focus on the role of these hormones in modulating glucose control in diabetes. The cellular mechanisms for the insulin-antagonistic effect of catecholamines and other hormones acting through the cAMP system, such as glucagon, will also be elucidated. Unfortunately, very little is known about the mechanisms for the interaction between insulin and cortisol or growth hormone.

New Concepts in the Pathogenesis of NIDDM, Edited by
C. G. Östenson *et al.*, Plenum Press, New York, 1993

1. INFLUENCE OF THE COUNTERREGULATORY HORMONES ON GLUCOSE CONTROL IN DIABETES

In general, the counterregulatory hormones switch metabolism from a normal anabolic state to a catabolic state. This is achieved by antagonizing the effect of the most important anabolic hormone, insulin. For glucose homeostasis, this is achieved by inhibiting both the ability of insulin to stimulate glucose uptake in peripheral tissues and to exert a suppressive effect on hepatic glucose production. Although these effects can be achieved by all counterregulatory hormones, they differ in terms of times of onset, organ selectivity and duration of effects. The body is, thus, equipped with a powerful system consisting of several hormones with different profiles acting in concert to increase the glucose levels in the blood by inducing insulin resistance. A corollary to this concept is that the failure of the body to produce any given counterregulatory hormone would also lead to metabolic consequences since the remaining hormones do not have the same profile. That this prediction is upheld and also leads to important clinical consequences has been particularly well demonstrated in diabetics in conjunction with a hypoglycemia (1).

Table I summarizes the characteristics of the counterregulatory hormones. The effects of catecholamines and glucagon are fast in onset and can, thereby, rapidly influence glucose control. This makes them important as "emergency" hormones since they can rapidly increase hepatic glucose production and induce a peripheral insulin resistance leading to an attenuated glucose uptake in insulin-regulated organs. Glucose is, thus, conserved for organs, such as the brain, which are dependent on a normal glucose level for their substrate extraction. Glucagon, at least in our hands, seems to act exclusively on the liver (2) where it transiently increases glucose production (2, 3). The catecholamines also elicit a transient increase in hepatic glucose production while the inhibitory effect on peripheral glucose disposal is maintained for at least several hours (2).

As discussed by Cherrington and coworkers elsewhere in this issue catecholamines, but not glucagon, are also able to sustain gluconeogenesis by supplying the liver with important substrates. Thus, these two emergency hormones complement rather than mimic each other.

Table I. Counterregulatory hormones and glucose turn-over

<u>Emergency hormones</u>

<u>Glucagon</u> — levels can change ~ 5-fold
- acts exclusively on the liver
- elicits a rapid but transient increase in the glucose production (glycogenolysis)
- does not sustain gluconeogenesis alone since gluconeogenic precursors are not supplied
- attenuated increase in response to hypo-glycemia seen in diabetics

<u>Catecholamines</u>
- levels can change ~ 100-fold
- elicit a sustained impairment of insulin-stimulated glucose disposal. This effect is not dependent on elevated FFA levels/oxidation but on direct cellular effects of cAMP on insu-lin action
- elicit a rapid but transient increase in hepatic glucose production (glycogenolysis)
- sustain gluconeogenesis by supplying gluconeo-genic precursors
- increase in response to hypoglycemia frequently impaired in diabetics

<u>Hormones with delayed onset (1-2 hours)</u>

<u>Cortisol</u> — antagonizes effect of insulin in suppressing glucose production and stimulating glucose disposal
- effect on gluconeogenesis unclear and dependent on supply of gluconeogenic precursors

<u>Growth hormone</u>
- released intermittently
- antagonizes effect of insulin in suppressing glucose production and stimulating glucose disposal
- normal peak levels can elicit a prolonged (up to 8 hrs) impairment of insulin action
- effect on gluconeogenesis is unclear and de-pendent on supply of gluconeogenic precursors

Cortisol and growth hormone have a delayed time of onset (1-2 hours) and, thus, are of less significance during an emergency situation such as a pronounced hypoglycemia (1, 4). However, by virtue of having a long effect duration they can induce a prolonged insulin resistance and elicit a significant deteriora-

Table II. Role of counterregulatory hormones in diabetes

Contribute to the recovery from hypoglycemia

Elicit prolonged insulin resistance after hypoglycemia
(Somogyi effect)

Elicit insulin resistance with increased insulin
requirement in conjunction with:

- infections, trauma, surgery
- smoking
- impaired glucose control?

Contribute to the dawn phenomenon

tion of the glucose control in diabetics. Table II outlines
various conditions that are important in diabetes and where the
counterregulatory hormones play a significant role.

Hypoglycemia. Hypoglycemia elicits the most pronounced
stressor response known and is associated with huge elevations
in the counterregulatory hormones. In normal individuals, this
leads to a rapid increase in blood glucose levels and recovery
from hypoglycemia. Diabetics, particularly type I, have an
impaired recovery from hypoglycemia. This defect may be
extremely pronounced and lead to hypoglycemic unawareness and
sudden unconsciounsness (1). To the best of our knowledge, the
first report that type I diabetes is associated with an impaired
recovery from insulin-induced hypoglycemia was published in 1979
by Lager et al (5). This was, at the time, an unexpected finding
although it had previosly been demonstrated that the increase
in glucagon in response to hypoglycemia was attenuated in
diabetics (6). It has subsequently been shown that not only
glucagon but also the release of catecholamines and/or the other
counterregulatory hormones may be impaired in certain diabetics
during hypoglycemia (1). The clinical consequences of this are
an impaired recovery as well as attenuated hypoglycemic
symptoms.

Detailed experimental studies have shown that hypoglycemia
is followed by a prolonged insulin resistance (6-8 hours) (7,
8). Catecholamines, through β_2-adrenergic stimulation, elicit an
insulin resistance which is rapid in onset and lasts 3-4 hours,

i.e., prevails even after the hormone and glucose levels have become normalized (8). Cortisol and growth hormone act in concert to gradually (1-2 hours) produce an insulin resistance which prevails after the waning of the catecholamine effect (7).

Although an insulin-antagonistic effect is advantageous for the recovery from hypoglycemia, a prolonged posthypoglycemic insulin resistance may lead to a rebound hyperglycemia. This phenomenon, the Somogyi effect, has been the subject of much controversy including the denial of its existence (9). However, those results may be attributable to the inclusion of relatively poorly controlled diabetics (9). A poor metabolic control leads to insulin resistance as discussed below. Inclusion of initially insulin-resistant subjects makes it difficult to evaluate the effect of a further attenuation of the insulin sensitivity. The counterregulatory hormones impair the insulin-sensitive glucose uptake but not uptake due to the mass action effect of glucose.

Studying relatively well-controlled type 1 diabetics under carefully supervised conditions, Fowelin et al (10) could clearly show a rebound hyperglycemia after an insulin-induced hypoglycemia (Fig. 1). Interestingly, this effect lasted around 8 hours which also is the expected duration based on the experimental studies referred to (7, 8). The magnitude of this effect varied between different individuals and was related to the peak growth hormone levels achieved during the hypoglycemia (10). In some individuals, the glucose levels increased more than 5 mmol/l in spite of unchanged insulin levels (Fig. 1). These increases are of suffcient magnitude to be clinically important, not least since most diabetics with currently used treatment regimens have one or several weekly hypoglycemic episodes of varying intensity.

Other stressor effects. A variety of common conditions (Table 2) lead to the release of counterregulatory hormones and, concomitantly, insulin resistance and impaired glucose control. Smoking has recently been shown to elicit insulin resistance (11), probably through the release of catecholamines and growth hormone (11, 12). Chronic smokers also exhibit the various characteristics of the insulin-resistance syndrome (13) and this may be one important reason for the associated increased risk for cardiovascular disease.

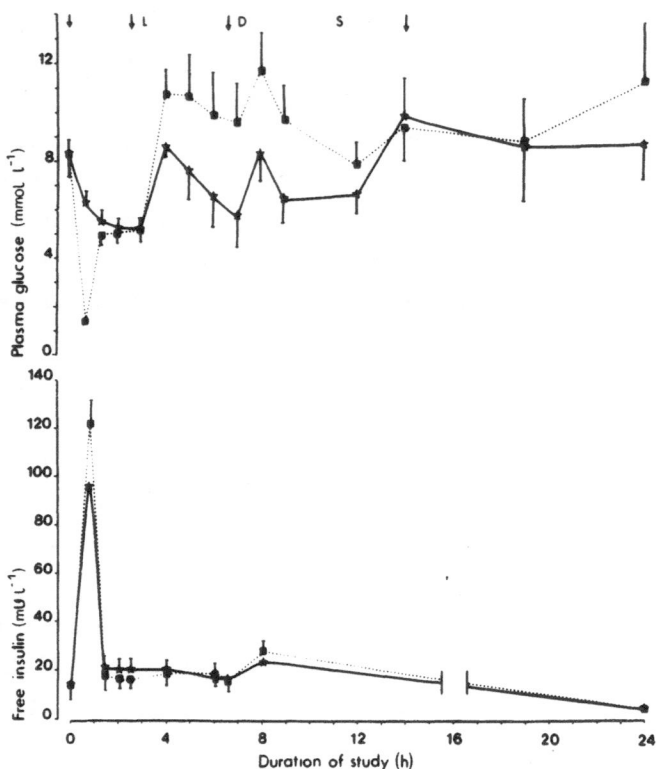

Fig. 1. Diurnal blood glucose and insulin levels in type 1 diabetics before or following an insulin-induced hypoglycemia. The arrows denote food intake. Reproduced from ref. 10 by permission.

Severely hyperglycemic and ketotic diabetics have elevated levels of growth hormone and other counterregulatory hormones. Since hyperglycemia leads to an impaired insulin sensitivity in diabetics (14, 15), these perturbations in hormone levels could play a role. However, in a detailed study of type 1 diabetics, experimental hyperglycemia for 44 hours (15-20 mmol/l) in the face of an unchanged degree of insulinization did not lead to significant elevations in the counterregulatory hormones (15). Thus, it is likely that the elevated hormone levels seen in poorly controlled diabetics are due to other effects associated with severe hyperglycemia such as hypovolemia and dehydration.

<u>Dawn phenomenon</u>. The importance of the dawn phenomenon in diabetes, i.e., rising glucose levels during the early morning, is controversial. However, most studies have shown that the postabsorptive glucose levels have a diurnal rhythm. They tend to be highest in the early morning when also the degree of insulin resistance is most pronounced. There is much evidence to support a role of growth hormone for the dawn phenomenon (1).

2. CELLULAR MECHANISMS FOR THE INSULIN-ANTAGONISTIC EFFECT OF CATECHOLAMINES AND ELEVATED cAMP LEVELS

Catecholamines and elevated cAMP levels produce a marked cellular insulin resistance. Most studies aimed at elucidating the mechanisms for this have used the isolated adipocyte as a model cell. However, the effect of catecholamines in eliciting insulin resistance in man in vivo, both in the liver and in peripheral tissues, is well-established and has already been referred to above (also see ref. 8).

The insulin resistance at the cellular level is seen both as an attenuated insulin sensitivity, i.e., increased EC_{50} for all metabolic effects of insulin measured, and attenuated responsiveness (16-22). These findings are in agreement with perturbations that have been demonstrated both at the receptor and postreceptor levels.

Insulin receptor level. Catecholamines and elevated cAMP levels elicit a rapid reduction (~ 30 %) in the number of insulin receptors in the plasma membrane of isolated fat cells (23, 24). This reduction is not sufficient, by itself, to account for the marked attenuation in insulin sensitivity (EC_{50} increases 5-10-fold). However, the reduction measured under normal receptor-binding assay conditions is underestimated since it does not take into account the ability of insulin to rapidly uncover additional binding sites in the plasma membrane (25, 26). This novel finding was disclosed in experiments where isolated fat cells were preincubated with insulin for 20 min. Receptor cycling was then stopped by energy-depleting the cells with KCN and cold insulin removed by careful washing. Labelled insulin was then bound or cross-linked to the receptors. Extensive studies by Eriksson et al (25-27) using this methodology have provided evidence for a large "hidden" pool of receptors associated with the plasma membrane but not appropriately inserted and capable of binding insulin. Insulin and the insulin-mimicker, vanadate, are both able to rapidly uncover receptors from this pool (25-27). The magnitude of this effect is closely related to the insulin sensitivity of the cells (26). Thus, this novel mechanism may be an important regulator of insulin sensitivity.

Catecholamines and elevated cAMP levels markedly attenuate the ability of insulin to enhance its receptor binding (27). As described below, the concept that receptors may be associated with the plasma membrane but not appropriately fused and/or inserted seems to be more general and include at least two other integral plasma membrane proteins, i.e., the IGF II receptor and the glucose transporting protein. Taken together, when measured in a non-recycling system, catecholamines and elevated cAMP levels do, in fact, reduce insulin binding in the plasma membrane to such an extent that it may account for the accompanying pronounced insulin resistance (27). In addition, there have been reports that catecholamines reduce insulin receptor tyrosine kinase activity, possibly through serine phosphorylation (19). In our hands, this effect is only seen after prolonged exposure to high catecholamine levels (27), a condition which is not required to elicit the insulin resistance.

Postreceptor level. In addition to reducing the cellular insulin sensitivity catecholamines and elevated cAMP levels, particularly in the absence of agonists linked to the inhibitory GTP-binding protein (16-18, 20. 21), also attenuate insulin responsiveness.

A key mechanism for the glucose-regulating effect of insulin is its ability to stimulate glucose transport over the cell membrane. This effect is achieved by the translocation of specific glucose transporting proteins from a large intracellular pool to the plasma membrane (see ref. 28 for review). The glucose transporters fuse with the plasma membrane through unclear mechanisms and, thereby, increase glucose uptake. Several studies have shown that the size of the intracellular pool of transporting proteins can be regulated and that it is frequently reduced in insulin-resistant states (28). As a consequence, the number of glucose transporters translocated by insulin to the plasma membrane is reduced. Also other integral membrane proteins, such as the IGF II receptor, undergo translocation in a similar manner in response to insulin.

A reduction in the effect of a supramaximal insulin concentration can either be due to an altered intrinsic activity of the proteins, to a reduction in the size of the intracellular pool or to a perturbation at the various steps of the translocation process. Current evidence favour that catecholamines and elevated cAMP levels can influence both latter processes.

However, chronically elevated levels are required to alter the size of the intracellular pool of membrane proteins (29) while the acute effect seems to be exerted at the translocation process.

Lönnroth et al (20) could show that the ability of insulin to translocate IGF II receptors from the large intracellular pool to the plasma membrane was essentially normal in the presence of catecholamine and that the "missing" receptors could be recovered in the plasma membrane upon solubilization. However, they were obviously not appropriately fused and/or inserted in the plasma membrane since they did not fully bind the ligand in the intact cell (~ 50 % reduction). A similar finding has recently been obtained with the glucose transporters in the presence of both insulin and catecholamines (21). Thus, glucose transport activity was markedly reduced and so was the number of appropriately fused glucose transporting proteins. However, the effect of insulin to translocate the proteins was apparently normal since the "missing" transporters were recovered in the plasma membrane upon solubilization and immuno-precipitation (21).

Taken together, these results indicate that at least three integral membrane proteins, i.e., the insulin receptor, IGF II receptor and glucose transporter, undergo a regulated step (S) upon insulin-stimulated translocation and cycling, whereby they bind (docking protein?) and fuse with the plasma membrane. This process is influenced by catecholamines and elevated cAMP levels so that the membrane proteins still seem to be associated with the plasma membrane but are not appropriately docked and/or fused. Fig. 2 schematically illustrates this process as it is currently understood. Much work clearly remains to elucidate the mechanisms and regulation of these processes.

However, this novel mechanism may well prove to be of fundamental importance in regulating both insulin sensitivity and action. It may, for instance, provide a reason why several recent studies have found a normal number of glucose trans-porters as well as m RNA levels in cells from type 2 diabetic individuals while the glucose transporting activity in response to insulin is markedly attenuated. In addition, the ability of insulin to uncover binding sites in the plasma membrane seems to be closely related the insulin sensitivity of the cells. It should also be emphasized that chronically elevated cAMP levels seem to regulate the synthesis of key proteins for insulin action such as the glucose transporting proteins (29). Thus, the counterregulatory hormones can influence insulin action, both acutely and chronically, by modulating processes extending from gene transcription of key proteins to their insertion and fusion with the plasma membrane of target cells.

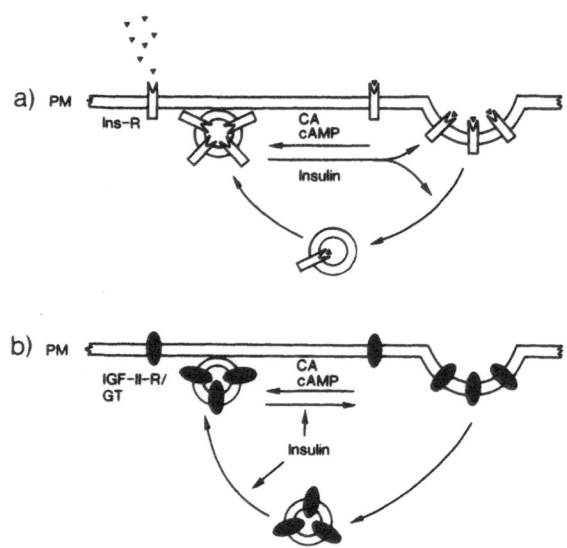

Fig. 2. Schematic illustration of the docking, insertion and fusion of the insulin receptor (a) and glucose transporter and IGF II receptor (b) with the plasma membrane (PM) in the presence of insulin and catecholamines (CA) or other agents that elevate the cAMP levels. From ref. 22.

ACKNOWLEDGEMENTS

The studies from the authors laboratory were supported by the Swedish Medical Research Council.

REFERENCES

1. J. Gerich, M. Mokau, T. Veneman, M. Korytkowski and A. Mitrakon. Hypoglycemia unawareness. Endocr. Rev. 12:356-71 (1991).
2. S. Attvall, J. Fowelin, H. von Schenck, U. Smith and I. Lager. Insulin-antagonistic effects of pulsatile and continuous glucagon infusions in man - a comparsion with the effect of adrenaline. J. Clin. Endocrinol. Metab. 74:1110-15 (1992).
3. G. Paolisso, S. Buonocore, S. Gentile, S. Sgambato, M. Varrichio,, A. Sheen, FD Onofrio and P. Lefebvre.

Pulsatile glucagon has greater hyperglycaemic, lipolytic and ketogenic effects than continuous hormone delivery in man: The effect of age. Diabetologia 32:272-77 (1990).

4. J. Fowelin, S. Attvall, H. von Schenck, U. Smith, and I. Lager. Characterization of the insulin-antagonistic effect of growth hormone in man. Diabetologia 34:500-6 (1991).

5. I. Lager, G. Blohmé, and U. Smith. Effect of cardioselective and non-selective β-blockade on the hypoglycaemic response in insulin-dependent diabetics. Lancet I:458-62 (1979).

6. J.E. Gerich, M. Langlois, C. Noacco, J.H. Karam, and P.H. Forsham. Lack of glucagon response to hypoglycemia in diabetes. Science 182:171-73 (1973).

7. J. Fowelin, S. Attvall, H. von Schenck, U. Smith, and I. Lager. Combined effect of growth hormone and cortisol on late posthypoglycemic insulin resistance in humans. Diabetes 38:1357-64 (1989).

8. S. Attvall, B-M. Eriksson, J. Fowelin, H. von Schenck, I. Lager, and U. Smith. Early posthypoglycemic insulin resistance in man is mainly an effect of β-adrenergic stimulation. J. Clin. Invest. 80:437-42 (1987).

9. K.M. Tordjman, C.E. Havlin, C.A. Levandovski, N.H. White, J. Santiago, and P.E. Cryer. Failure of nocturnal hypoglycemia to cause fasting hyperglycemia in patients with IDDM. N. Engl. J. Med. 317:1552-59 (1987).

10. J. Fowelin, S. Attvall, H. von Schenk, U. Smith, and I. Lager. Postprandial hyperglycaemia following a morning hypoglycaemia in type I diabetes mellitus. Diabetic Medicine 7:156-61 (1990).

11. S. Attvall, J. Fowelin, I. Lager, and U. Smith. Smoking induces insulin resistance - a potential link with the insulin resistance syndrome. J. Intern. Med. In press (1993).

12. P. Cryer, M. Haymond, J. Santiago, and S. Shah. Norepinephrine and epinephrine release and adrenergic mediation of smoking - associated hemodynamic and metabolic events. N. Engl. J. Med. 295:573-77 (1976).

13. F. Facchini, C. Hollenbeck, J. Jeppesen, I. Chen, and G. Reaven. Insulin resistance and smoking. Lancet 339:1128-30 (1992).

14. H. Yki-Järvinen, E. Helve, and V. Koivisto. Hyperglycemia decreases glucose uptake in type 1 diabetes. Diabetes 36:892-96 (1987).

15. J. Fowelin, S. Attvall, H. von Schenk, B-Å. Bengtsson. U. Smith, and I. Lager. Effect of prolonged hyperglycemia on growth hormone levels and insulin sensitivity in insulin-dependent diabetes mellitus. Metabolism - in press (1993).

16. U. Smith, M. Kuroda, and I.A. Simpson. Counter-regulation of insulin-stimulated glucose transport by catecholamines in the isolated rat adipose cell. J. Biol. Chem. 259:8758-63 (1984).

17. P. Lönnroth, I. Davis, I. Lönnroth, and U. Smith. The interaction between the adenylate cyclase system and insulin-stimulated glucose transport. Biochem. J. 243:789-95 (1987).

18. M. Kuroda, R.C. Honnor, S.W. Cushman, C. Londos, and I.A. Simpson. Regulation of insulin-stimulated glucose transport in the isolated rat adipocyte. <u>J. Biol. Chem.</u> 262:245-53 (1987).
19. H. Häring, D. Kirsch, B. Obermaier, B. Ermel, and F. Machicao. Decreased tyrosine kinase activity of insulin receptors isolated from rat adipocytes rendered insulin-resistant by catecholamine treatment in vitro. <u>Biochem. J.</u> 234:59-66 (1986).
20. P. Lönnroth, K.C. Appell, C. Wesslau, S. Cushman, I. Simpson, and U. Smith. Insulin-induced subcellular redistribution of insulinlike growth factor II receptors in the rat adipose cell. Counterregulatory effects of isoproterenol, adenosine, and cAMP analogues. <u>J. Biol. Chem.</u> 263:15386-91 (1988).
21. S. Vanucci, H. Nishimura, S. Satoh, S.W. Cushman, G.D. Holman, and I.A. Simpson. Cell surface accessibility of GLUT 4 glucose transporters in insulin-stimulated rat adipose cells. Modulation by isoproterenol and adenosine. <u>Biochem. J.</u> 288:325-30 (1992).
22. C. Wesslau. The role of cAMP and G_i protein in modulating the actions of insulin and beta-adrenergic agonists. Academic dissertation, University of Göteborg (1992).
23. P. Lönnroth, and U. Smith. ß-adrenergic dependent down-regulation of insulin binding in rat adipocytes. <u>Biochem. Biophys. Res. Commun.</u> 112:972-79 (1983).
24. J.E. Pessin, W. Gitomer, Y. Oka, C.L. Oppenheimer, and M.P. Czech. ß-adrenergic regulation of insulin and epidermal growth factor receptors in rat adipocytes. <u>J. Biol. Chem.</u> 258:7386-94 (1983).
25. J. Eriksson, P. Lönnroth, and U. Smith. Insulin can rapidly increase cell surface insulin binding capacity in rat adipocytes. <u>Diabetes</u> 41:707-14(1992).
26. J.W. Eriksson, P. Lönnroth, and U. Smith. Vanadate increases cell surface insulin binding and improves insulin sensitivity in both normal and insulin-resistant rat adipocytes. <u>Diabetologia</u> 35:510-11 (1992).
27. J.W. Eriksson, P. Lönnroth, and U. Smith. cAMP impairs the rapid effect of insulin to enhance cell surface insulin binding capacity in rat adipocytes. <u>Biochem. J.</u> 288:625-29 (1992).
28. I. A. Simpson, and S.W. Cushman. Hormonal regulation of mammalian glucose transport. <u>Ann. Rev. Biochem.</u> 55:1059-89 (1986).
29. T.M. Kaestner, J.R. Flores-Riveros, J.C. McLenithan, M. Janicot, and M.D. Lane. Transcriptional regression of the mouse insulin-responsive glucose transporter (GLUT4) gene by cAMP. <u>Proc. Natl. Acad. Sci. USA.</u> 88: 1933-37 (1991).

ON INSULIN ACTION IN VIVO: THE SINGLE GATEWAY HYPOTHESIS

Richard N. Bergman, David C. Bradley and Marilyn Ader

Department of Physiology and Biophysics
University of Southern California School of Medicine
1333 San Pablo Street
Los Angeles, CA 90033   USA

INTRODUCTION

It has become increasingly clear that insulin resistance plays an important role in the pathogenesis of chronic diseases of Westernized societies. Compelling evidence exists not only that insulin resistance is a risk factor for diabetes mellitus (1,2) but this metabolic characteristic is also associated with hypertension and atherosclerosis (3,4). In addition, there is evidence from the Pima Indians (5) and from high-risk offspring of two NIDDM parents (6) that insulin resistance is an inherited trait. Given that there must be one or more specific underlying mechanistic defects which are responsible for insulin resistance, and given its potential importance in the pathogenesis of chronic disease, it is incumbent upon us to continue to try to identify the cellular or molecular defects which underlie this condition.

It is beyond the scope of this review to cover the myriad of studies attempting to elucidate insulin action defects in obesity, NIDDM and other states. A common paradigm has been to pinpoint insulin resistance at the level of the intact individual, using the glucose clamp or other methods, and then examine biopsied tissues to uncover the cellular defect responsible (7). Much is known regarding the cascade of insulin action in isolated tissues. To act, insulin must bind to and activate its specific target cell receptor, and presumably will phosphorylate a "second messenger" (8) which results in the translocation of (and possibly, activation of) insulin dependent glucose transporters, GLUT4, from intracellular sequestration to the cell membrane. Glucose may then enter the cell to be stored or oxidized. It has generally been assumed that insulin resistance is caused by defects in one or more of these processes, and, indeed, there are reports of defects at virtually all levels in this cascade in NIDDM (9-11).

One problem with the scenario by which defects at the cellular level are related to overall insulin resistance in NIDDM (the "*in vivo/in vitro*" paradigm) is that it is generally unclear whether the observed defect is the *cause* or the *result* of the disease (6,12,13). This has resulted in a shift of emphasis from studying individuals *with* the disease to those at *high risk.* Fortunately, it is possible to find such individuals using various clinical and genetic tools (13,14). A second problem with the *in vivo/in vitro* approach is that there is a distinct *temporal disparity* between the effects of insulin in the intact organism and at the level of the cell. *In vitro, insulin acts rapidly.* For example, when added to adipocytes in culture, glucose uptake increases almost immediately, achieving steady-state within 10 minutes. However, *in vivo insulin acts slowly.* This is most clearly seen with the euglycemic clamp protocol. Infusing insulin achieves a hyperinsulinemic steady-state within a few minutes, whereas it takes at least 180 minutes to achieve a steady-state rate of increased glucose utilization provoked by the sustained hyperinsulinemia (Figure 1; ref. 15). In insulin resistant states this delay appears to be increased (16). Clearly, if one is to understand insulin action in the intact organism, it will be necessary to understand clearly why the hormone acts so slowly, and to examine the importance of the reason for the extensive delay to insulin action under normal and pathological conditions.

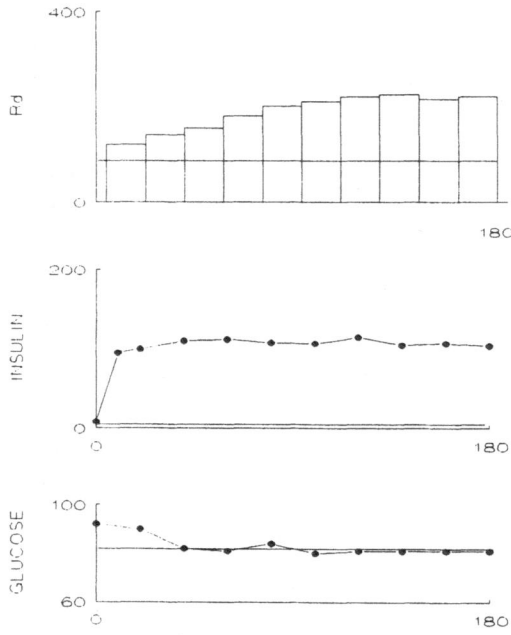

Figure 1. Plasma glucose, insulin, and R_d from a typical euglycemic glucose clamp. Insulin is administered as a primed infusion to rapidly raise concentrations to ~100 μU/ml (adapted from ref. 24).

INSULIN EFFECT ON GLUCOSE UPTAKE

How may we account for the dynamic disparity between insulin's effects in vitro and in vivo? An important hint came from studies of Scow and his colleagues, who demonstrated that insulin acted more slowly when perfused into isolated fat pads than when added to adipocytes (17). This leads to the hypothesis that it is the time required to pass *from the circulation to the cells* which is the cause of the delay when insulin is given *in vivo*. Additional evidence was gleaned by Rasio and his colleagues, who measured insulin in lymph, and showed that there was a dynamic disparity between insulin concentrations in plasma and lymph following insulin injection (18,19). They reasoned that if lymph insulin represented interstitial insulin, a substantial delay would be incurred during the passage from the vascular to the interstitial compartment. Finally, Sherwin et al. had suggested that insulin acted from a compartment "remote" from plasma (20,21), and we incorporated this same concept in a minimal model of insulin-dependent glucose metabolism (22-25).

Lymph Studies - Insulin and Inulin Dynamics

We have recently examined the concept that the delay in insulin action *in vivo*, compared to *in vitro* may be explained by the time necessary for insulin to move from the vascular to the interstitial space, across the endothelial barrier. We performed euglycemic clamps in conscious dogs, and followed Rasio's lead in measuring insulin in thoracic duct lymph (26). We infused insulin and ^{14}C-inulin, the latter being an inert molecule with a molecular weight similar to insulin, but which, in contrast to insulin, is not taken up by cells, but is excreted exclusively by the kidneys (27).

We found a sharp distinction between the dynamics of the inert inulin and insulin itself. When infused (Figure 2) *inulin* increased slowly, consistent with its ability to cross the endothelial barrier between plasma and interstitium. By 3 hours of constant infusion, inulin achieved an equilibrium between plasma and interstitium, such that its concentration was equal in the two compartments. The equilibrium achieved by inulin is consistent with no significant disappearance of inulin from the interstitial compartment, as well as no dilution or concentration of the interstitial fluid as it enters the lymphatic system.[1]

---

[1] The possibility exists, of course, that insulin is degraded in the interstitium, thus lowering its concentration in that compartment, and then the interstitium is concentrated as it passes through lymph nodes, again increasing its concentration. However, it seems very unlikely that the concentration should be exactly equivalent and opposite to the degradation, which is the only way the lymph concentration could be equal to plasma. It seems much more likely that the molecule is degraded *only* from plasma, and that there is no significant concentration or dilution in the lymph system.

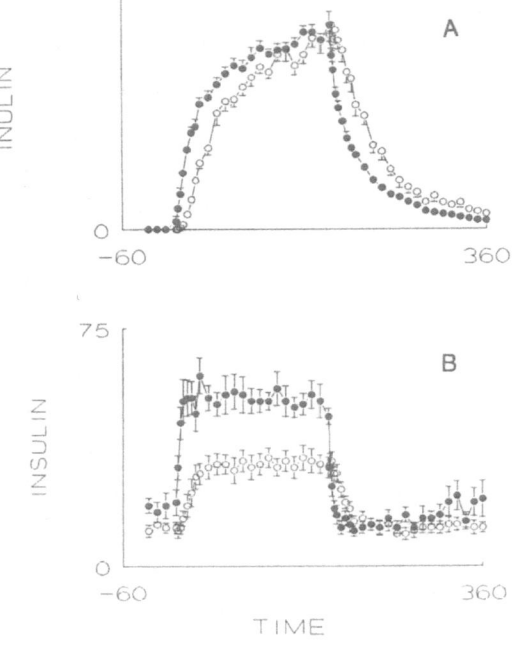

Figure 2. Concentrations of (A) inulin and (B) insulin measured during clamps in dogs. Insulin and inulin were infused from 0 to 180 min during euglycemia, after which infusions were terminated, and euglycemia maintained for an additional 180 min. Samples were drawn simultaneously from mixed venous blood (•) and thoracic duct lymph (O) (Adapted from ref. 26).

Insulin dynamics were very different. First, under basal fasting conditions, lymph insulin concentration is only about *two-thirds* the plasma insulin concentration. Second, unlike insulin, even at steady-state, after 3 hours of insulin infusion, this 3:2 gradient between plasma and lymph insulin was maintained. Thus we observed an *attenuation* of the insulin signal between plasma and the interstitial compartment. Finally, while plasma insulin rose rapidly upon infusion, to a hyperinsulinemic steady-state, lymph insulin rose considerably more slowly. Thus, we observed a significant *delay* in the time it took for the insulin concentration to achieve steady-state. Attenuation and delay of the insulin signal was also observed at higher rates of insulin infusion (28).

What is the explanation of the attenuation and delay of the insulin signal? We envision the passage of molecular species across the endothelium as observed in Figure 3. Hypothetically, molecules may either pass easily across the barrier, or be restricted; once in the interstitial space, molecules may be bound, internalized and degraded by cells, or not affected. Inulin, the large, inert molecule, is potentially restricted in transport, but not taken up by cells. This is consistent with the dynamics we observed, i.e., slow approach to a true equilibrium between the plasma and interstitial compartment (Figure 3a). Small molecules, such as glucose, should pass easily from plasma to interstitium; even if taken up by cells, a concentration gradient between plasma and interstitium will not develop due to the permeability of the endothelial barrier (Figure 3b). We found no significant decrement in concentration between the two compartments for glucose (ref. 28; data not shown). For insulin however, we hypothesize *both* a transport barrier, *and* cellular binding, uptake and degradation (26,28).

We therefore propose that our studies of the dynamic relationship between plasma and lymph insulin are consistent with the model of Figure 3c, that is, 1) that the capillary endothelium provides a significant *barrier* to the transport of insulin, and 2) that insulin, once in the interstitial fluid is bound and degraded by cells at a finite rate. This model is consistent with the attenuation and delay of the insulin signal.

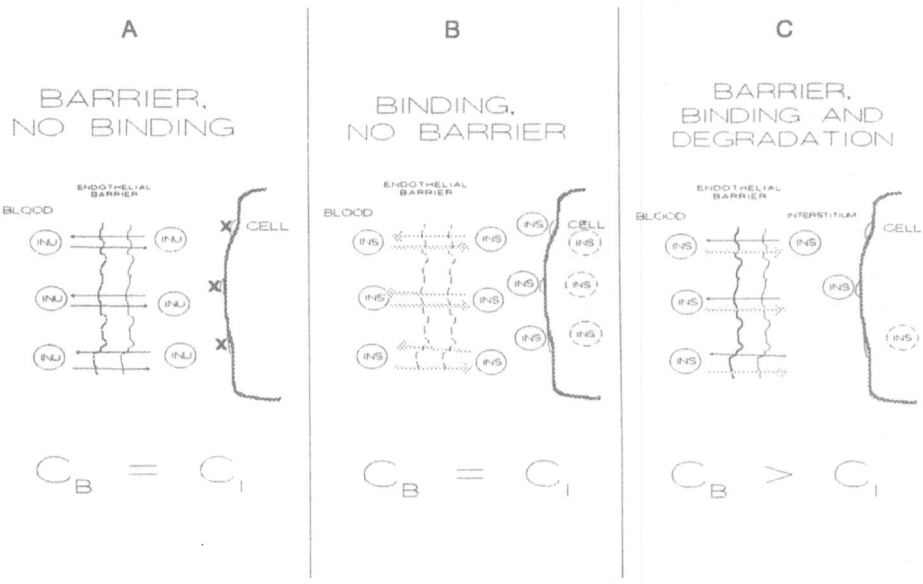

Figure 3. Possible fates of molecular substances upon passage across the capillary endothelium. (A) Compound may be restricted across endothelial barrier, but not be bound from interstitial space by cells. Such a substance (e.g. inulin) would exhibit equal concentrations in blood (C_B) and interstitial fluid (C_I). (B) Compound may pass freely across endothelial barrier, but not be sequestered by cells. These compounds (e.g. glucose) would also achieve equal concentrations in blood and interstitial fluid. (C) Some species may be restricted in their movement across endothelial barrier, and upon entering interstitium, they are bound, internalized, and degraded by cells. Such substances (e.g. insulin) would exhibit a concentration gradient between vascular and interstitial compartments such that $C_B > C_I$.

Lymph Studies - Glucose Kinetics

Of particular interest is the extent to which the delay in insulin action *in vivo* my be attributed to restricted trans-endothelial passage. To answer this question we examined the relationship between insulin concentration in plasma or lymph, and the rate of glucose uptake, measured using accurate tracer dilution methodology (29,30). We found (Figure 4) that during clamps, there is a very strong relationship between the rate of glucose uptake and the concentration of insulin in thoracic duct lymph (i.e., interstitium), even while lymph insulin and glucose uptake are changing (during the initial 180 minutes). In contrast, there is only a relatively weak relationship between plasma insulin and glucose uptake during the same period. This strong relationship held over the entire physiological range of insulin action (Figure 4; ref. 28). An even stronger relationship was observed in animals in which we compared lymph draining insulin-sensitive tissue (hindlimb lymph in anesthetized dogs) with glucose uptake (31).

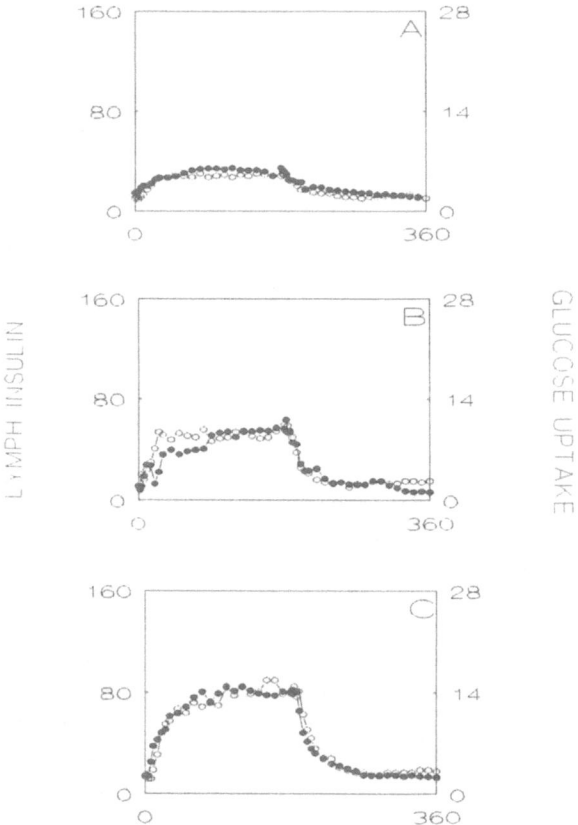

Figure 4. Time courses of thoracic duct lymph insulin (O) and glucose uptake (•) during glucose clamps with insulin infused at (A) 0.6, (B) 0.9, and (C) 1.2 mU/min per kg (adapted from ref. 28).

Thus we observed during the dynamic phases of the glucose clamp that plasma insulin changes more rapidly than glucose utilization, whereas the rates of increase and decrease of lymph insulin and glucose utilization are virtually identical. Several conclusions may be drawn from this result. First, that lymph insulin and glucose utilization are in temporal synchrony during clamps provides compelling evidence that thoracic duct insulin is a useful indicator of insulin in interstitial fluid. If it were not, it would be fortuitous indeed that at several rates of insulin infusion, interstitial insulin would be very strongly related to glucose utilization measured using tracer dilution. A second conclusion is that the synchrony between interstitial insulin (but not plasma insulin) and glucose utilization indicates that *the step in the insulin cascade which determines the rate at which insulin acts in vivo is transendothelial insulin transport ("TET") (Figure 5).*

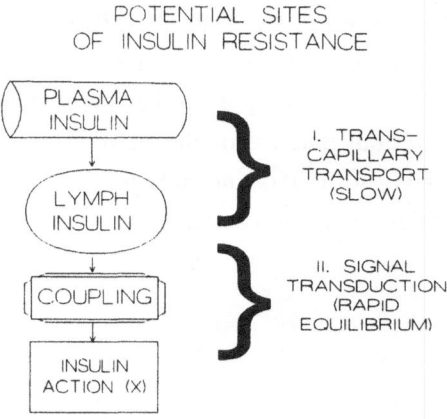

Figure 5. Schematic illustrating major steps in insulin action which may be involved in the pathogenesis of insulin resistance.

Significance of Transendothelial Insulin Transport

What is the significance of the demonstration that transport of insulin across the capillary determines how rapidly the hormone acts? The significance is not yet clear, but several speculations may be entertained.

A burning question is whether alteration in "TET" is responsible for alterations in insulin action seen with environmental changes (diet, adiposity, fitness, aging), and/or in pathological states (hypertension, NIDDM). We have failed to find any change in TET in animals rendered insulin resistant with chronic nicotinic acid therapy (32). However, Prager and Olefsky and their colleagues have reported that the rate of activation of insulin action is retarded in obese subjects (16), and it will be of interest whether this retardation is caused by alteration in insulin transport, or action at the cellular level. Also, changes in TET could well be a secondary effect of capillary changes associated with complications in long-term obese NIDDM patients.

It will be particularly interesting to examine the interaction between TET, insulin action, and hemodynamics. Insulin resistance is well documented in Caucasian hypertensives (4,33). It is possible to hypothesize that changes in peripheral blood flow in essential hypertension could effect alterations in TET due to reduction in peripheral blood flow. In ongoing studies we are examining the interaction between blood pressure regulation, TET, and insulin action in animal models.

Regardless of the pathogenic importance of TET, the demonstrated delay between plasma and interstitial insulin levels has interesting consequences regarding physiological regulation of the blood glucose level. Clearly, a substantial delay in TET will attenuate rapid changes in plasma insulin concentration. Mutiphasic insulin secretion, demonstrated in normal and reduced in NIDDM subjects, will be much "damped" at the level of the cell. In fact, one may hypothesize that the existence of first-phase insulin secretion would have the effect of overcoming the normal TET delay, and providing a "feed-forward" signal to the cells to rapidly store carbohydrate (Figure 6). Is it possible that the increased first phase in obesity (23) acts to overcome the increased delay in insulin transport in that state? Clearly, further studies of the relationship between various patterns of change in plasma insulin, TET, and glucose utilization are necessary to understand fully the physiological significance of a very substantial temporal delay in insulin action.

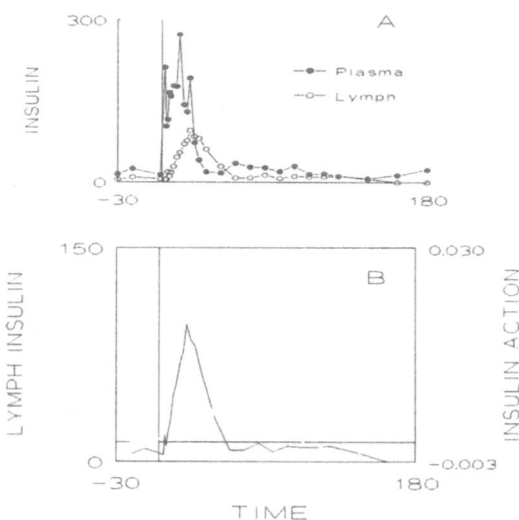

Figure 6. Effect of glucose-stimulated insulin response on (A) insulin concentrations in lymph and (B) insulin action ("X"), as quantified by the minimal model. Glucose was injected intravenously (0.3 g/kg) in a conscious dog, and samples were drawn at frequent intervals from right atrial blood (•) and thoracic duct lymph (O).

INSULIN EFFECT ON GLUCOSE PRODUCTION

In the fasting state, glucose needs of the body, and particularly the central nervous system, are provided primarily by the liver. Upon the ingestion of a carbohydrate meal, endogenously produced glucose is presumably no longer needed. It has been an article of faith that upon the administration of carbohydrate, hepatic glucose output (HGO) is suppressed directly by insulin, to be replaced by hepatic carbohydrate storage. In fact, rapid and total suppression of HGO has been reported during insulin administration (34).

The presumed rapidity of insulin's action to suppress HGO can be seriously questioned in view of two considerations. First, in the absence of elevated glucagon, suppression *in vitro* of glucose production by liver has been stubbornly difficult to demonstrate. Second, methodology for measurement of endogenous HGO *in vivo* was plagued by systematic artifacts for many years, artifacts that have only recently been eliminated (c.f., 29,30). Because of these considerations, we felt it would be fruitful to apply validated methodology for assessment of HGO *in vivo*, and reexamine the effects of insulin on liver in the conscious dog. In particular, we wished to test the hypothesis that insulin had a rapid effect to suppress glucose output under basal, fasting conditions.

Dose-response effects of insulin on HGO

To examine the effects of insulin infusion on HGO, it was necessary to suppress endogenous insulin and glucagon infusion with somatostatin, and replace basal hormone levels by infusion. We found that insulin replacement at 200 $\mu U/min$ per kg, and glucagon replacement at 2 ng/min per kg were adequate for re-establishment of basal rates of glucose turnover and plasma hormone levels (35). Glucose clamps were performed, at fasting glycemia, with variable rates of insulin infusion (Figure 7). Glucose production was measured by two methods in these studies -- by the so-called "HOT GINF" methodology, in which exogenously infused glucose contains ^3H-glucose so as to achieve relatively constant specific activity (29,30) -- and by direct arteriovenous glucose difference across the liver. We found the results from these two methods to be very strongly correlated (r = 0.97; ref. 36).

As expected, the sensitivity to insulin of liver glucose production (ED_{50} = 23 $\mu U/ml$) was considerably greater than the sensitivity to insulin of glucose uptake (ED_{50} = 74 $\mu U/ml$). However, we found that the effect of insulin to suppress hepatic glucose output, to be surprisingly slow, with a half-time for total suppression of 43 minutes. This slow suppression was revealed by tracer-determined as well as arte-

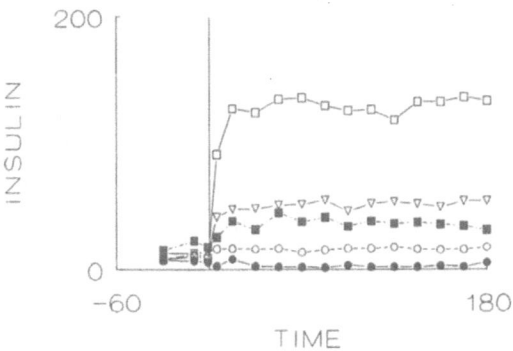

Figure 7. Insulin concentrations during euglycemic clamps with insulin infused systemically at various rates into dogs (•) 0, (○) 0.5, (■) 1.0, (▽) 1.67, or (□) 3.33 mU/min per kg (adapted from ref. 36).

Table 1. Half-times (in minutes) for insulin action to accelerate glucose uptake and suppress hepatic glucose production at multiple insulin doses (adapted from ref. 36).

Insulin Dose (mU/min per kg)	Glucose Uptake	Glucose Production	
		HGO	NHGO
1.00	39±9	53±12	29±7
1.67	58±13	37±5	52±26
3.33	40±4	41±9	43±8
overall	45±5	43±5	42±9
n	17	17	14

riovenous difference-assessed glucose output.[2] Of particular interest was the result that the half-time for insulin to suppress fasting glucose production was *virtually identical* to the half-time for insulin to increase glucose utilization -- 45 versus 43 minutes (Figure 8, Table 1).

Significance of the Dynamics of Insulin Action on Liver

Having measured glucose production by two accurate methods, we feel confident that we have directly assessed the temporal effects of insulin on glucose output.

---

[2] Note from Figure 8 that *net* hepatic glucose output (NHGO) during insulin infusion changes from a positive value (net production) to a negative value (net uptake), and that this value is approximately 1.5 mg/min per kg more negative that tracer-determined HGO. The reason for this constant difference (HGO - NHGO) is presumably due to an approximately constant rate of glucose *uptake* of 1.5 mg/min per kg. Since NHGO is HGO minus glucose uptake, the absolute value of NHGO will be lower.

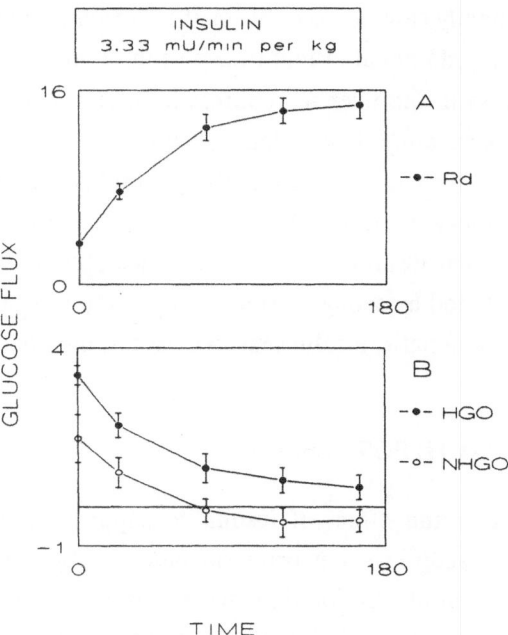

Figure 8. Time courses for the (A) stimulation of R_d and (B) suppression of glucose production by insulin during euglycemic clamps. Glucose production was measured both by tracer methodology (HGO) and hepatic A-V difference techniques (NHGO), as described in text (adapted from ref. 36).

What, then, is the meaning of the remarkable similarity of the dynamics of insulin to enhance uptake and suppress production? Several possibilities come to mind.

Assuming that measurement artifact is ruled out, it could simply be that the mechanism by which insulin performs these two actions could be similar at liver versus periphery. However, as we have seen above, it is the transendothelial transport of insulin which determines the rate at which glucose uptake increases. Is this a possible rate limiting step in the liver?

Insulin sensitive tissues, namely muscle and fat, have tight endothelial barriers. Thus it makes sense that this barrier would restrict insulin transport and, as we have seen, allow for only a slow increase in the rate of glucose uptake. However, there is no such barrier in liver. In fact, capillaries in liver are fenestrated. When labeled insulin is injected, it is sequestered within 2-4 minutes in the liver (37), consistent with the substantial degradation of insulin by that organ. Thus, insulin seems to be taken up by the liver according to a time course which is much too fast to account for the slow, 43 minute half-time seen in our studies. Therefore, we find it unlikely that a similar transendothelial transport step, which determines the rate of insulin action in muscle, accounts for the delay at the level of the liver.

A second possibility is that insulin is taken up rapidly by liver, but the secondary coupling of insulin to glucose production mechanisms (glycogenolysis and gluconeogenesis) is sufficiently slow to account for the 43 minute half-time. This would require that 1) the transendothelial insulin transport in the periphery, and 2) the post-binding effects of insulin in the liver to suppress both glycogen breakdown and glu-coneogenesis would *fortuitously* have identical time courses. While this seems improbable, at this time we cannot rule our the possibility that the similarity in the time courses is just a lucky accident. However, there is a third possibility that we consider more likely -- that insulin's effects to enhance glucose uptake and suppress glucose output are mediated by a single primary step -- the transport of insulin across the endothelium of extra-hepatic, insulin sensitive tissues, muscle and adipose.

THE SINGLE GATEWAY HYPOTHESIS

We hypothesize that the effect of insulin to suppress glucose output, in the absence of elevated glucagon levels, is primarily *indirect*. By this concept (Figure 9) the rate limiting step for insulin action on glucose metabolism is the transport of the hormone across the capillaries perfusing muscle (and possibly adipose) beds. Once insulin is present in the interstitial fluid of these tissues, it will bind to cells and

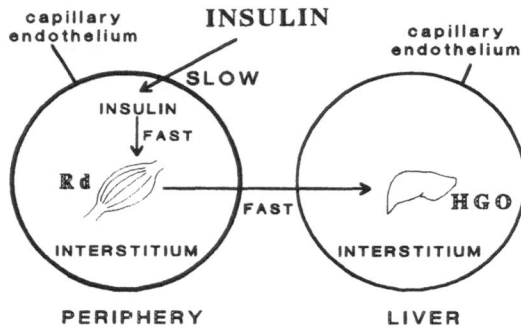

Figure 9. "Single Gateway" hypothesis of insulin action. The rate-determining step in both peripheral and hepatic actions of insulin is believed to be transendothelial insulin transport, which is slow. Once insulin enters peripheral interstitium, it is bound and elicits a rapid cascade of insulin action resulting in increased R_d. Interstitial insulin will inhibit release of a blood-borne substance from the periphery; suppression of this release by insulin results in inhibition of HGO.

enhance glucose uptake. In addition, we suggest that insulin will inhibit a blood-borne signal originating in muscle and/or fat which normally supports hepatic glucose production. Reduction in this signal reduces HGO. Thus, by this hypothesis transen-dothelial insulin transport at the level of muscle and adipose tissue is the rate limiting step *not only for glucose uptake, but also for glucose production.* Thus, by this conceptual framework, there is a *single gateway for insulin action in vivo,* and that single gateway is the capillary endothelium of extra-hepatic insulin sensitive tissues.

The single gateway hypothesis is consistent with the data reviewed above which demonstrates that 1) insulin acts slowly to stimulate glucose uptake; 2) the rate of glucose uptake is proportional to *interstitial* but not *plasma* insulin, indicating transendothelial transport is rate-limiting for glucose uptake; 3) insulin also acts slowly to suppress fasting glucose output, and 4) the time courses of insulin's effects on glucose output and glucose uptake are virtually identical.

Indirect Effect of Insulin on Hepatic Glucose Production

Implicit in the single gateway concept is the notion that insulin's effect to suppress HGO is indirect, via peripheral tissues rather than direct, on the liver itself. However, this concept collides with the fact that insulin is secreted directly into the portal vein, and is expected to inhibit the directly liver via the portal route. If that were the case, then the single gateway idea would fall.

In fact, there is compelling evidence that the suppression of glucose output is mediated indirectly, rather than directly. This question has been addressed multiple times, usually by the administration of similar doses of insulin via portal vein or peripheral vein, with comparison of hepatic glucose output (38,39) between different routes of hormone administration. However, because the liver degrades approximately half of the insulin administered portally on each pass through the organ, the same rate of insulin infusion via the two routes will match *neither portal nor peripheral insulin levels (Figure 10a).*

The reason is the following: infusion of insulin via the portal route at rate "R" will result in a large increment in portal insulin (R divided by portal plasma flow), but an infusion rate of only **R/2** into the peripheral circulation (as half the insulin is degraded by the liver on first pass). Alternatively, systemic insulin infusion at rate R will cause approximately equal increments in insulin in portal and peripheral plasma; increments which are lower than the portal increment and greater than the peripheral increment. Thus, infusion of insulin at equal rates in portal versus peripheral blood will match *neither portal nor peripheral concentrations.*

Ader and her colleagues, in this laboratory, introduced a new protocol which allows for matched peripheral but not portal insulin concrentrations (Figure 10b; ref. 40). This protocol involved the infusion of insulin at one rate into the portal vein, and infusion at *half this rate* into the peripheral circulation. Under such a protocol, portal insulin levels are very different, but systemic insulin levels are matched (Figure 11).

We reported that infusion of insulin systemically was equally potent in suppressing hepatic glucose output as infusion of insulin portally at twice the rate. In other words, *the degree of suppression of hepatic glucose production depended upon the*

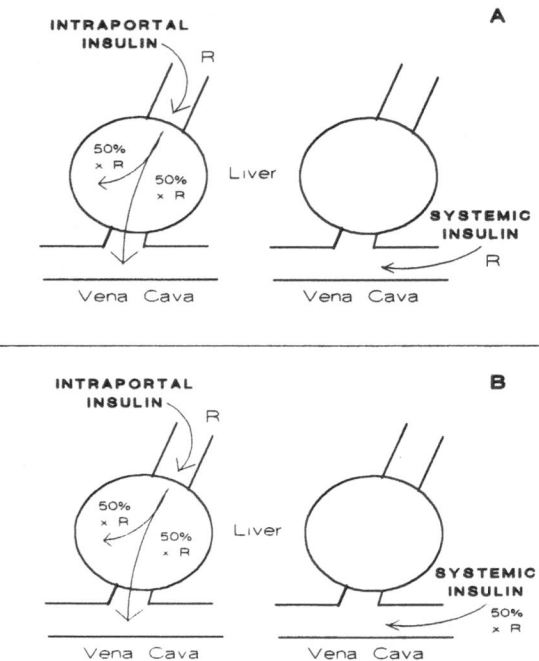

Figure 10. Experimental designs to examine the relative importance of portal vs peripheral insulin to suppress hepatic glucose production. (A) Insulin is administered intraportally or peripherally at rate "R". After ~50% hepatic extraction, (50% x R) would reach systemic circulation during intraportal infusion, but full infusion rate R would enter systemic circulation during peripheral infusion. (B) Insulin is administered intraportally at rate "R", and systemically at rate (50% x R) to effectively match the amount of insulin arriving in systemic circulation during intraportal infusion.

peripheral but not the portal insulin level (Figure 11). This surprising result suggested that the effect of insulin to suppress glucose output by the liver is mediated by a peripheral, rather than a portal effect. Recently, Giacca and Vranic and their colleagues have substantiated this result, although in the diabetic dog model (41). Taken together, our results and the results of the Toronto group support the concept that the primary effect of insulin to suppress hepatic glucose output is indirect, via the systemic circulation. This concept is supportive of the primary gateway hypothesis, enunciated above, which proposes that the suppression of glucose output occurs secondarily to transcapillary transport of insulin into a peripheral tissue.

UNANSWERED QUESTIONS

The single gateway hypothesis was proposed based upon circumstantial evidence: evidence that the dynamics of lymph insulin are similar to the dynamic changes in glucose turnover *in vivo*, and evidence that the rate of increase of glucose utilization

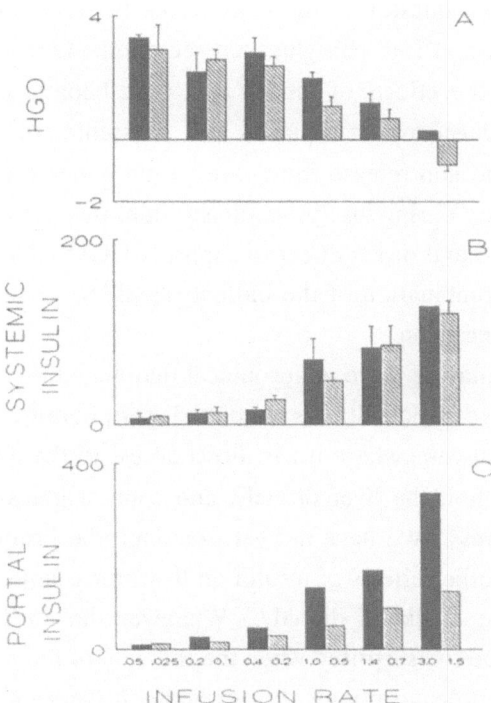

Figure 11. Importance of systemic, rather than portal, insulin concentrations in suppression of hepatic glucose production (HGO). Insulin was infused at various rates intraportally (solid bars) or systemically at half the intraportal rate (hatched bars; c.f. Figure 10b). (A) Decline in HGO followed systemic insulin (B) rather than portal levels (C). In other words, despite higher portal insulinemia during intraportal infusion, HGO was no further suppressed than during systemic infusion (adapted from ref. 40).

during clamps is virtually identical to the rate of suppression of endogenous glucose output. However, important questions remain to be investigated before this concept can be accepted.

What is the putative "signal" by which peripheral tissues may signal the liver to squelch glucose output? One possible answer is free fatty acids. This metabolite is suppressed by insulin, and this suppression is known to be significantly more sensitive to insulin than is the stimulation of glucose uptake (36,42). In addition, FFA have been demonstrated to be a potent stimulus to gluconeogenesis (43). Thus, one can imagine a scenario in which insulin suppresses FFA release from adipose tissue, and this results in a reduction in liver glucose output. Additional possible signals include gluconeogenic precursors, including alanine, the release of which are insulin-suppressible (44,45). Finally, the possiblity exists that the central nervous system plays some role in the putative signalling of the liver by peripheral tissues. Further studies in our laboratory are addressing this important issue.

A second important question is the relationship between indirect suppression of fasting glucose output and islet cell glucagon secretion. Our studies were carefully designed to examine the effects of insulin *per se* on hepatic glucose output under conditions in which glucagon was replaced at a constant rate. However, insulin is known to suppress glucagon release from α-cells, and under conditions in which glucagon is allowed to vary during insulin administration, this insulin effect may play an important, quantitative and direct effect to suppress HGO. Thus, the overall effects of insulin may be a combination of the indirect signalling from liver and direct suppression of glucagon secretion.

A third question may be more philosophical than scientific: "if the effect of insulin to regulate glucose production is mediated peripherally, then why is insulin secreted into the portal vein, where it has direct access to the liver?" One possibility is that insulin may inhibit the liver directly, and counter glucagon and epinephrine effects on HGO in stress. We have not yet examined the insulin effects in stressful circumstances. Also, other effects of insulin on liver, for example stimulation of protein synthesis, may be mediated directly. Whatever the answer to this teleogical query, we should not be too surprised when the function of the metabolic system does not coincide with our preconceptions. For example, it is now clear that much of the glycogen synthesized in liver after carbohydrate ingestion originates not in glucose entering the portal vein directly, but from 3-carbon moieties emanating from peripheral tissues, primarily muscle. Is it possible that the so-called "indirect" route of glycogen deposition, which also violates our preconceptions, represents a paradigm which is similar to the importance of the indirect effects of insulin on hepatic glucose output?

These questions, and others, form the basis for future testing of the single gateway paradigm.

CONCLUSIONS

Regardless of the support or lack thereof for the single gateway concept, we believe that it underscores the limitations of extrapolating *in vitro* results to the *in vivo* situation. No one can minimize the enormous impact of cellular and molecular studies of insulin action. However, there is a necessary role for studies in the intact organism, which allow for a careful consideration of the dynamic regulation of metabolic physiology, and to examine carefully signalling between different tissues. Clearly, important questions remain with regard to this signalling process which molecular studies will not elucidate. Also, considering the importance of elevated glucose production in the etiology of fasting hyperglycemia (2), the present studies will likely have important ramifications for understanding the pathogenesis of diabetes mellitus. The role of the single gateway idea in diabetes deserves further investigation.

REFERENCES

1. M.F. Saad, W.C. Knowler, D.J. Pettitt, R.G. Nelson, D.M. Mott, and P.H. Bennett, The natural history of impaired glucose tolerance in the Pima Indians, *N Engl J Med* 319:1500 (1988).
2. R.A. DeFronzo, The triumvirate: B-cell, muscle, liver. A collusion responsible for NIDDM, *Diabetes* 37:667 (1987).
3. M. Laakso, H. Sarlund, and L. Mykkanen, Insulin resistance is associated with lipid and lipoprotein abnormalities in subjects with varying degrees of glucose tolerance, *Arteriosclerosis* 10:223 (1990).
4. E. Ferrannini, G. Buzzigoli, R. Bonadonna, M.A. Giorico, M. Oleggini, L. Graziadei, R. Pedrinelli, L. Brandi, and S. Bevilacqua, Insulin resistance in essential hypertension, *N Engl J Med* 317:350 (1987).
5. C. Bogardus, S. Lillioja, B.L. Nyomba, F. Zurlo, B.A. Swinburn, A. Esposito-del Puente, W.C. Knowler, E. Ravussin, D.M. Mott, and P.H. Bennett, Distribution of in vivo insulin action in Pima Indians as mixture of three normal distributions, *Diabetes* 38:1423 (1989).
6. B.C. Martin, J.H. Warram, A.S. Krolewski, R.N. Bergman, J.S. Soeldner, and C.R. Kahn, Role of glucose and insulin resistance in development of type 2 diabetes mellitus: results of a 25-year follow-up study, *Lancet* 340:925 (1992).
7. T.P. Ciaraldi, J.M. Molina, and J.M. Olefsky, Insulin action kinetics in adipocytes from obese and noninsulin-dependent diabetes mellitus subjects: identification of multiple cellular defects in glucose transport, *J Clin Endocrinol Metab* 72:876 (1991).
8. A.R. Saltiel, Second messengers of insulin action, *Diab Care* 13:244 (1990).
9. O.G. Kolterman, J. Insel, M. Saekow, and J.M. Olefsky, Mechanisms of insulin resistance in human obesity: evidence for receptor and postreceptor defects, *J Clin Invest* 65:1272 (1980).
10. J.M. Olefsky, W.T. Garvey, R.R. Henry, D. Brillon, S. Matthaei, and G.R. Freidenberg, Cellular mechanisms of insulin resistance in non-insulin- dependent (type II) diabetes, *Am J Med* 85:86 (1988).
11. L.J. Mandarino, Regulation of skeletal muscle pyruvate dehydrogenase and glycogen synthase in man, *Diab Metab Rev* 5:475 (1989).
12. J. Eriksson, A. Franssila-Kallunki, A. Ekstrand, C. Saloranta, E. Widen, C. Schalin, and L. Groop, Early metabolic defects in persons at increased risk for non-insulin-dependent diabetes mellitus, *N Engl J Med* 321:337 (1989).
13. S.S. Rich, Mapping genes in diabetes, *Diabetes* 39:1315 (1990).
14. J.I. Rotter, C.E. Anderson, and D.L. Rimoin, Genetics of diabetes mellitus, *in:* "Diabetes Mellitus: Theory and Practice," M. Ellenberg and H. Rifkin, eds., Medical Examination Publishing Co., New York (1983).
15. L. Doberne, M.S. Greenfield, B. Schulz, and G.M. Reaven, Enhanced glucose utilization during prolonged glucose clamp studies, *Diabetes* 30:829 (1981).
16. R. Prager, P. Wallace, and J.M. Olefsky, In vivo kinetics of insulin action on peripheral glucose disposal and hepatic glucose output in normal and obese subjects, *J Clin Invest* 78:472 (1986).
17. S.S. Chernick, R.J. Gardiner, and R.O. Scow, Restricted passage of insulin across capillary endothelium in perfused rat adipose tissue, *Am J Physiol* 253:E475 (1987).
18. E.A. Rasio, E. Mack, R.H. Egdahl, and M.G. Herrera, Passage of insulin and inulin across vascular membranes in the dog, *Diabetes* 17:668 (1968).
19. E.A. Rasio, C.L. Hampers, J.S. Soeldner, and G.F. Cahill, Jr., Diffusion of glucose, insulin, inulin, and Evans blue protein into thoracic duct lymph of man, *J Clin Invest* 46:903 (1967).
20. R.S. Sherwin, K.J. Kramer, J.D. Tobin, P.A. Insel, J.E. Liljenquist, M. Berman, and R. Andres, A model of the kinetics of insulin in man, *J Clin Invest* 53:1481 (1974).
21. P.A. Insel, J.E. Liljenquist, J.D. Tobin, R.S. Sherwin, P. Watkins, R. Andres, and M. Berman, Insulin control of glucose metabolism in man, *J Clin Invest* 55:1057 (1975).
22. R.N. Bergman, Y.Z. Ider, C.R. Bowden, and C. Cobelli, Quantitative estimation of insulin sensitivity, *Am J Physiol* 236:E667 (1979).
23. R.N. Bergman, L.S. Phillips, and C. Cobelli, Physiologic evaluation of factors controlling glucose tolerance in man: measurement of insulin sensitivity and B-cell glucose sensitivity from the response to intravenous glucose, *J Clin Invest* 68:1456 (1981).
24. M. Ader and R.N. Bergman, Insulin sensitivity in the intact organism, *in:* "Bailliere's Clinics in Endocrinology and Metabolism," K.G.M.M. Alberti, P.D. Home, and R. Taylor, eds., Bailliere Tindall, London (1987).
25. R.N. Bergman, Toward physiological understanding of glucose tolerance: minimal- model approach, *Diabetes* 38:1512 (1989).
26. Y.J. Yang, I.D. Hope, M. Ader, and R.N. Bergman, Insulin transport across capillaries is rate limiting for insulin action in dogs, *J Clin Invest* 84:1620 (1989).

27. M. Gaudino, Kinetics of distribution of inulin between two body water compartments, *Proc Soc Exp Biol Med* 70:672 (1949).

28. M. Ader, R.A. Poulin, Y.J. Yang, and R.N. Bergman, Dose response relationship between lymph insulin and glucose uptake reveals enhanced insulin sensitivity of peripheral tissues, *Diabetes* 41:241 (1992).

29. D.T. Finegood, R.N. Bergman, and M. Vranic, Estimation of endogenous glucose production during hyperinsulinemic-euglycemic glucose clamps: comparison of unlabeled and labeled exogenous glucose infusates, *Diabetes* 36:914 (1987).

30. D.T. Finegood, R. N. Bergman, and M. Vranic, Modelling error and apparent isotope discrimination confound estimation of endogenous glucose production during euglycemic glucose clamps, *Diabetes* 37:1025 (1988).

31. R.A. Poulin, G. Steil, D. Banks, H. Xiang, M. Ader, and R.N. Bergman, Transcapillary insulin transport (TCIT) is rate-limiting in insulin-sensitive tissues, *FASEB J* 5:A653 (1991).

32. M. Ader, R.A. Poulin, G.M. Steil, and R.N. Bergman, Nicotinic acid induces insulin resistance at a post-capillary transport locus as demonstrated by simultaneous measures of hindlimb lymph insulin and glucose uptake, *Clin Res* (1992)

33. M.F. Saad, S. Lillioja, B.L. Nyomba, C. Castillo, R. Ferraro, M. De Gregario, E. Ravussin, W.C. Knowler, P.H. Bennett, B.V. Howard, and C. Bogardus, Racial differences in the relation between blood pressure and insulin resistance, *N Engl J Med* 324:733 (1991).

34. R. Prager, P. Wallace, and J.M. Olefsky, Direct and indirect effects of insulin to inhibit hepatic glucose output in obese subjects,*Diabetes* 36:607 (1987).

35. D.C. Bradley and R.N. Bergman, Restoration of stable metabolic conditions during islet suppression in the dog,*Am J Physiol* 262:E532 (1992).

36. D.C. Bradley, R.A. Poulin, and R.N. Bergman, Dynamics of hepatic and peripheral insulin effects suggest common rate-limiting step in vivo, *Diabetes* 42:296 (1993).

37. I. Jensen, V. Kruse, and U.D. Larsen, Scintigraphic studies in rats: kinetics of insulin analogues covering wide range of receptor affinities, *Diabetes* 40:628 (1991).

38. T. Ishida, Z. Chap, J. Chou, R.M. Lewis, C.J. Hartley, M.L. Entman, and J.B. Field, Effects of portal and peripheral venous insulin infusion on glucose production and utilization in depancreatized, conscious dogs, *Diabetes* 33:984 (1984).

39. R.W. Stevenson, J.A. Parsons, and K.G.M.M. Alberti, Comparison of the metabolic responses to portal and peripheral infusions of insulin in diabetic dogs, *Metabolism* 30:745 (1981).

40. M. Ader and R.N. Bergman, Peripheral effects of insulin dominate suppression of fasting hepatic glucose production, *Am J Physiol* 258:E1020 (1990).

41. A. Giacca, S. Fisher, Z.Q. Shi, R. Gupta, H.L.A. Lickley, and M. Vranic, Importance of peripheral insulin levels for insulin-induced suppression of glucose production in depancreatized dogs, *J Clin Invest* 90:1769 (1992).

42. K.L. Zierler and D. Rabinowitz, Effect of very small concentrations of insulin on forearm metabolism: persistence of its action on potassium and free fatty acids without its effect on glucose, *J Clin Invest* 43:950 (1964).

43. A.D. Cherrington and M. Vranic, Hormonal control of gluconeogenesis in vivo, *in:* "Hormonal Regulation in Gluconeogenesis," N. Kraus-Friedman, ed., CRC Press, Boca Raton, FL (1986).

44. N.K. Fukagawa, K.L. Minaker, V.R. Young, and J.W. Rowe, Insulin dose-dependent reductions in plasma amino acids in man, *Am J Physiol* 250:E13 (1986).

45. N. Nurjhan, P.J. Campbell, F.P. Kennedy, J.M. Miles, and J.E. Gerich, Insulin dose-response characteristics for suppression of glycerol release and conversion to glucose in humans, *Diabetes* 35:1326 (1986).

ACUTE HORMONAL REGULATION OF GLUCONEOGENESIS IN THE CONSCIOUS DOG

Alan D. Cherrington, Ralph W. Stevenson, Kurt E. Steiner,
Cynthia C. Connolly, Masahiko Wada and Richard E. Goldstein

Vanderbilt University School of Medicine
Department of Molecular Physiology and Biophysics
710 Medical Research Building
Nashville, TN 37232-0615 U.S.A.

INTRODUCTION

The glucose level in blood is precisely controlled by the coordinated regulation of both glucose production and utilization. The former involves two processes, glycogenolysis, the break down of stored glycogen, and gluconeogenesis, the conversion of gluconeogenic substrates into glucose. Gluconeogenesis is carried out both by the liver and the kidneys, with the former being dominant and usually accounting for the majority of gluconeogenic glucose production.

Hepatic gluconeogenesis is controlled at three sites 1) within tissues which supply gluconeogenic precursors to the liver, 2) transport proteins which facilitate the entry of the precursors into the liver, and 3) enzymatic steps within the gluconeogenic and glycolytic pathways of the liver per se.[1] This complexity has made it difficult to quantitate gluconeogenesis in vivo. In addition to multiple sites of regulation, there are multiple precursors (alanine, serine, glycine, threonine, glutamine, lactate, pyruvate and glycerol) involved. Thus, administration of a given labelled gluconeogenic precursor and subsequent estimation of its rate of conversion into glucose reflects only a portion of the entire response. In addition certain intermediate pools in the liver are common to gluconeogenesis and other metabolic pathways so that loss of the isotopic molecule can occur simply as the result of exchange.[2] For that reason, estimation of gluconeogenesis by calculating the rate of conversion of a precursor into glucose (i.e. [14]C glucose production rate ÷ [14]C precursor SA in blood entering the liver) yields a minimum estimate of the conversion rate of that precursor into glucose.[2]

The A-V difference technique can also be used to estimate the gluconeogenic rate. The net hepatic uptake of each gluconeogenic precursor can be determined and one can then simply assume that there is quantitative conversion of the carbon atoms to glucose. This approach of course yields a maximal estimate of the rate of gluconeogenesis from circulating precursors. Combination of the tracer and A-V difference approaches currently provides the most useful assessment of gluconeogenesis in vivo. By administering a labelled gluconeogenic precursor, determining the rate of uptake of all labelled gluconeogenic precursors by the liver and calculating the fraction of the extracted precursors which are converted to glucose one can estimate the efficiency of the gluconeogenic process. Once again this represents a minimal estimate for the reasons alluded to above. The product of efficiency and the maximal gluconeogenic rate then yields a minimal estimate of the gluconeogenic process. In this way the true gluconeogenic rate can be bracketed between a minimal and a maximal estimate. Currently attempts are being made to use data obtained

New Concepts in the Pathogenesis of NIDDM, Edited by
C. G. Östenson *et al.*, Plenum Press, New York. 1993

from liver biopsies and NMR methodology to more precisely evaluate gluconeogenesis in vivo, but further validation studies are required before these newer approaches can be broadly implemented.

Yet another reason for the difficulty in quantifying gluconeogenesis in vivo is the number of hormones which alter the process. It is the purpose of the present paper to summarize our results in this area and to evaluate the ability of physiologic changes in the plasma levels of some of the hormones involved in the acute regulation of the glucose level to stimulate gluconeogenesis in vivo.

GLUCAGON

Glucagon concentrations in plasma can fall by as much as 75% in response to carbohydrate loading or rise by as much as 5 fold in response to stress.[3] In a study carried out a number of years ago[4], we used the pancreatic clamp technique to control the endocrine pancreas so as to bring about a selective 4-fold increase in plasma glucagon. Somatostatin was infused to inhibit the endocrine pancreas and basal intraportal replacement infusions of insulin and glucagon were given to replace endogenous secretion. The glucagon infusion rate was fixed, the glucose level was monitored every 5 min, and the insulin infusion rate was then adjusted so as to maintain euglycemia. After the hormonal titration was complete the animal was allowed to stand calmly in the Pavlov harness for 30 min prior to initiating the control period of the experiment. After a 40 min control period the glucagon infusion rate was increased four fold (Fig. 1). The arterial glucagon level thus rose from 101±10 to 366±47 pg/ml while the arterial insulin, epinephrine, norepinephrine, and cortisol levels remained basal and unchanged. The selective rise in glucagon caused glucose production to rise rapidly to 7.7±0.5 mg/kg-min. As a result it exceeded glucose utilization and the plasma glucose level quickly doubled. Glucose utilization subsequently rose as a consequence of the increase in the plasma glucose level. After about 30 min glucose production began to fall such that by the end of 3h the glucagon-induced rise was reduced by 3.6 mg/kg-min.

Control experiments in which the glucagon-induced glycemic excursion was mimicked by glucose infusion in the presence of a pancreatic clamp (but no extra glucagon) indicated that glucose production would have fallen by 50% as a result of the hyperglycemia per se. Thus there was still a substantial effect of glucagon on glucose production (~2.6 mg/kg-min) at 3h. The control data also demonstrated that glucagon did not alter peripheral glucose uptake because the rise in glucose utilization under the hyperglycemic condition was similar whether glucagon was present or not.

Glucagon did not have an appreciable effect on the rate of delivery of gluconeogenic substrates to the liver but did augment the fractional extraction of alanine (and the other gluconeogenic amino acids) by the liver, causing it to rise from 0.31±0.05 to 0.66±0.10. This did not result in a significant increase in amino acid uptake by the liver, however, since the amino acid level in plasma fell until it offset the augmented transport process. Control data indicated that hyperglycemia per se caused a slight increase in peripheral alanine production so that unlike the case for the other amino acids alanine uptake by the liver did increase slightly (1.5 µmol/kg-min). This effect was indirect, however, was mediated by the rise in glucose rather than glucagon per se.

The efficiency with which the liver converted gluconeogenic precursors to glucose tripled (relative to the control period rate) by the third hour of hyperglucagonemia. If anything, this underestimates the effect of glucagon since hyperglycemia of the magnitude observed decreased gluconeogenic efficiency by about 40% in control studiies. The minimal gluconeogenic rate thus increased from 0.2±0.1 to 0.6±0.2 mg/kg-min while the maximal gluconeogenic rate rose only slightly from 0.6±0.2 to 0.8±0.2 mg/kg-min. Thus by the end of the third hour of elevated glucagon the gluconeogenic rate was between 0.6 and 0.8 mg/kg-min and the glycogenolytic rate (glucose production - gluconeogenic rate) was 3.3 to 3.5 mg/kg-min. As such gluconeogenesis was accounting for only 15 to 20% of overall glucose production. Since glucagon's glycogenolytic effect was very potent and since its gluconeogenic effect on the liver was unaccompanied by a push of substrate from the periphery the latter process was always a minor contributor to hepatic glucose release. Furthermore since no change in any gluconeogenic parameter had occurred by 15 min, the initial response to glucagon can be attributed exclusively to glycogenolysis.

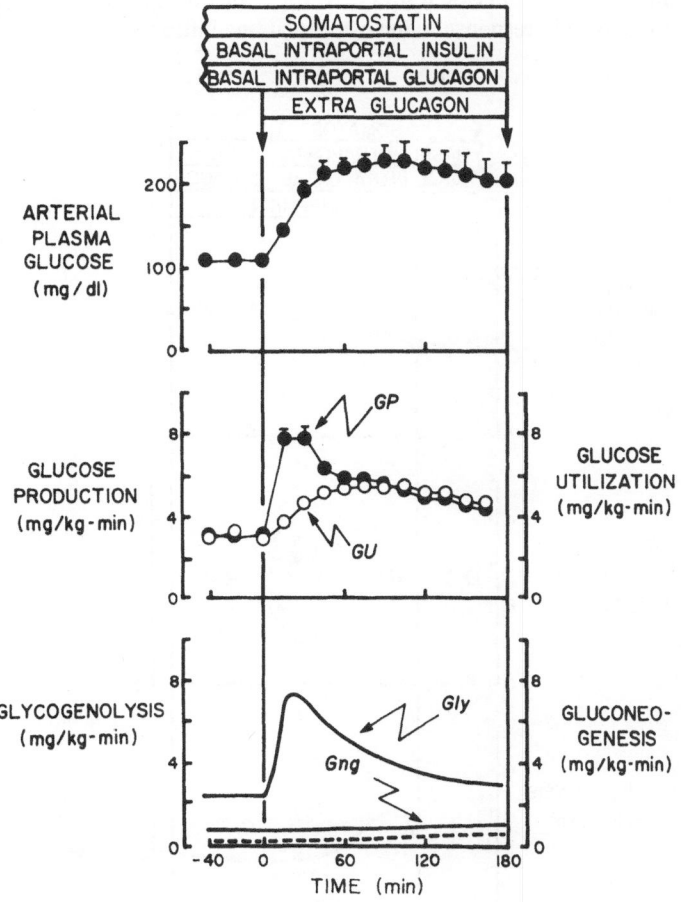

Figure 1. The effects of a selective rise in glucagon brought about during a pancreatic clamp on the arterial plasma glucose level, glucose turnover assessed using ^{3}H-3 glucose infusion, maximal (—) and minimal (---) estimates of gluconeogenesis (see introduction) and minimal glycogenolysis in overnight-fasted conscious dogs. Redrawn from ref. 4 & 7.

Thus a physiologic increase in glucagon can bring about a rapid and marked increment in glycogenolysis which, however, wanes substantially with time. Elevated glucagon progressively increases gluconeogenic efficiency within the liver, but brings about a quantitatively small increase in gluconeogenically derived glucose production since it is unable to provide extra substrate for conversion to glucose. By the end of 3h of hyperglucagonemia glucose production remains somewhat elevated and the gluconeogenic/glycogenolytic mix is slightly shifted toward the former.

EPINEPHRINE

Plasma epinephrine concentrations are basally 50 to 100 pg/ml but can rise as much as 80 fold in respond to certain stimuli, as discussed elsewhere.[5] In the study depicted in Figure 2 the pancreatic clamp was again used to fix the insulin and glucagon levels at basal values so as to allow a selective increase in epinephrine to be brought about. Epinephrine was infused through a peripheral vein at a rate of 0.04 μg/kg-min and its arterial level increased from 62±19 to 424±48 pg/ml. Neither insulin, glucagon, norepinephrine nor cortisol levels changed in response to the catecholamine infusion.

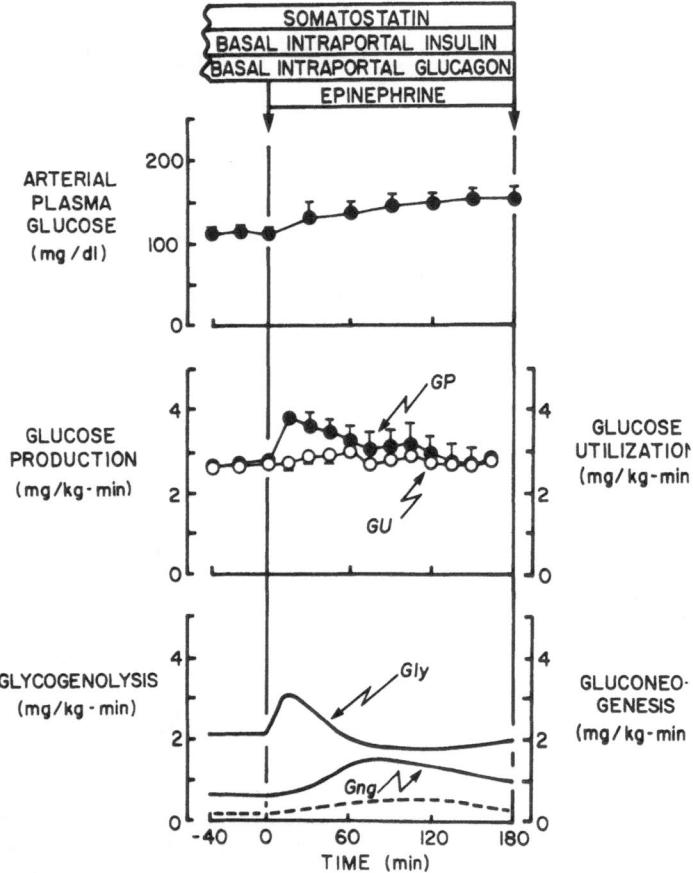

Figure 2. The effects of a selective rise in epinephrine brought about during a pancreatic clamp on the arterial plasma glucose level, glucose turnover assessed using ³H-3 glucose infusion, maximal (—) and minimal (---) estimates of gluconeogenesis (see introduction) and minimal glycogenolysis in overnight-fasted conscious dogs. Redrawn from ref. 5 & 6.

The selective rise in epinephrine caused glucose production to increase by 1.1±0.3 mg/kg-min (Fig. 2) and as a consequence glucose production exceeded glucose utilization and the glucose level rose.[6] As with glucagon the increase in glucose production waned with time and eventually the rates of glucose utilization and production were again equal and the glucose level plateaued (160±16 mg/dl). In this protocol, as in the former, hyperglycemia confounds data interpretation, necessitating a control study in which the rise in glucose is brought about in isolation.[5] When the rise in glucose is considered it is clear that the catecholamine, like glucagon, is still stimulating glucose production at 3h although its effect is reduced relative to the peak response. In addition, the slight increase in glucose utilization which occurred during epinephrine infusion was considerably less than that which would normally occur in response to the hyperglycemia observed, indicating that epinephrine can impair peripheral glucose uptake.

If one examines the gluconeogenic data it becomes apparent that the effect of epinephrine on gluconeogenesis begins slowly, peaks at 90 min and then begins to fall. The increase in the maximal rate of gluconeogenesis at 90 min (0.6 ± 0.1 to 1.4 ± 0.4 mg/kg-min) resulted primarily from an increase in the uptake of lactate by the liver secondary to a marked rise in the blood lactate level. The latter in turn resulted from increased lactate production by non hepatic tissues (muscle). By the end of 3h net hepatic lactate uptake had returned to zero but hepatic uptake of the gluconeogenic amino acids and glycerol was still increased. The latter responses resulted from increased arterial levels of both substrates secondary to increased production by non-hepatic tissues (muscle & fat). By the end of 3h of increased epinephrine the maximal gluconeogenic rate was 0.9 ± 0.2 mg/kg-min. Gluconeogenic efficiency increased by only 84% in response to epinephrine but it must be remembered that hyperglycemia reduces gluconeogenic efficiency. The rise in gluconeogenic efficiency that normally occurs over time during saline infusion[4] was offset by the inhibitory action of hyperglycemia in control studies. The minimal gluconeogenic rate thus rose from a control value of 0.2 ± 0.1 to 0.7 ± 0.2 mg/kg-min at 90 min and 0.4 ± 0.1 mg/kg-min at 180 min. By the end of 3h of a selective increase in epinephrine the gluconeogenic rate was between 0.4 and 0.9 mg/kg-min and thus gluconeogenesis comprised 14 to 30% of the glucose production rate.

If one examines the epinephrine-induced change in glucose production it is clear that the initial (15 min) increase was primarily (~75%) glycogenolytic in origin since the gluconeogenic parameters changed minimally. A peripheral effect of the catecholamine on muscle and fat quickly became evident such that lipolysis, as well as the Cori and Alanine cycles, increased. As a result there was a push of substrate from muscle and fat to the liver, and consequently hepatic gluconeogenic precursor uptake increased. Since the uptake was accompanied by a relatively small increase in gluconeogenic efficiency, precursors accumulated in the liver and eventually lactate uptake in particular began to decrease. This increase in epinephrine had little residual effect on glucose production by 3h because its inhibition of glucose clearance caused the increase in glycemia to offset the increment in glucose output. Nevertheless, the gluconeogenic/glycogenolytic mix of the liver was shifted slightly in favor of gluconeogenesis.

Comparison of the effects of the rise in glucagon to the rise in epinephrine indicates that glucagon (on a fold basis) is much more effective (5 to 10 times) than the catecholamine in stimulating glucose production in vivo. Since on a molar basis the basal epinephrine level is 25 times greater than the basal glucagon level a molecule of glucagon is 125 to 250 times as effective as an epinephrine molecule in causing glucose output by the liver. Interestingly, the catecholamine can stimulate both glycogenolysis and gluconeogenesis within the liver and in fact it brings about a slightly greater gluconeogenic/glycogenolytic mix than does glucagon. When the marked extra-hepatic effects of the catecholamine on gluconeogenic precursor supply are added to its hepatic gluconeogenic action its gluconeogenic potential becomes quite prominent. Interestingly, when the catecholamine was infused at 0.24 μg/kg-min so that the plasma epinephrine level increased by ~40 fold the initial (15 min) increase in glycogenolysis reached almost 3.0 mg/kg-min. Three hours later glucose production (~3.3 ± 0.6 mg/kg-min) was almost exclusively gluconeogenic (75 to 100%). This was the result of a marked increase in the push of gluconeogenic substrate to the liver, and a 136% in gluconeogenic efficiency per se. Thus, unlike glucagon which continues to be primarily glycogenolytic 3h after its level is raised, epinephrine at a dose which is able to bring about a glycogenolytic response equal to that of a 3-fold increase in glucagon, soon switches the liver to a predominantly gluconeogenic organ. Since the epinephrine level can change up to 80 fold there are undoubtedly certain stress conditions in which epinephrine plays an important role in shifting the liver to a predominantly gluconeogenic state.

NOREPINEPHRINE

Under most circumstances the basal arterial norepinephrine level (100-150 pg/ml) does not increase more than 10 fold. Since, however, the level of this catecholamine in plasma is primarily derived from nerve endings its action at target tissues may by associated with levels far in excess of those seen in the blood. Figure 3 contains data from a set of experiments in which the pancreatic clamp was again used to fix the level of insulin and glucagon at basal values so that a selective increase in norepinephrine (0.32 μg/kg-min) could be brought about. This infusion rate resulted in norepinephrine in blood of ~3000 pg/ml, an amount assumed to exist in the synaptic cleft.[8] Insulin, glucagon, epinephrine and cortisol levels remained unchanged at basal values throughout the study.

Norepinephrine increased glucose production from 2.8±0.2 to 3.8±5 mg/kg-min by 15 min. As a result of the increase in glucose production the rate of glucose entry into plasma exceeded its rate of exit and the glucose level rose progressively over 3h from 108±4

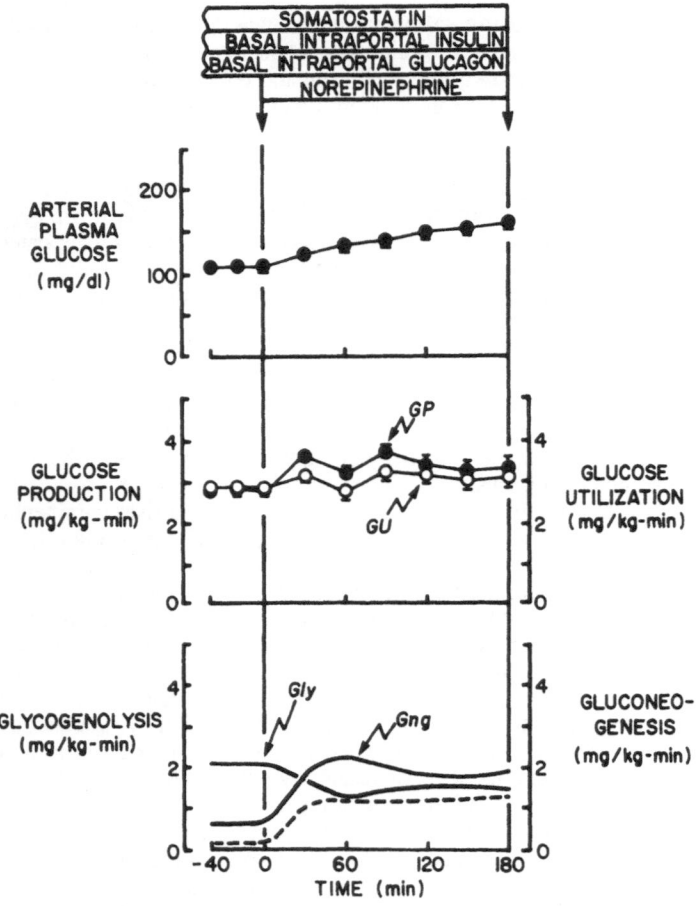

Figure 3. The effects of a selective rise in norepinephrine brought about during a pancreatic clamp on the arterial plasma glucose level, glucose turnover assessed using ³H-3 glucose infusion, maximal (—) and minimal (---) estimates of gluconeogenesis (see introduction) and minimal glycogenolysis in overnight-fasted conscious dogs. Redrawn from ref. 8.

to 159±15 mg/dl. As with epinephrine, consideration of the effects of the accompanying hyperglycemia leads to the conclusion that the effect of norepinephrine on glucose production in fact does not decline with time.   In addition, it becomes clear that norepinephrine also inhibits glucose utilization. Norepinephrine caused a marked increase in the maximal gluconeogenic rate which peaked at 60 min. This increase resulted from a rapid effect on muscle and fat which increased the supply of alanine, lactate and glycerol reaching the liver. In addition the catecholamine caused the gluconeogenic efficiency to increase by over 200%; thus the minimal gluconeogenic rate also showed a marked increase in response to norepinephrine.

Clearly norepinephrine failed to increase glycogen breakdown and may have actually reduced it. By the end of 3h of increased norepinephrine glucose production (3.4±0.4 mg/kg-min) was derived equally from glycogenolysis and gluconeogenesis.   Thus norepinephrine can stimulate gluconeogenesis in the absence of a significant effect on glycogenolysis. This raises the possibility that neural input to the liver may determine the gluconeogenic set point of the organ.

It should also be noted that a selective increment in norepinephrine similar in magnitude (310 pg/ml) to the aforementioned increase in epinephrine (362 pg/ml), had no effect on overall glucose production, slightly increased the supply of gluconeogenic precursors reaching the liver but increased the efficiency of the gluconeogenic process within the liver by $\sim$300%. After 3h of this small increase in norepinephrine gluconeogenesis was responsible for 25 and 40% of overall glucose production (0.7 to 1.1 mg/kg-min as opposed to 0.6 to 0.8 mg/kg-min with glucagon). The two hormones had a reasonably similar effect on hepatic gluconeogenesis per se, but norepinephrine also had an effect on fat and muscle and thus was able to provide extra substrate for the gluconeogenic process. As such the gluconeogenic effect of a four fold increase of norepinephrine was larger than that of a four fold rise in glucagon.

CORTISOL

The basal plasma cortisol level in the dog (0.5 to 3.0 µg/dl) can rise to as high as 18 µg/dl in response to maximal stress. Figure 4 contains data from a study in which the acute effects of a selective physiologic increase in cortisol were examined. The plasma cortisol level rose from 2.2±0.4 to 7.3±1.1 µg/dl. Once again the pancreatic clamp was used to ensure that changes in insulin and glucagon did not occur. Similarly, neither the epinephrine level nor the norepinephrine level changed in response to the rise in cortisol.[13]

The metabolic consequences of the acute rise in cortisol are indistinguishable from those obtained in a control protocol[4] in which infusion of saline during a pancreatic clamp caused a slight decrease in the plasma glucose level and a 10% fall in glucose production. Cortisol caused a small decrease in the glucose level, and little or no change in glucose turnover. There was no difference in the increase in gluconeogenic efficiency (53±26 and 65±43%) seen with saline or cortisol infusion. Cortisol had little or no effect on the uptake of lactate or glycerol by the liver but did augment the uptake of the gluconeogenic amino acids slightly by virtue of an effect ($\sim$40%) on their fractional extraction by the liver. Therefore the overall hepatic uptake of gluconeogenic precursors increased and with that the maximal gluconeogenic rate rose (0.8 to 1.1 mg/kg-min) in response to cortisol. By the end of 3h of hypercortisolemia the liver had shifted to a slightly more gluconeogenic state. At 3h gluconeogenesis was responsible for between 20 and 44% of glucose production in the cortisol group as opposed to 12 to 32% in the saline infused group. When one considers the effect of the selective increase in cortisol relative to the effects of glucagon and the catecholamines, it becomes evident that the glucocorticoid by itself is without significant effect. It is important to now determine the ability of an increase in cortisol to potentiate the action of increases in the levels of the previously discussed agonists.

INSULIN

The aforementioned hormones are all positive regulators of glucose production by the liver, while insulin is a well known inhibitor of hepatic glucose output. In the study shown in Figure 5 a selective four fold rise (10±21 to 35±14 µU/ml) in insulin was brought about in conscious overnight-fasted dogs. The plasma glucagon level remained unchanged and although epinephrine, norepinephrine and cortisol were not measured subsequent studies[9] have shown that they would not change under such conditions.

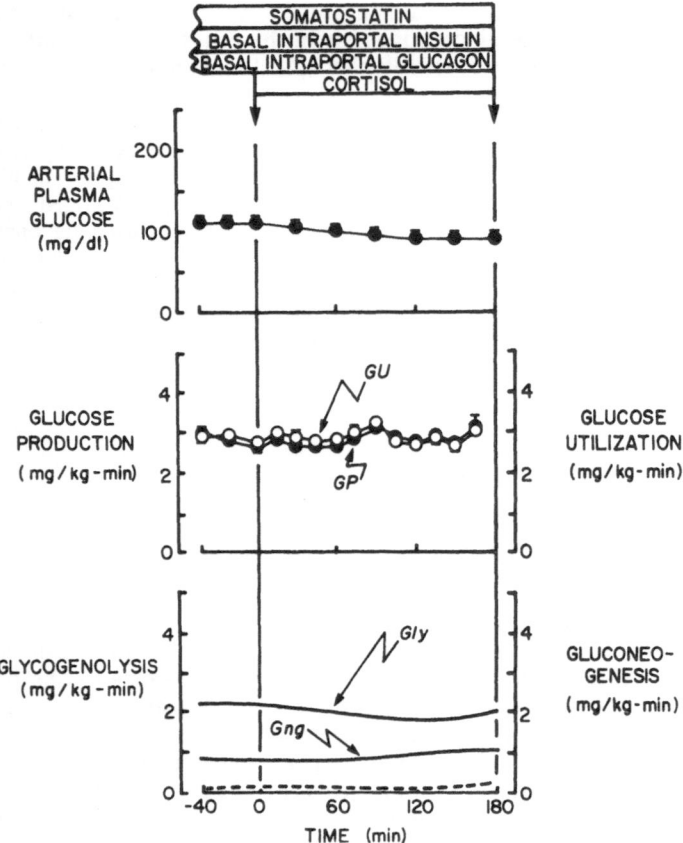

Figure 4. The effects of a selective rise in cortisol brought about during a pancreatic clamp on the arterial plasma glucose level, glucose turnover assessed using ^{3}H-3 glucose infusion, maximal (——) and minimal (---) estimates of gluconeogenesis (see introduction) and minimal glycogenolysis in overnight-fasted conscious dogs. Redrawn from ref. 104.

In response to the rise in insulin glucose production fell over the course of an hour and was eventually reduced from 2.8±0.2 to 1.6±0.8 mg/kg-min by 3h. Glucose utilization, on the other hand, rose from 2.7±0.3 to 8.1±1.1 mg/kg-min by 3h. The glucose level (112±2 µg/dl) was maintained by a peripheral glucose infusion adequate to keep total glucose appearance equivalent to total glucose disappearance.

Separation of glucose production into its components reveals that the fall in hepatic glucose output was exclusively a function of reduced glycogenolysis. There was a slight fall in hepatic gluconeogenic precursor uptake secondary to an inhibition of lipolysis and thus a fall in net hepatic glycerol uptake. The maximal gluconeogenic rate fell trivially over 3h. Gluconeogenic efficiency rose somewhat less than the rise which occurred in response to saline infusion.[4] The small decrease, however, resulted from the diversion of labelled glucose into hepatic glycogen rather than a true change in gluconeogenic efficiency. The minimal gluconeogenic rate, like the maximal rate, was thus virtually unaffected by the rise in insulin. Since gluconeogenesis was not appreciably reduced by the increment in insulin it can be concluded that the fall in glucose production was due to a potent reduction in glycogen breakdown.

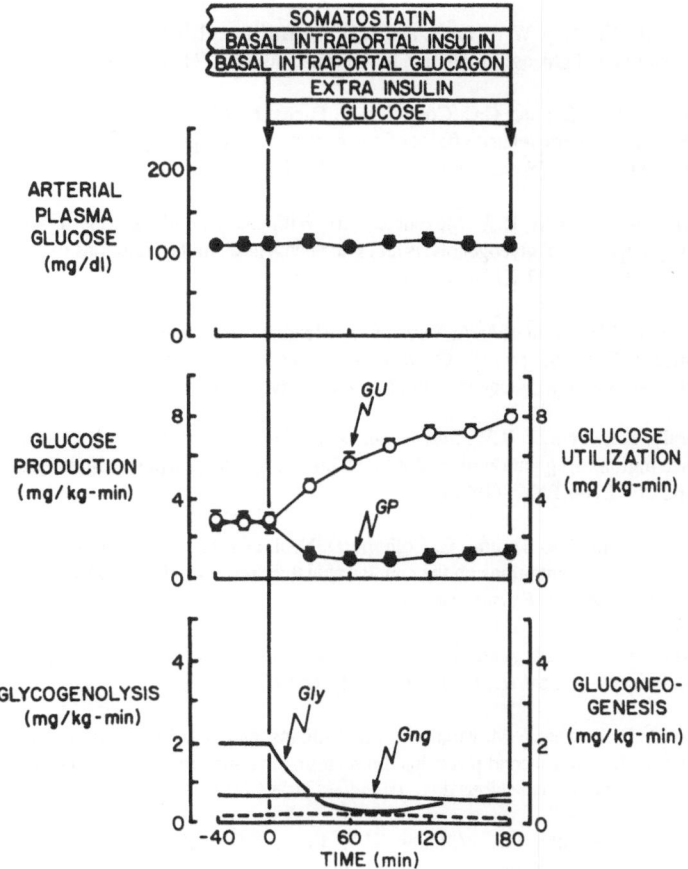

Figure 5. The effects of a selective rise in insulin brought about during a pancreatic clamp on the arterial plasma glucose level, glucose turnover assessed using ^{3}H-3 glucose infusion, maximal (—) and minimal (---) estimates of gluconeogenesis (see introduction) and minimal glycogenolysis in overnight-fasted conscious dogs. Redrawn from ref. 4.

These data should not be interpreted to mean that insulin does not regulate the gluconeogenic process in vivo. Clearly acute removal of the basal level of insulin results in an increase in both glycogenolysis and gluconeogenesis.[10] Quantitative assessment of this condition is less complete because of the absence of A-V difference data. It should also be noted that apart from glucagon[4,11,12] the interaction of insulin with the catecholamines and cortisol deserves further study.

REFERENCES

1. A.D. Cherrington. Gluconeogenesis:its regulation by insulin and glucagon, in: "Diabetes Mellitus," M.Brownlee, ed., Garland Press, New York. (1981).

2. J.L. Chiasson, J.E. Liljenquist, W.W. Lacy, A.S.Jennings, and A.D. Cherrington. Gluconeogenesis: Methodological approaches in vivo. *Fed. Proc.* 36:229- 235 (1977).

3. A.D. Cherrington, P.E. Williams, G.I. Shulman, and W.W. Lacy. Differential time course of glucagon's effect on glycogenolysis and gluconeogenesis in the conscious dog. *Diabetes* 30:180-187 (1981).

4. K.E. Steiner, P.E.Williams, W.W. Lacy, and A.D. Cherrington. Effects of insulin on glucagon-stimulated glucose production in the conscious dog. *Metabolism* 39:1325-1333 (1990).

5. R.W. Stevenson, K.E. Steiner, C.C. Connolly, H. Fuchs, K.G.M.M. Alberti, P.E. Williams, and A.D. Cherrington. Dose related effects of epinephrine on glycogenolysis and gluconeogenesis in the conscious dog. *Am. J. Physiol.* 260:E363-370 (1991).

6. A.D. Cherrington, H. Fuchs, R.W. Stevenson, P.E. Williams, K.G.M.M. Alberti, and K.E. Steiner. Effect of epinephrine on glycogenolysis and gluconeogenesis in the conscious overnight fasted dog. *Am. J. Physiol.* 247:E137-E144 (1984).

7. R.W. Stevenson, K.E. Steiner, M.A. Davis, G.K. Hendrick, P.E. Williams, W.W. Lacy, L. Brown, P. Donahue, D.C. Lacy, and A.D. Cherrington. Similar dose responsiveness of hepatic glycogenolysis and gluconeogenesis to glucagon in vivo. *Diabetes* 36:382-389 (1987).

8. C.C. Connolly, K.E. Steiner, R.W. Stevenson, D.W. Neal, P.E. Williams, K.G.M.M. Alberti, and A.D. Cherrington. Regulation of glucose metabolism by norepinephrine in conscious dogs. *Am. J. Physiol.* 261:E764-E772, (1992).

9. S.N. Davis, R. Dobbins, R. Tarumi, C. Colburn, D. Neal, and A.D. Cherrington. The effects of differing insulin concentrations on the counterregulatory response to equivalent hypoglycemia in conscious dogs. *Am. J. Physiol.* (in press).

10. A.D. Cherrington, W.W. Lacy, and J.L. Chiasson. Effect of glucagon on glucose production during insulin deficiency in the dog. *J. Clin. Invest.* 62:664-677 (1978).

11. K.E. Steiner, C.R. Bowles, S.M. Mouton, P.E. Williams, and A.D.Cherrington. The relative importance of first and second phase insulin secretion in countering the action of glucagon on glucose turnover in the conscious dog. *Diabetes* 31:964-972 (1982).

12. K.E. Steiner, S.M. Mouton, P.E. Williams, W.W. Lacy, and A.D.Cherrington. Relative importance of first- and second-phase insulin secretion in glucose homeostasis in the conscious dog. II. Effects of gluconeogenesis. *Diabetes* 35:776-784 (1986).

13. R.E. Goldstein, G.W. Reed, D.H. Wasserman, P.E. Williams, D.B. Lacy, R. Buckspan, N.N. Abumrad, and A.D. Cherrington. The effects of acute elevations in plasma cortisol on alanine metabolism in the conscious dog. *Metabolism* (in press).

ESTIMATING GLUCONEOGENIC RATES IN NIDDM

Bernard R. Landau

Department of Medicine
School of Medicine
Case Western Reserve University
Cleveland, OH 44106

INTRODUCTION

Lactate and alanine are the major precursors of glucose formed in liver by gluconeogenesis. However, incorporation of ^{14}C, when those precursors are labeled with ^{14}C, cannot be used to quantitate rates of gluconeogenesis. That is because an intermediate in the formation of the glucose is oxaloacetate, i.e. alanine and lactate → pyruvate → oxaloacetate → glucose, and oxaloacetate is also an intermediate in the tricarboxylic acid cycle(Figure l). Consequently, labeled carbon in oxaloacetate, formed from the labeled precursors, exchanges with unlabeled carbon in oxaloacetate formed in the cycle from acetyl-CoA.

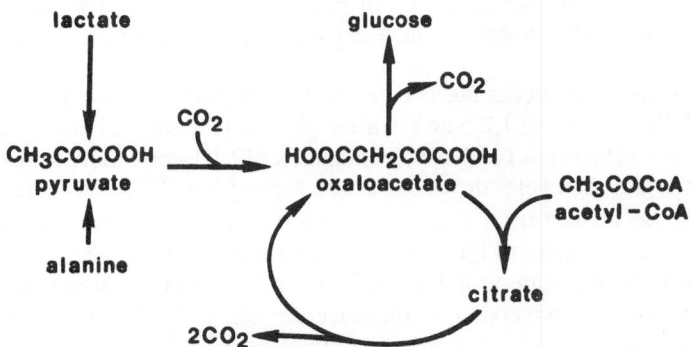

Figure 1. Oxaloacetate as an intermediate in gluconeogenesis and the tricarboxylic acid cycle.

Chaikoff and his associates[1,2] introduced an approach for estimating in in vitro liver studies rates of gluconeogenesis relative to the rates of tricarboxylic acid cycle flux. Hetenyi[3] applied and Katz[4] refined and extended that approach in order to correct for exchange in in vivo circumstances. Correction depends upon the ratio of the incorporation of ^{14}C from [2-^{14}C]acetate into carbons 1 to 3 of oxaloacetate. Carbon 1 of oxaloacetate is the precursor of carbon 1 of phosphoenolpyruvate(PEP) and hence carbons 3 and 4 of glucose formed by gluconeogenesis(Figure 2). Carbon 3 of oxaloacetate is the precursor of carbons 3 of PEP and hence carbons 1 and 6 of the glucose. Thus, when [2-^{14}C]acetate is metabolized via the tricarboxylic acid cycle, i.e. converted to acetyl-CoA and the acetyl

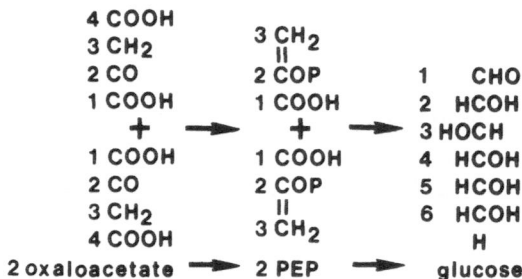

Figure 2. Nomenclature of carbons in the conversion of oxaloacetate to glucose.

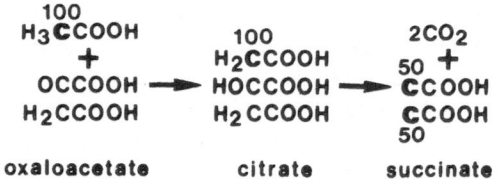

Figure 3. Succinate formation via the tricarboxylic acid cycle.

CoA condenses with oxaloacetate to form citrate, [2,3-^{14}C]succinate and hence [2,3-^{14}C]oxaloacetate are formed(Figure 3). In Figure 3 the specific activity of carbon 2 of acetyl-CoA is set to 100. Succinate then has equal specific activities, 50, in its carbons 2 and 3.

If that (2,3-^{14}C]oxaloacetate is converted to glucose, it forms glucose, via [2,3-^{14}C]PEP, with ^{14}C in carbons 1,2,5 and 6 of the glucose (Figure 4). However, to the extent the labeled oxaloacetate first condenses with another [2-[14]]acetyl-CoA to form citrate and that citrate traverses the cycle, the oxaloacetate formed has ^{14}C in all its carbons(Figures 5). Then, glucose formed from that oxaloacetate will have ^{14}C in its carbons 3 and 4 as well as its other carbons(from Figure 5 the C1/C3 ratio will be 75/25). The more the rate of conversion of oxaloacetate to glucose relative to the rate of oxaloacetate's conversion to citrate and citrate's conversion to oxaloacetate and CO_2, the more ^{14}C in C-1 will exceed that in C-3(Figure 6). It can be readily shown[2,4] that for the reactions, depicted in Figure 6, the ratio of incorporation of ^{14}C into carbon 1 to carbon 3 of glucose, i.e. C1/C3, will equal $2 + 2y$, where y is the rate of gluconeogenesis relative to tricarboxylic acid cycle flux, here designated V_6/V_3.

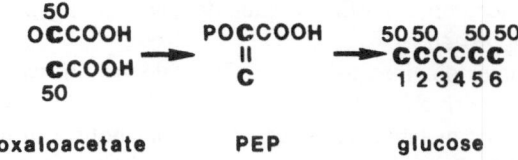

Figure 4. Conversion of [2,3-^{14}C]oxaloacetate to [1,2,5,6-^{14}C]glucose.

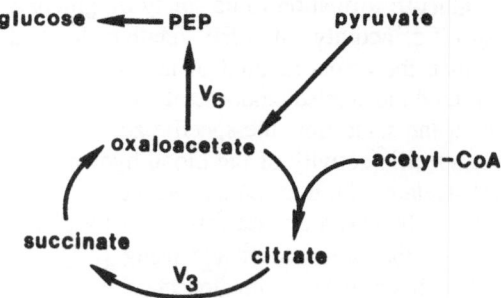

Figure 5. Conversion of citrate formed from [2-^{14}C]acetate (Figure 2) and oxaloacetate (Figure 4) to oxaloacetate via the tricarboxylic acid cycle.

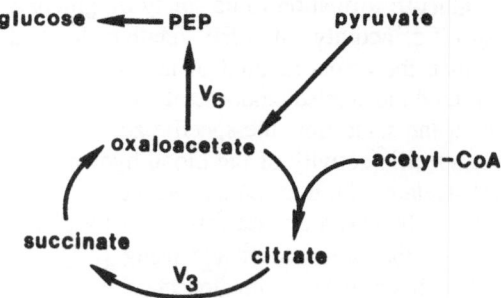

Figure 6. Reactions of gluconeogenesis relative to the tricarboxylic acid cycle.

USE OF [2-^{14}C]ACETATE

Consoli et al., giving [2-^{14}C]acetate to normal subjects[5] and noninsulin dependent diabetics[6,7], and obtaining C1/C3 ratios of about 3, calculated y values of about 0.5, i.e. rates of gluconeogenesis half the rates of cycle flux. From those y values the specific activity of PEP relative to acetyl-CoA was calculated, since the exchange of ^{14}C with ^{12}C in oxaloacetate is a simple function of y. The specific activity of hydroxybutyrate in blood was taken as the measure of the specific activity of hepatic acetyl-CoA, as suggested by Katz[4], since hydroxybutyrate is formed in liver from acetyl-CoA, i.e. 2 acetyl-CoA → acetoacetyl-CoA → hydroxybutyrate. In normal individuals fasted 60 h and infused with [2-^{14}C]acetate, the calculated specific activity of PEP was half the specific activity of blood glucose. In individuals fasted 60 h all of blood glucose would be expected to arise by gluconeogenesis and hence, since 2 PEP → glucose, that finding was taken as evidence for the validity of the approach.

However, the estimates obtained using [2-^{14}C]acetate are not valid. That is because acetate is metabolized extensively in tissues other than liver[8,9]. The incorporation of ^{14}C in glucose then reflects that metabolism as well as its metabolism in liver. Indeed, Katz and Chaikoff[10], over thirty-five years ago, from a comparison of acetate's metabolism in the intact animal and in liver slices, concluded that acetate's metabolism in vivo might show little resemblance to acetate's metabolism in liver. In the extreme, suppose [2-^{14}C]acetate were only metabolized in other than liver to yield [14]CO_2. That [14]CO_2 would be fixed by pyruvate in liver to form [4-^{14}C]oxaloacetate and following equilibration with fumarate(Figure 7), [1,4-^{14}C]oxaloacetate. The [1,4-^{14}C] oxaloacetate would yield [1-^{14}C]PEP and hence glucose labeled only in its carbons 3 and 4.

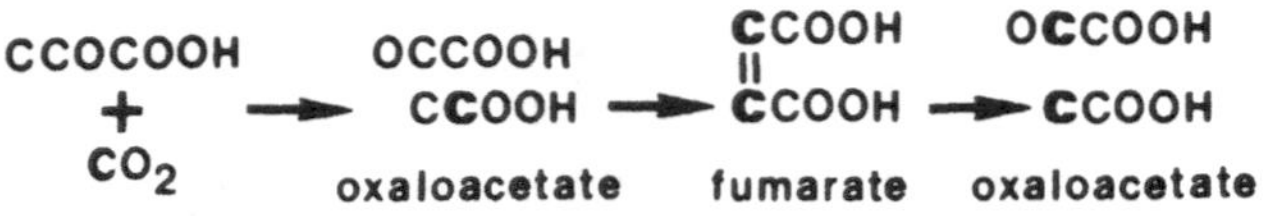

CCOCOOH   OCCOOH   CCOOH   OCCOOH
   +   →   ‖   →
  CO₂   CCOOH   CCOOH   →   CCOOH

CCOCOOH + CO₂	OCCOOH CCOOH	CCOOH‖CCOOH	OCCOOH CCOOH
	oxaloacetate	fumarate	oxaloacetate

Figure 7. Formation of oxaloacetate from pyruvate and [14]C.

Thus, to the extent [2-^{14}C]acetate is oxidized to [14]CO_2 and ^{14}C from it is incorporated into carbons 3 and 4 of glucose, the Cl/C3 ratio will be lowered and hence y underestimated. We[11] believe it was fortuitous that in the 60 h fasted individual, Consoli et al[5,7] calculated blood glucose formation to be solely by gluconeogenesis. Since y was underestimated, the specific activity of PEP relative to that of acetyl-CoA was overestimated, i.e. less than the actual amount of unlabeled oxaloacetate from pyruvate was calculated to have diluted the labeled oxaloacetate formed from the [2-^{14}C]acetyl-CoA in the cycle. However, at the same time the specific activity of hepatic acetyl-CoA was underestimated from the specific activity of the blood hydroxybutyrate. That is because exchange of labeled acetoacetate with unlabeled acetoacetate formed in the periphery will dilute the specific activity of hydroxybutyrate formed in liver[12].

There is other evidence that estimates made using [2-^{14}C]acetate are invalid. [3-^{14}C]Lactate is converted in liver to [3-^{14}C]oxaloacetate via pyruvate's fixation of CO_2. Relative to its other fates, since there is extensive equilibration of oxaloacetate with fumarate in liver, [2,3-^{14}C]oxaloacetate is formed. Whether [2,3-^{14}C]oxaloacetate is formed from [3-^{14}C]lactate or [2-^{14}C]acetate, if the metabolism of both is in liver, the Cl/C3 ratios of ^{14}C in glucose should be the same for both and hence the estimate of y. In fact, when [2-^{14}C]acetate is given to animals and humans, the Cl/C3 in glucose is less than when [3-^{14}C]lactate is given[11,13]. That is also the case on comparing data from Consoli et al.[4,5,14,15]. When [2-^{14}C]acetate was incubated with liver slices[16], the Cl/C3 ratio was that found on administering [3-^{14}C]lactate. Thus, when [2-^{14}C]acetate is metabolized in liver, it gives the distribution in glucose observed in vivo with [3-^{14}C] lactate.

Since isotopic equilibration of [3-^{14}C]oxaloacetate from [3-^{14}C]lactate with fumarate is not quite complete, carbon 3 of oxaloacetate will at steady state still have somewhat more ^{14}C than carbon 2 of oxaloacetate. Therefore, carbon 1 of glucose will have somewhat more ^{14}C than carbon 2 of glucose. This contrasts with [2-^{14}C]acetate where carbons 1 and 2 will have the same amount of ^{14}C, irrespective of the extent of equilibration, since the ^{14}C incorporated into oxaloacetate arises from fumarate. Therefore, for estimates of y using [3-^{14}C]lactate, also as suggested by Katz[4], and for comparison with distributions obtained with [2-^{14}C]acetate, the ratio [(Cl+C2)/2]C3 rather than Cl/C3 is used[4,13].

The ratio of specific activities of blood glucose to urinary urea, when individuals are fasted for 60 h and given [2-^{14}C]acetate, is 19[11] and when given [3-^{14}C]lactate the ratio is 133[13]. This must also mean that compared to lactate, acetate is extensively oxidized to CO_2 in tissues other than liver(Figure 8). That is relative to incorporation of ^{14}C into glucose in liver, ^{14}C from CO_2 formed from [2-^{14}C]acetate has equilibrated with hepatic bicarbonate, resulting in a higher specific in urea, whose carbon source is bicarbonate, than from [3-^{14}C]lactate. Incorporations of ^{14}C from [2-^{14}C]acetate then occurs into glucose carbons 1,2,5,6, but not to a large extent into carbons 3 and 4, by metabolism of the [2-^{14}C]acetate in liver. Also invalid are interpretations of liver metabolism requiring that the production of labeled CO_2 from labeled acetate occurs primarily in liver[8,17].

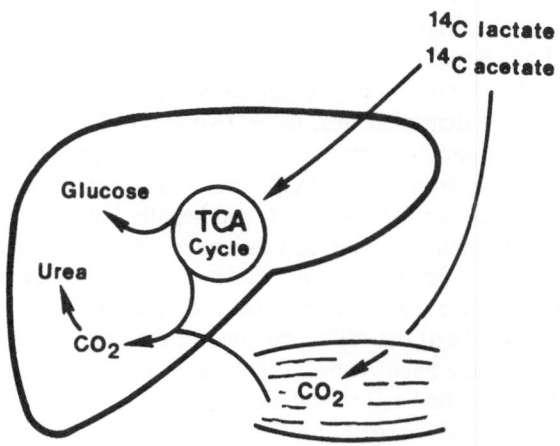

Figure 8. Sources of ^{14}C in hepatic glucose and urea.

CORRECTION FOR CO_2 FIXATION

Consoli et al., having obtained higher Cl/C3 ratios in glucose on administering [3-^{14}C]lactate than [2-^{14}]acetate to 14 h fasted normal individuals[14], calculated y values from [3-^{14}C]lactate of about 0.9 rather than 0.5. However, they did not correct for $^{14}CO_2$ that was formed from the [3-^{14}C]lactate and then incorporated into carbons 3 and 4 of glucose(Footnote 1). We, having administered [2-^{14}C]acetate and [3-^{14}C]lactate to individuals fasted for 60 h, did make the correction. In the fasted state, ^{14}C is incorporated in equal amounts into carbons 3 and 4. We gave NaH^{14}CO$_3$ to individuals fasted 60 h and determined the specific activity of blood glucose to urinary urea. Since ^{14}C from the bicarbonate is only incorporated into carbons 3 and 4 and via fixation by pyruvate in liver, we then had a measure of how much ^{14}C is incorporated into carbon 3 of glucose for a given specific activity of urea. From the specific activity of urea under the same conditions, when the [2-^{14}C] acetate was given, we calculated that 60% of the specific activity in carbon 3 of glucose was due to fixation of $^{14}CO_2$ and not [2-^{14}C]acetate's passage through the tricarboxylic acid cycle[11] and when [3-^{14}C]lactate was given 30%[13]. That is:

$$\frac{\text{Specific activity in carbons 3 and 4 of glucose due to fixation of } ^{14}CO_2 \text{ from } [2\text{-}^{14}C]\text{acetate or } [3\text{-}^{14}C]\text{lactate}}{\text{Specific activity of urea from } [2\text{-}^{14}C]\text{acetate or } [3\text{-}^{14}C]\text{lactate}} = \frac{\text{Specific activity in carbons 3 and 4 of glucose due to fixation of } ^{14}CO_2 \text{ from } NaH^{14}CO_3}{\text{Specific activity of urea from } NaH^{14}CO_3}$$

Making corrections for fixation gave y values about two or more, i.e. for example, when Cl/C3$=4=2+2$y, y$=1.0$, but when incorporation into C3 is reduced by one-third, Cl/C3$=6=2+2$y, y$=2$. Of note, Hetenyi et al.[19], using the distribution of ^{14}C in glucose from [2-^{14}C]acetate given to normal and diabetic dogs to estimate y, concluded that fixation of $^{14}CO_2$ was not significant. They concluded that because on giving $NaH^{14}CO_3$, relative to the specific activity of urea, the specific activity of glucose was only 5% as much as from [2-^{14}C]acetate. What they failed to note was that only about 14% of the ^{14}C in glucose from the [2-^{14}C]acetate was in carbons 3 and 4 of the glucose, i.e. the relative distribution of ^{14}C in glucose was Cl$=22$, C2$=22$, C3$=7$, C4$=7$, C5$=22$, C6$=22$[20], while all the ^{14}C from $NaH^{14}CO_3$ was in carbons 3 and 4. Thus, about one-third of the incorporation in carbons 3 and 4 in their study, i.e. 5%/14% was due to fixation.

Consoli et al.[14,15,21] gave $NaH^{14}CO_3$ at the same time as [3-^{14}C]lactate to normal as well as noninsulin dependent diabetics, fasted overnight. The giving of the $NaH^{14}CO_3$ would further favor incorporation into carbons 3 and 4 of glucose through fixation of $^{14}CO_2$ and not through ^{14}C from the [3-^{14}C]lactate traversing the cycle. Since there is extensive equilibration between expired $^{14}CO_2$ and hepatic bicarbonate, from the specific activity of the $^{14}CO_2$ their subjects expired, I estimated a y value of 1.9[22] rather than their 0.9 value. Consoli et al.[21] concluded that a y value of about 2 is not possible because the rate of ATP generation in liver would provide less ATP than required for gluconeogenesis. However, while they calculated the ATP generated in the oxidation of acetyl-CoA in the tricarboxylic acid cycle, they neglected ATP generated in the conversion of fatty acid to the acetyl-CoA and ketone bodies[23,24] and that y is the relative rate of oxaloacetate's conversion to PEP and to the extent the PEP is recycled, i.e. PEP → pyruvate → oxaloacetate → PEP, only one-third as much ATP is expended as when the oxaloacetate is converted to glucose. Furthermore, in the 14 h fasted individual they assumed ATP was required for a rate of formation of urea about twice the actual rate[25]. When a balance for ATP is made with those corrections, using the y values we estimated, sufficient ATP is generated to provide for gluconeogenesis as well as for other needs of the liver(unpublished observations).

We extended the model of gluconeogenesis and tricarboxylic acid cycle reactions to include the rates of equilibration of oxaloacetate with fumarate and pyruvate cycling(Figure 9). To estimate the rates of the reactions in that model more information is required than is available from the distribution of label in the carbons of glucose. Sufficient information is obtained from the distribution of label in α-ketoglutarate. To sample hepatic α-ketoglutarate noninvasively, we[11,13] gave subjects phenylacetate which is conjugated in liver to glutamine. Glutamine is formed from α-ketoglutarate via glutamate, with the carbon skeleton unchanged, so the distribution of label in glutamine is that in the α-ketoglutarate(Figure 10). Carbons 1,2 and 3 of α-ketoglutarate are formed in the cycle respectively from carbons 4,3 and 2 of oxaloacetate. Carbons 4 and 5 of α-ketoglutarate are formed from the acetyl-CoA that condenses with the oxaloacetate to form citrate and hence α-ketoglutarate.

When [3-^{14}C]lactate was given to normal subjects fasted 60 h the relative distribution of ^{14}C in the five carbons of the glutamine, corrected for the fixation of CO_2, averaged C1$=7$, C2$=48$, C3$=36$, C4$=7$, C5$=2$[13]. The relative labeling in carbons 4,3

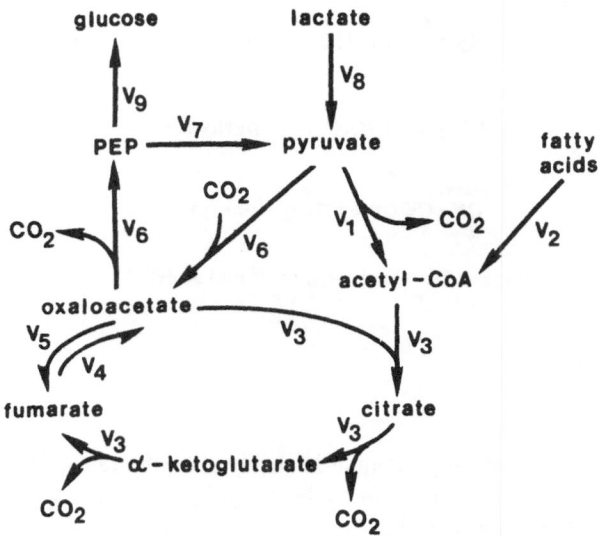

Figure 9. Model of gluconeogenesis and tricarboxylic acid cycle metabolism. V's indicate rates in molecules/unit time.

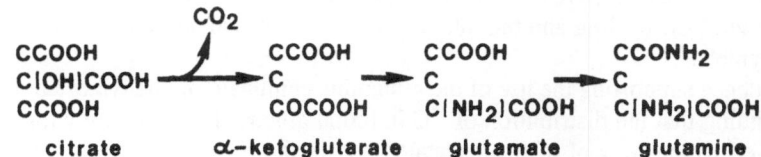

Figure 10. Retention of carbon skeleton in conversion of α-ketoglutarate from citrate to glutamine.

and 2 of oxaloacetate was therefore 7:48:36. From the distribution in glutamine, the relative distribution of ^{14}C that was in the four carbons of oxaloacetate that condensed with acetyl-CoA can be calculated to be 7:48:36:8.[13] The ratio of ^{14}C in carbon 1 to carbons 2 and 3 of glutamine then provides a measure of the rate of oxaloacetate's conversion to PEP relative to tricarboxylic acid cycle flux. The difference between the amount of ^{14}C in carbon 2 and 3 is the measure the extent of equilibration of oxaloacetate with fumarate. The incorporation of ^{14}C in carbon 4 is a measure of the rate of pyruvate's oxidation to acetyl-CoA, used in the cycle, and CO_2, i.e. [3-^{14}C]pyruvate → [2-^{14}C]acetyl-CoA → + CO_2, relative to its fixation of CO_2 to form oxaloacetate. The relative amount of ^{14}C in carbon 5 is the measure of pyruvate cycling, since ^{14}C from [3-^{14}C]lactate can only be incorporated into carbon 1 of acetyl-CoA via the conversion of PEP to pyruvate(Figure 11).

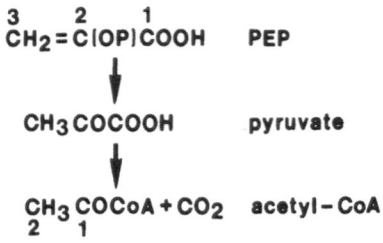

Figure 11. Conversion of PEP to acetyl-CoA.

From the distribution in glutamine at steady state the rates, as given in Figure 9, relative to the rate of tricarboxylic acid cycle flux, V_3, set to 1.0, can then be calculated: $V_1=0.1$, $V_2=0.9$, $V_4=12$, $V_5=11$, $V_6=3$, $V_7=2.0$, $V_8=1.4$ and $V_9=1.3$[13]. Thus, in the 60 h fasted normal individual the rate of carboxylation of pyruvate to oxaloacetate is about thirty times the rate of its decarboxylation to acetyl-CoA metabolized in the tricarboxylic acid cycle, the rate of oxaloacetate to PEP is about three times the rate of tricarboxylic acid cycle flux and the rate of pyruvate cycling appears to be more than 50% the rate of conversion of oxaloacetate to PEP. The estimate of pyruvate cycling depends on measuring only very small incorporations into carbon 5 and therefore can be subject to considerable error[13]. Furthermore, the rate of pyruvate cycling not only includes the rate of conversion of PEP to pyruvate as shown in Figure 9, but also the conversion of PEP to pyruvate via Cori cycling and the rate of conversion of malate to pyruvate catalyzed by malate enzyme[13].

Evidence supporting the use of the glutamine conjugate of phenylacetate is obtained from the finding that the distribution of ^{14}C in blood glucose is in keeping with its carbons being derived from the pool of oxaloacetate used in the formation of the glutamine, i.e. in the above example, the relative distribution of ^{14}C in carbons 1,2,3 of the glucose was about 48,36,8, the incorporation into carbon 3 of the glucose also having been corrected for fixation of $^{14}CO_2$[13]. Also, when livers isolated from 24 h starved rats were perfused with [3-^{13}C]lactate, the distribution of ^{13}C in glutamate from liver extracts (DiDonato et al. unpublished observation) was similar to that in glutamine from the conjugate excreted by the 60 h fasted individuals given [3-^{14}C]lactate.

USE OF CARBON LABELED SODIUM BICARBONATE

From the relative rates, the absolute rate of gluconeogenesis can be calculated if the absolute specific activity of one of the intermediates in the gluconeogenic process on giving a ^{14}C-labeled substrate is known. We used $^{14}CO_2$ as that intermediate, measuring the specific activity of expired CO_2 and blood glucose on giving $NaH^{14}CO_3$ to the 60 h fasted individuals[26]. There is good evidence that the specific activity of expired $^{14}CO_2$ is a good measure of the specific activity of $^{14}CO_2$ in hepatic mitochondria. [4-^{14}C]Oxaloacetate formed by $^{14}CO_2$ fixation of pyruvate will equilibrate with fumarate to form [1,4-^{14}C] oxaloacetate and hence [3,4-^{14}C]glucose. However, if that [1,4-^{14}C]oxaloacetate is instead metabolized in the tricarboxylic acid cycle, an equivalent amount of unlabeled oxaloacetate will be formed, i.e. [1,4-^{14}C]oxaloacetate → [1,6-^{14}C] citrate → $2^{14}CO_2$ + succinate → oxaloacetate. If the [1-^{14}C]PEP formed from the [1,4-^{14}C]oxaloacetate is converted to [1-^{14}C]pyruvate, the fixation of $^{14}CO_2$ will yield [1,4-^{14}C]oxaloacetate.

Hence, the specific activity of PEP formed on fixation of $^{14}CO_2$ by pyruvate in the liver will depend on the rates of equilibration, V_4 and V_5, the rate of oxaloacetate's conversion to PEP, V_6, and the rate of pyruvate's cycling, V_7, all relative to the rate of the tricarboxylic acid cycle, V_3. Using the relative rates obtained from the distribution of ^{14}C in glutamine on giving [3-^{14}C]lactate to the 60 h fasted subjects, and taking the specific activity of liver $^{14}CO_2$ as that they expired when given $NaH^{14}CO_3$, we estimated the specific activity of hepatic PEP. The specific activity of the PEP was about half that of the blood glucose on giving the $NaH^{14}CO_3$, in accord with essentially all the glucose in the blood having been formed by gluconeogenesis, i.e. 2 PEP → glucose.

Recently we tested the above approach by giving [3-^{14}C]lactate and $NaH^{13}CO_3$ in one sitting to a normal individual fasted 60 h. The distribution of ^{14}C in glutamine from the phenylacetate conjugate was corrected for $^{14}CO_2$ fixation from the ratio of the ^{13}C enrichments in carbon 1 of the glutamate and in urea and the ^{14}C specific activity of the urea. The relative rates were then calculated from the corrected distribution. From the enrichment of ^{13}C in the expired CO_2 the enrichment of ^{13}C in PEP was estimated. That enrichment was about half the enrichment in carbons 3 plus 4 of glucose, again in keeping with all the blood glucose arising by gluconeogenesis. Thus, that approach may be applicable to estimates of gluconeogenic rates in NIDDM. However, despite a single sitting the analyses remain time consuming and there is the uncertainty in the estimate of pyruvate cycling.

OTHER APPROACHES

Three other approaches have been introduced or suggested which could allow estimates of gluconeogenic rates in NIDDM.

Rothman et al.[27] measured by ^{13}C-NMR spectroscopy the concentration of glycogen in liver of normal subjects during fasting. By magnetic resonance imaging, they determined liver volume and therefore were able to calculate the rate of decrease of liver glycogen with time of fasting. They also measured hepatic glucose production using tritiated glucose. Rates of gluconeogenesis during fasting were then calculated as the difference between the rates of hepatic glucose production and glycogen depletion. While the equipment required is expensive, this is a relatively simple noninvasive approach. It depends upon the reliability of the measurement of hepatic glucose production, but those methods for measurement seem established. It also depends on the measurement of the liver glycogen concentration. There has been a question raised with regard to the visibility to NMR of glycogen in liver[28], despite evidence for its visibility[29]. Further, as with all such estimates, the liver is assumed to function as a homogenous system. Thus, if in one portion of the liver glycogen was being glycolyzed to lactate and in another part, glucose was being formed from the lactate, the extent glucose in the blood arose by gluconeogenesis would be underestimated.

The incorporation of deuterium from ^2H_2O into carbon 6 of glucose may also be used to measure gluconeogenesis[30]. The approach is based on the extensive incorporation of the label from ^3H_2O into the hydrogens of pyruvate during the process of gluconeogenesis. The approach has been used to measure in the rat the indirect pathway, i.e. the gluconeogenic pathway, of glycogen formation[31]. One need only determine at steady state the enrichment of deuterium in body water relative to that in the hydrogens bound to carbon 6 of the circulating glucose. However, toxicity of ^2H_2O limits the amount that can be given and a suitable method for measuring deuterium content at carbon 6 has yet to be established. Furthermore, incorporation at carbon 6 while extensive is not complete. It is also possible for deuterium from ^2H_2O to be incorporated at carbon 6 of glucose by the transaldolase exchange reaction, i.e. fructose-6-P + [3-^2H]glyceraldehyde-3-P ⇌ [6-^2H]fructose-6-P + glyceraldehyde-3-P[32]. Then glucose formed from glycogen via glucose-6-P, to the extent it equilibrates with fructose-6-P, would also bear label at carbon 6. Gluconeogenesis from glycerol would not be included in the estimate.

Isotopomer patterns in glucose on administering either ^{13}C-labeled glycerol or lactate also have been proposed for use in measuring rates of gluconeogenesis[33,34]. Interpretations of the patterns can also be compromised by the existence of heterogeneity of cellular function in the liver and the operation of the transaldolase exchange reaction. Furthermore, quantitation depends upon the extent of isotopic equilibration of dihydroxyacetone-3-P with glyceraldehyde-3-P formed from the labeled substrate.

SUMMARY

To measure the rate of gluconeogenesis in humans directly, one must administer and determine the specific activity or the enrichment in an intermediate in the gluconeogenic process and in the glucose formed, thus obtaining the fraction of the glucose formed by gluconeogenesis. By a separate determination of the rate of hepatic glucose production, the rate of gluconeogenesis can then be calculated. The closer the intermediate is to glucose-6-P, the more complete will be the measurement of the rate. Thus, if the intermediate is below the level of the triose phosphates, gluconeogenesis from glycerol will not be included in the estimate. Estimates of rates of gluconeogenesis from estimates of PEP enrichment or specific activity require a measure of the extent of exchange of label at the level of oxaloacetate. By using ^{14}C or ^{13}C labeled CO_2 as the intermediate and estimating the relative rates of the reactions of the tricarboxylic acid cycle relative to gluconeogenesis from the distribution of ^{14}C from [3-^{14}C]lactate in glutamine from the glutamine conjugate of phenylacetate, the enrichment or specific activity of PEP has been estimated. Correction must be made for the incorporation into the glutamine of $^{14}CO_2$ formed from the [3-^{14}C]lactate. Data support the validity of this approach toward estimating gluconeogenesis in NIDDM, but the approach is complex, time consuming and with uncertainties. Estimates that have been made using [2-^{14}C] acetate are invalid because of the extensive metabolism of [2-^{14}C]acetate in other than liver. Other approaches have promise, but technical problems may exist in their use and other problems, such as hepatic zonation and exchange reactions, may compromise their application.

FOOTNOTE

1. In subjects infused with [3-^{14}C]lactate, Consoli et al.[14,15] also determined net balance and specific activities of arterial and venous lactate across muscle. From the arterial and venous concentrations of lactate and the fractional extraction of labeled lactate, they concluded muscle is a much larger source of lactate for hepatic gluconeogenesis than measured by net balance. However, because the uptake of ^{14}C-lactate proceeds through an exchange reaction by which ^{14}C-lactate is replaced with ^{12}C-lactate, their conclusion is untenable[18].

ACKNOWLEDGEMENTS

This work has been supported by Grant DK 14507 from the National Institutes of Health. Studies described[11,13,26] have been done with colleagues whose collaborations are gratefully acknowledged.

REFERENCES

1. E.H. Strisower, G.D.Kohler, and I.L. Chaikoff. Incorporation of acetate carbon into glucose by liver slices from normal and alloxan-diabetic rats. J. Biol. Chem. 198:115-120(1952).

2. E.O. Weinman, E.H. Strisower, and I.L. Chaikoff. Conversion of fatty acids to carbohydrate. Application of isotopes to this problem and role of the Krebs cycle as a synthetic pathway. Physiol. Rev. 37:252-272(1957).

3. G. Hetenyi, Jr. Correction for the metabolic exchange of ^{14}C from ^{12}C atoms in the pathway of gluconeogenesis in vivo. Fed. Proc. 41:104-109(1982).

4. J. Katz. Determination of gluconeogenesis in vivo with ^{14}C-labeled substrates. Am. J. Physiol.(Regulatory Integrative Comp. Physiol. 17)248:R391-399(1985).

5. A. Consoli, F. Kennedy, J. Miles, and J. Gerich. Determination of Krebs cycle carbon exchange in vivo and its use to estimate the individual contributions of gluconeogenesis and glycogenolysis to overall glucose output in man. J. Clin. Invest. 80:1303-1310(1987).

6. A. Consoli, N. Nurjhan, F. Capani, and J. Gerich. Predominant role of gluconeogenesis in increased hepatic glucose production in NIDDM. Diabetes 38:550-557(1989).

7. A. Consoli, and N. Nurjhan. Contribution of gluconeogenesis to overall glucose output in diabetic and non-diabetic men. Ann. Med. 22:191-195(1990).

8. B.R. Landau. Acetate's metabolism, CO_2 production and the TCA cycle. Am. J. Clin. Nutr. 53:981(1991).

9. B. Bleiberg, T.R. Beers, M. Persson, and J.M. Miles. Systemic and regional acetate kinetics in dogs. Am. J. Physiol. 262(Endocrinol. Metab. 25):E197-202(1992).

10. J. Katz, and I.L. Chaikoff. Synthesis via the Krebs cycle in the utilization of acetate by rat liver slices. Biochim. Biophys. Acta.18:87-101(1955).

11. W.C. Schumann, I. Magnusson, V. Chandramouli, K. Kumaran, J. Wahren, and B.R. Landau. Metabolism of [2-^{14}C]acetate and its use in assessing hepatic Krebs cycle activity and gluconeogenesis. J. Biol. Chem. 266:6985-6990(1991).

12. C. Des Rosiers, J.A. Montgomery, M. Garneau, F. David, O.A. Mamer, P. Daloze, G. Toffolo, C. Cobelli, B.R. Landau, and H. Brunengraber. Pseudoketogenesis in hepatectomized dogs. Am. J. Physiol. 258(Endocrinol. Metab. 21)E519-528(1990).

13. I. Magnusson, W.C. Schumann, G.E. Bartsch, V. Chandramouli, K. Kumaran, J. Wahren, and B.R. Landau. Noninvasive tracing of Krebs cycle metabolism in liver. J. Biol. Chem. 266:6975-6984(1991).

14. A. Consoli, N. Nurjhan, J.J. Reilly, Jr., D.M. Bier, and J.E. Gerich. Contribution of liver and skeletal muscle to alanine and lactate metabolism in humans. Am. J. Physiol. 259(Endocrinol. Metab. 22):E677-684(1990).

15. A. Consoli, N. Nurjhan, J.J. Reilly, Jr., D.M. Bier, and J.E. Gerich. Mechanism of increased gluconeogenesis in noninsulin dependent diabetes mellitus: Role of alterations in systemic, hepatic and muscle lactate and alanine metabolism. J. Clin. Invest. 86:2038-2045(1990).

16. W. Kam, K. Kumaran, and B.R. Landau. Contribution of ω-oxidation to fatty acid oxidation by liver of rat and monkey. J. Lipid Res. 19:591-600(1978).

17. R.R. Wolfe. Reply to B.R. Landau. Am. J. Clin. Nutr. 53:982(1992).

18. B.R. Landau and J. Wahren. Nonproductive exchanges: The use of isotopes gone astray. Metabolism 41:457-459(1992).

19. G. Hetenyi Jr., B. Lussier, C. Ferrarotto, and J. Radziuk. Calculation of the rate

of gluconeogenesis from the incorporation of ^{14}C atoms from labeled bicarbonate or acetate. Can. J. Physiol. Pharmacol. 60:1603-1609(1982).

20. G. Hetenyi, Jr. Correction factor for estimation of plasma glucose synthesis from the transfer of ^{14}C-atoms from labeled substrate in vivo: A preliminary report. Can. J. Physiol. Pharmacol. 57:767-770(1979).

21. A. Consoli, N. Nurjhan, D. Bier, and J.E. Gerich. Reply. Am. J. Physiol. 261(Endocrinol. Metab. 24):E675-676(1991).

22. B.R. Landau. Correction of tricarboxylic acid cycle exchange in gluconeogenesis: Why the y's are wrong? Am. J. Physiol. 261(Endocrinol. Metab. 24):E673-674(1991).

23. J. Wahren, S. Efendic, R. Luft, L. Hagenfeldt, O. Bjorkman, and P. Felig. Influence of somatomedin on splanchnic glucose metabolism in postabsorptive and 60-hour fasted humans. J. Clin. Invest. 59:299-307(1977).

24. R. Kibler, W. Taylor, and J. Myers. The effect of glucagon on net splanchnic balances of glucose, amino acid, nitrogen, urea, ketones and oxygen in man. J. Clin. Invest. 43:904-915(1964).

25. D.E. Matthews and R.S. Downey. Measurements of urea kinetics in humans: a validation of stable isotope tracer methods. Am. J. Physiol. 246(Endocrinol. Metab. 9):E519-E529(1984).

26. E. Esenmo, V. Chandramouli, W.C. Schumann, K. Kumaran, J. Wahren, and B.R. Landau. Use of $^{14}CO_2$ in estimating rates of hepatic gluconeogenesis. Am. J. Physiol. 263(Endocrinol. Metab. 26):E36-41(1992).

27. D.L. Rothman, I. Magnusson, L.D. Katz, R.G. Shulman and G.I. Shulman. Quantitation of hepatic glycogenolysis and gluconeogenesis in fasting humans with ^{13}C-NMR. Science 54:573-576(1991).

28. B. Kunnecke and J. Seelig. Glycogen metabolism as detected by in vivo and in vitro ^{13}C-NMR spectroscopy using $[1,2\text{-}^{13}C_2]$glucose as substrate. Biochim. Biophys. Acta. 1095:103-113(1991).

29. L.O. Sillerud and R.G. Shulman. Structure and metabolism of mammalian liver glycogen monitored by carbon 13 nuclear magnetic resonance. Biochemistry 22:1087-1094(1983).

30. R. Rognstad. Estimation of gluconeogenesis and glycogenolysis in vivo using tritiated water. Biochem. J. 279:911(1991).

31. M. Kuwajima, S. Golden, J. Katz, R.H. Unger, D.W. Foster and J.D. McGarry. Active hepatic glycogen synthesis from gluconeogenic precursors despite high tissue levels of fructose 2,6-bisphosphate. J. Biol. Chem. 261:2632-2637(1986).

32. L. Ljungdahl, H.G. Wood, E. Racker, and D. Couri. Formation of unequally labeled fructose-6-phosphate by an exchange reaction catalyzed by transaldolase. J. Biol. Chem. 236:1622-1625(1961).

33. D. Faix, R. Neese, and M.K. Hellerstein. Measurement of gluconeogenesis in vivo using mass isotopomer distribution analysis. FASEB(Abstract)6:A1788(1992).

34. J.K. Kelleher, and A.L. Holleran. Model equations estimating gluconeogensis and glycogenolysis as components of hepatic glucose output using ^{13}C tracers. FASEB(Abstract)6:A3167(1992).

GLUCOSE METABOLISM DURING PHYSICAL EXERCISE IN PATIENTS WITH NONINSULIN-DEPENDENT (TYPE II) DIABETES

Iva K. Martin[1] and John Wahren[2]

[1]Department of Chemistry and Biology
Victoria University of Technology
Footscray Victoria 3011, Australia
[2]Department of Clinical Physiology
Karolinska Hospital
S-104 01 Stockholm, Sweden

INTRODUCTION

The beneficial effects of physical exercise on glucose metabolism in diabetic patients were documented already before the discovery of insulin. Thus, in a treatise on dietary regulation in the treatment of diabetes, Allen and colleagues (1919) described an association between exercise and reduced levels of sugar in blood and urine, as well as increased carbohydrate tolerance. Subsequent studies have shown that exercise also enhances the hypoglycaemic effect of injected insulin (Lawrence, 1926) and increases the rate of insulin-stimulated glucose disposal in diabetic patients (Wallberg-Henriksson et al., 1982; Koivisto and DeFronzo, 1984; Krotkiewski et al., 1985; Devlin et al., 1987). In addition, the insulin response to ingested glucose in diabetic patients with endogenous insulin secretion has been shown to be blunted following physical training (Rogers et al., 1988; Eriksson and Lindgärde, 1991).

Despite the well-documented improvements in insulin action that are associated with exercise training in both insulin-dependent (Type I, IDDM) and noninsulin-dependent (Type II, NIDDM) diabetic patients, conclusive evidence for improved long-term glycaemic control has been presented only for the latter group of patients. However, because exercise has additional beneficial effects on factors such as cardiorespiratory fitness, psychological well-being, blood lipid profile, adiposity and blood pressure, it has achieved wide-spread recognition as an important component of the treatment of all patients with diabetes mellitus (American Diabetes Association, 1990a,b).

Studies of the metabolic response to exercise of patients with NIDDM are of considerable interest because not only is this group of patients more numerous, but it is also likely to have better metabolic control and, therefore, be less susceptible to the risks that might make exercise inadvisable in poorly-controlled IDDM patients (Horton, 1988). Thus, it would seem that patients with NIDDM could potentially derive more benefits from exercise than patients with IDDM.

Although the beneficial actions of exercise to lower the blood glucose concentration and improve insulin sensitivity in NIDDM patients are apparent after a single bout of high-intensity exercise (Devlin et al., 1987), improvements of glucose tolerance following more moderate exercise may not become evident until after several days of exercise training (Rogers et al., 1988). Positive cumulative effects of exercise training on insulin sensitivity and glucose tolerance of NIDDM patients have been demonstrated by some (Krotkiewski et al., 1985; Holloszy et al., 1986; Rogers et al., 1988; Eriksson and Lindgärde, 1991), but not all studies (Ruderman et al., 1979; Segal et al., 1991).

Relatively little information is available regarding glucose metabolism in NIDDM patients during short-term exercise. Interpretation of the findings of the available short-term exercise studies in patients with NIDDM (Minuk et al., 1981; Jenkins et al., 1988; Kjaer et al., 1990) is complicated by the use of methods that may underestimate true rates of glucose production and utilization (Coggan, 1991), or by the choice of NIDDM patients who are also obese (Minuk et al., 1981; Kjaer et al., 1990); obesity is characterized by metabolic derangements such as hyperinsulinaemia and insulin resistance that are difficult to separate from those associated with noninsulin-dependent diabetes *per se*. The purpose of this communication is to briefly review the findings of studies that have examined glucose metabolism during short-term exercise in NIDDM patients and to present an outline of one such recently completed study of nonobese NIDDM patients.

Blood Glucose Concentration

During exercise, as well as at rest, blood glucose concentration is determined by the balance between hepatic glucose production and peripheral glucose utilization. In healthy subjects, the increase in glucose utilization by working muscle during exercise is matched by a corresponding increase in glucose production; unless the exercise is prolonged or intense, the blood glucose concentration is maintained at or near the resting value. In patients with IDDM, the balance between glucose utilization and production during such exercise appears to be strongly influenced by the degree of insulinisation. In severely insulin-deficient patients with marked resting hyperglycaemia (18 mmol/l and above) and mild to moderate ketonaemia (2-3 mmol/l), exercise is known to cause further increases in blood glucose and ketone body levels, while in insulin-treated patients with mild to moderate resting hyperglycaemia (up to 15 mmol/l) exercise results in a lowering of the blood glucose levels (Wahren et al., 1975; Berger et al., 1977); over-insulinised patients are vulnerable to developing hypoglycaemia during and after exercise (Wasserman and Abumrad, 1989).

Short-term exercise can also bring about changes in the blood glucose levels of patients with NIDDM. Thus, a modest decline in blood glucose levels during exercise at low to moderately high work intensities was demonstrated in NIDDM patients who were treated with diet alone or in conjunction with oral hypoglycaemic agents and who displayed mild to moderate resting hyperglycaemia (Minuk et al., 1981; Karamanos et al., 1982; Koivisto and DeFronzo, 1984;

Hübinger et al., 1987; Paternostro-Bayles et al., 1989). In another study, marked heterogeneity among the plasma glucose responses of nonobese NIDDM patients to exercise was noted; the plasma glucose levels in 3 of 7 subjects rose above and then slowly declined towards the resting level during 60 min of exercise at 50% of estimated maximal pulmonary oxygen uptake ($\dot{V}O_{2max}$), while plasma glucose in the remaining subjects fell, at various rates, below the resting level (Jenkins et al., 1988). Exercise at 70-75% of $\dot{V}O_{2max}$ in obese men with NIDDM also caused a minor, non-significant decline of about 0.7 mmol/l over 30 min (Schneider et al., 1987).

The combined observations from the above studies thus suggest that in patients with NIDDM, short-term exercise at low to moderately high intensities (up to 75% of $\dot{V}O_{2max}$) elicits a minor decline in blood glucose levels; the magnitude is largely unrelated to the exercise intensity and is generally insufficient to cause hypoglycaemia. However, it is conceivable that normoglycaemic NIDDM patients would be at increased risk of developing hypoglycaemia if they exercised at 1-2 hours after ingesting sulphonylureas; it has been demonstrated that moderate exercise fails to attenuate sulphonylurea-stimulated insulin secretion and accelerates the accompanying fall of blood glucose levels in normoglycaemic healthy subjects (Kemmer et al., 1987).

Only limited information appears to be available regarding changes in blood glucose levels during high-intensity exercise. In a study protocol that included 7 min exercise at 60%, 3 min exercise at 100% and 2 min at 110% $\dot{V}O_{2max}$, a minor increase in plasma glucose levels, from 8.0±1.0 to 8.6±1.5 (P<0.05), was observed during the exercise in obese NIDDM subjects but not in weight-matched controls (Kjaer et al., 1990). Previous observations in trained nonobese healthy subjects have shown a similar minor increase in plasma glucose levels during such high-intensity exercise with no changes in plasma glucose being observed in untrained controls (Kjaer et al., 1986).

Glucose Utilization

Although glucose utilization by resting skeletal muscle is small (Andres et al., 1956), within minutes of onset of exercise it rises progressively to meet the increased energy requirements of this tissue (Jorfeldt and Wahren, 1970). The degree to which glucose utilization is augmented by exercise is determined by many factors including the intensity, duration and mode of exercise, availability of energy substrates such as muscle glycogen and plasma free fatty acids, levels of glucoregulatory hormones, preceding carbohydrate intake, the degree of cardiorespiratory fitness, blood flow and oxygen availability. The exercise-induced increase in glucose utilization is usually matched by a corresponding rise in hepatic glucose production and the blood glucose concentration remains constant (Björkman and Wahren, 1988).

Glucose uptake by the leg tissues in healthy men, determined directly from measurements of the arterio-venous glucose difference and blood flow, rose 7- to 10-fold after 40 min cycling exercise at mild to moderately high work intensities, while with more strenuous cycling the rise was 20-fold (Wahren et al., 1971). Even greater increments of glucose uptake, up to 35 times basal, were shown for exercising forearm muscles (Jorfeldt and Wahren, 1970). The rate of whole-body glucose disappearance (R_d), a derived measure obtained from estimates of isotopic tracer dilution in plasma, was doubled in healthy controls after 45-60 min of cycling exercise at moderately high intensity (Zinman et al., 1977; Minuk et al., 1981; Jenkins et al., 1988). Even greater increments of R_d after a brief bout of high intensity running exercise have been reported (Kjaer et al., 1986), but the validity of estimates such as these might be questioned given the very short time available for equilibration of the infused tracer within the glucose distribution volume and other limitations of isotopic tracer methods (Wolfe, 1990; Coggan, 1991).

In studies of IDDM patients with mild to moderate hyperglycaemia that employed the arterio-venous glucose balance method and were performed 24 hours after the last injection of insulin, glucose uptake by the leg tissues at rest was similar to that in healthy controls; during exercise, it rose at least as much as that in controls (Wahren et al., 1975; Lyngsoe et al., 1978). Isotopic tracer studies of IDDM patients receiving insulin showed normal or elevated resting R_d, depending on the degree of insulinisation. During moderate exercise, the increment of R_d in well- or over-insulinised patients was comparable to that in controls (Zinman et al., 1977).

Observations on glucose utilization in exercising NIDDM patients are few and have been limited to studies employing isotopic tracer techniques. In a study of obese NIDDM subjects, R_d at rest was elevated compared with that in weight-matched controls (220±25 vs 140±12 mg/min, P<0.001), but rose by a comparable amount during 45 min of moderate exercise (Minuk et al., 1981). In another study, R_d at rest was similar in obese patients and controls and was augmented during a brief bout of intense exercise to a similar degree (Kjaer et al., 1990). In a group of nonobese NIDDM patients R_d at rest was also similar to that in healthy controls; however, a small but significant attenuation of the increment of R_d (which reached 1.6±0.2 vs 1.9±0.3 mmol/min, P<0.05) was seen in the NIDDM patients after 60 min moderate exercise (Jenkins et al., 1988). Taken together, these observations suggest a minor reduction of the exercise-induced increment of R_d in some groups of NIDDM patients. This conclusion should be approached with caution, however, given the difficulty in interpreting isotopic tracer data and the tendency of the technique to underestimate the rate of whole-body utilization of glucose during exercise in non-steady state situations (Coggan, 1991).

Basal insulin-mediated glucose uptake by skeletal muscle of obese patients with NIDDM is likely to be reduced as a result of impaired glucose storage and oxidation (Kelley et al., 1992), although downregulation of glucose transporters by chronic hyperglycaemia may be a contributing factor. Total basal glucose uptake in patients with NIDDM could be normalized by the increased mass action effect of glucose (Vranic, 1992).

Glucose Production

In the overnight fasted state, the blood glucose concentration is maintained only by the liver, which produces approximately 0.8-1.0 mmol glucose/min. Most of the glucose derives from the breakdown of stored glycogen and a smaller proportion, about 25-35%, results from *de novo* synthesis from substrates such as lactate, pyruvate, glycerol and amino acids (Wahren et al., 1971). During exercise, the increased requirements of working muscle for blood glucose are met by an acceleration of hepatic glucose production. Both the glycogenolytic and gluconeogenic rates are increased and the relative contributions of these two metabolic processes to the overall rate of hepatic glucose production are influenced by factors such as the intensity, duration and mode of exercise, preceding carbohydrate intake, as well as the prevailing hormonal *milieu* (Coggan, 1991).

As in the case of glucose utilization, the rate of hepatic glucose production may be estimated by the catheter technique or isotopic tracer methods. Because of the difficulty in catheterising the portal vein in humans, the former technique may be employed to determine net glucose production by the splanchnic tissues (calculated as the product of hepatic blood flow and arterial-hepatic venous glucose difference) rather than the true hepatic glucose output. Splanchnic glucose output represents hepatic production minus glucose uptake by the intestine and other extrahepatic tissues. At rest splanchnic glucose output is thought to account for at least 95% of glucose production by the liver (Wahren et al., 1971). Estimates of R_a, the whole-body glucose appearance rate, by the isotopic tracer method are subject to the same limitations of interpretation as are those of R_d (Wolfe, 1990; Coggan, 1991).

Studies in healthy subjects employing the catheter technique have demonstrated that, after onset of exercise, there is a gradual increase in glucose output from the splanchnic area that keeps up with the augmented utilization of blood glucose by working muscle (Wahren et al., 1971). The increase in splanchnic glucose production ranges from 2- to 5-fold the basal rate, depending on the intensity of exercise (Wahren et al., 1971; Ahlborg et al., 1974). Studies employing the isotopic tracer method show a 2-fold increment of R_a during moderate exercise (Zinman et al., 1977; Minuk et al., 1981; Jenkins et al., 1988) and more than a 4-fold increment after a brief bout of very intense exercise in healthy controls (Kjaer et al., 1986).

In patients with IDDM, the basal rate of splanchnic glucose production 24 hours after withdrawal of insulin has been shown to be similar to that of controls (Wahren et al., 1972, 1975), unless marked hyperglycaemia and ketoacidosis were present; in such a situation, splanchnic glucose output was increased (Bondy et al., 1949). During exercise, the rise in splanchnic glucose production has been found to be similar in IDDM patients and controls, but the contribution of gluconeogenesis to the total glucose output, estimated from splanchnic uptake of gluconeogenic precursors, was greater in the diabetics (Wahren et al., 1975). In an isotopic tracer study of IDDM patients who were receiving insulin, basal R_a reflected increments of R_d with increasing insulinisation (Zinman et al., 1977). During exercise, the increments of R_a in well-insulinised patients were smaller than those in controls but still sufficient to maintain normoglycaemia. However, in over-insulinised patients, R_a actually fell, despite a concurrent increase in R_d, with a resulting drop in the plasma glucose concentration (Zinman et al., 1977).

The few data that are available for exercising NIDDM patients come from isotopic tracer studies. Obese NIDDM patients show a slightly elevated R_a at rest but only a minor increase in R_a during 40 min of moderate exercise despite a 70% increase in R_d (Minuk et al., 1981). In obese NIDDM patients performing a brief bout of very intense exercise, R_a rose earlier and to a similar degree as that in controls, exceeding R_d (Kjaer et al., 1990). In nonobese patients, R_a at rest was similar to that of healthy controls but failed to increase significantly, despite an increase in R_d, during 60 min of moderate exercise with a consequent decline of the plasma glucose concentration (Jenkins et al., 1988). From these combined studies of NIDDM patients it would appear that, during moderate exercise, there is an attenuation of the increment of glucose production that is responsible for the observed decline of the blood glucose levels. In contrast, during high-intensity exercise, the rate of glucose production rises normally, exceeding the rate of utilization and the plasma glucose concentration becomes even more elevated.

Hormonal Response to Exercise

Exercise initiates a complex series of neural and hormonal responses that serve to couple the increase in glucose utilization by working muscle with increased hepatic glucose production. Sympathetic adrenergic stimulation with accompanying increments of plasma noradrenaline and adrenaline, decreased insulin and raised glucagon levels are considered to be of primary importance in the regulation of hepatic glucose production, with cortisol and growth hormone playing secondary roles in this process.

Activation of the sympathetic nervous system during exercise is reflected in raised plasma levels of noradrenaline, resulting mainly from overflow to the blood stream from working skeletal muscle, and adrenaline, from increased adrenomedullary secretion (Galbo, 1986). These catecholamines have potent effects on metabolism, which they exert directly and through their influence on the actions and secretion of other hormones. Their direct metabolic effects include stimulation of lipolysis in adipose tissue and glycogenolysis in liver and muscle, although recent evidence suggests that, during exercise, glycogenolysis in rat skeletal muscle occurs independently of adrenaline (Coderre et al., 1991). Indirect actions of catecholamines include synergistic interactions with other insulin-counterregulatory hormones and antagonism of insulin action. Increments of catecholamine levels in exercising subjects may contribute to diminution of insulin secretion, although in humans this is chiefly attributed to alpha-adrenergic inhibition exerted by sympathetic nerves to the pancreas (Galbo, 1986). During very intense exercise, catecholamines may become regulators of hepatic glucose production (Marliss et al., 1991). In addition, catecholamines are critical to the prevention of hypoglycaemia during moderate exercise when changes in insulin and glucagon do not occur (Marker et al., 1991). Finally, stimulation of sympathetic nerves to the liver has been put forward as a possible primary mechanism responsible for increased hepatic glucose output during moderate exercise in humans (Hoelzer et al., 1986a, 1986b). This suggestion was partly based on studies in which adrenalectomized patients received somatostatin infusion with insulin and glucagon replacement during exercise and were able to maintain a normal plasma glucose concentration. However, it is possible that in these studies insulin was under-replaced and therefore a decline of the plasma glucose levels may have been prevented (Cryer, 1989). Relative hypoglycaemia, without accompanying changes in insulin and glucagon, may stimulate hepatic glucose production at rest (Wolfe et al., 1986b) as well as during exercise (Jenkins et al., 1986).

The decline of plasma insulin during exercise is accompanied by increases in glycogenolysis, gluconeogenesis and lipolysis, partly as a result of reduced antagonism of the stimulatory actions of catecholamines and glucagon on these processes. The decline of plasma insulin may not be necessary to maintain a normal blood glucose concentration, as long as glucagon is free to increase (Zinman et al., 1981). Glucagon itself may play a relatively minor role in the prevention of hypoglycaemia during moderate short-term exercise in humans. Its secretion tends not to increase substantially unless the exercise is of high intensity or long duration (Galbo, 1986); and deficiency of glucagon does not significantly impair the early hepatic glycogenolytic response to exercise (Björkman et al., 1981). In contrast to the uncertain role of changes in each of these hormones separately, an increase in the ratio of plasma glucagon to insulin has been suggested to be of crucial importance to adequate stimulation of hepatic glucose output during moderate exercise (Wolfe et al., 1986; Hirsch et al., 1991).

The role of hormones other than those discussed above in the maintenance of blood glucose homeostasis during moderate exercise has been questioned, given that hypoglycaemia develops when catecholamine actions are blocked and plasma insulin and glucagon do not change (Marker et al., 1991). Nevertheless, plasma levels of growth hormone and cortisol, for example, do increase during moderate exercise and both these hormones have well-documented insulin counterregulatory actions.

Regulation of glucose utilization during exercise is no less complex, due to synergistic and antagonistic interactions between the glucoregulatory hormones and other factors such as availability of fatty acids, glycogenolysis and level of phosphocreatine in muscle, blood flow and hypoxia. Insulin is known to stimulate glucose uptake by translocating glucose transporters from an intracellular microsomal pool to the plasma membrane of skeletal muscle cells; this effect is similar to that of contractile activity *per se* (Goodyear et al., 1991). However, leg glucose uptake in humans is strongly correlated with insulin delivery and increases synergistically when insulin infusion and exercise are superimposed (DeFronzo et al., 1981). Insulin and exercise also act synergistically in stimulating whole-body glucose disappearance and carbohydrate oxidation (Wasserman et al., 1991). In addition, insulin could enhance glucose utilization during exercise by counteracting the effects of other glucoregulatory hormones and factors that would interfere with glucose uptake by muscle, such as catecholamine-mediated stimulation of lipolysis (Vranic et al., 1990). Conversely, by stimulating lipolysis, the catecholamines could limit excessive utilization of blood glucose and thus help maintain blood glucose homeostasis (Wasserman and Abumrad, 1989).

Hormonal and metabolic responses to exercise in insulin-treated diabetic patients appear to be strongly influenced by the degree of insulinisation. In these patients the plasma insulin levels fail to decrease during exercise and may even increase due to enhanced insulin absorption from a subcutaneous injection site. As a result, the coupling between increased glucose utilization by working muscle and increased hepatic glucose production may be disrupted (Zinman et al., 1977). In moderately well-controlled patients with IDDM the marked fall in blood glucose is accompanied by increments of fatty acids, ketone bodies, glucagon, growth hormone and cortisol that are comparable to those of healthy controls (Berger et al., 1977). During mild to moderate exercise of short duration, the fall in blood glucose may be considered to be a beneficial effect but, during more prolonged exercise, clinical hypoglycaemia may result (Horton, 1988). Exercise under hypoglycaemic conditions has been shown to be associated with decreased increments of adrenaline and noradrenaline in some patients with IDDM (Schneider et al., 1991).

In contrast, in under-insulinised ketotic IDDM patients exercise leads to further elevations of blood glucose, fatty acids and ketone bodies (Berger et al., 1977). These adverse metabolic effects of exercise in poorly-controlled IDDM patients can be attributed to the prevailing insulinopenia, which could be expected to directly impair exercise-induced increases in muscle glucose uptake and oxidation and enhance hepatic glucose production (Wasserman et al., 1992). The concurrent excessive increases in glucagon, growth hormone, cortisol and catecholamines (Berger et al., 1977; Galbo, 1986) may also contribute to the metabolic alterations during exercise. Exercise-induced changes in catecholamines, growth hormone and cortisol in IDDM patients could be of secondary importance, however, given the finding that diabetics with autonomic neuropathy show an impaired response of these hormones to exercise when compared to those of diabetics without evidence of autonomic neuropathy, although the metabolic responses of both groups of patients to exercise are similar (Hilstead et al., 1980).

Very little information appears to be available regarding the hormonal response to exercise in patients with NIDDM. In contrast to patients with IDDM, this group of patients is characterised by impaired secretion and sensitivity to the actions of insulin, rather than a total insulin deficiency. Short-term moderate exercise leads to a gradual reduction in blood glucose that is unlikely to cause hypoglycaemia. This response may be related to a failure of plasma insulin to decline, as shown for some (Minuk et al., 1981; Schneider et al., 1987; Jenkins et al., 1988; Paternostro-Bayles et al., 1989; Kjaer et al., 1990) but not all (Karamanos et al., 1982; Koivisto and DeFronzo, 1984, Hübinger et al., 1987) groups of NIDDM patients.

Very few studies appear to have addressed the question of exercise-induced changes in the other glucoregulatory hormones. Thus, the few data that are available suggest that, as in healthy control, short-term exercise of moderate intensity is not associated with significant increments of plasma glucagon levels in NIDDM patients (Minuk et al., 1981). Increments of plasma levels of adrenaline during exercise may be higher in NIDDM patients than in controls, while noradrenaline levels may be similar to those of controls, or elevated, especially if the exercise is of high intensity (Schneider et al., 1987; Kjaer et al., 1990). There appears to be no difference between NIDDM patients and controls in the responses of growth hormone and cortisol to exercise (Jenkins et al., 1988; Kjaer et al., 1990). Clearly, further studies are needed to elucidate the role of glucoregulatory hormones in the control of metabolism during exercise in patients with NIDDM.

Recent Findings

In a recent study we examined glucose metabolism in a group of eight non-obese NIDDM patients at rest and during 40 min of bicycle exercise at 60% of maximal pulmonary oxygen uptake. A catheter technique was employed to determine leg and splanchnic glucose exchange. Muscle biopsies from the lateral portion of the quadriceps femoris muscle were obtained at rest and immediately after the exercise period. A needle biopsy technique was used. The findings for the diabetic patients were compared to those in a group of healthy controls matched for age, weight and sex. In the diabetic patients the arterial blood glucose level was 10.0±0.9 mmol/l in the basal state and declined gradually during exercise to 9.0±0.9 mmol/l (P<0.05) after 40 min. In contrast, the control subjects showed an increase in blood glucose concentration from 5.2±0.2 mmol/l at rest to 6.3±0.6 mmol/l (P<0.05) at the end of the exercise period, a response that agrees with previous observations (Wahren et al., 1971). Lactate concentrations were similar in the two groups at rest but rose more during exercise in the diabetic patients.

Leg uptake of glucose at rest was similar in the two groups and rose markedly during exercise (Fig. 1). At the end of the exercise period the leg glucose uptake was greater in the diabetic group (P<0.05), possibly as a consequence of the hyperglycaemia in the patients. Leg release of lactate in the basal state was no different in the two groups, but throughout the exercise period leg output of lactate was approximately twice as great in the diabetics as in the controls. Basal splanchnic glucose output was slightly lower in the diabetic patients (Fig. 1). During exercise there was a rapid 3-4 fold increase in

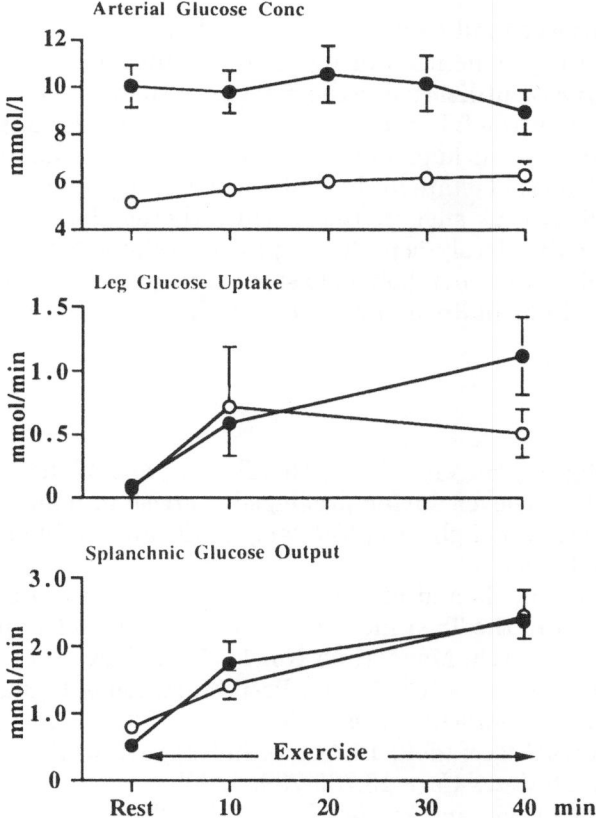

Fig. 1. Arterial glucose concentration, leg glucose uptake and splanchnic glucose output in the resting state and during exercise at 60% of max $\dot{V}O_2$. Solid symbols denote findings in type II diabetic patients, open symbols indicate results in healthy controls.

splanchnic glucose output of similar magnitude in both groups (Fig. 1).

The muscle content of ATP and phosphocreatine was similar in the basal state in the two groups; ATP remained unchanged during exercise and phosphocreatine decreased to the same extent in patients and controls. In contrast, muscle lactate content rose to higher levels in the patients (19.5±-1.7mmol/kg dw) as compared to the healthy controls (12.7±5.9 mmol/kg dw, $P<0.05$).

It is difficult to ascribe the observed differences in blood glucose metabolism in the present study to alterations in the levels of the major glucoregulatory hormones. The plasma levels of free insulin and C-peptide at rest were higher in the diabetics than in the controls, but during exercise the levels of free insulin and C-peptide did not change significantly in either group. The plasma levels of glucagon tended to be slightly lower in the diabetic patients than in the controls but in neither group was there a significant increase during exercise. Finally, the plasma levels of adrenaline and noradrenaline at rest were similar in both groups and rose comparably during exercise.

In summary, the above findings thus confirm previous observations that blood glucose levels tend to fall during exercise in NIDDM patients while the opposite is true for healthy controls exercising at the same work intensity.

However, the observed fall cannot be explained on the basis of a deficient rise in hepatic glucose production during exercise. Instead, a more pronounced increase in leg glucose utilization in the diabetic patients appears as the primary factor responsible for the fall in blood glucose. The latter finding is underscored by the observation of an augmented muscle content of lactate as well as an increased leg release of lactate in the diabetic patients compared to the controls. The current results thus suggest that besides the accelerated skeletal muscle metabolism along the glycolytic pathway, possibly related to the hyperglycaemia, the adaptation of glucose metabolism to short-term exercise in nonobese NIDDM patients appears to be quite similar to that in healthy subjects.

REFERENCES

Ahlborg, G., Felig, P., Hagenfeldt, L., Hendler, R., and Wahren, J., 1974, Substrate turnover during prolonged exercise in man: splanchnic and leg metabolism of glucose, free fatty acids, and amino acids. *J. Clin. Invest.* 53:1080.

Allen, F.M., Stillman, E., and Fitz, R., 1919, Exercise, *in*: Total Dietary Regulation in the Treatment of Diabetes, Rockefeller Institute for Medical Research, Monograph No. 11, New York.

American Diabetes Association, 1990a, Position statement: Diabetes mellitus and exercise. *Diabetes Care* 13:84.

American Diabetes Association, 1990b, Technical review: Exercise and NIDDM. *Diabetes Care* 13:785-789, 1990b.

Andres, R., G Cader, G., and Zierler, K.L., 1956, The quantitatively minor role of carbohydrate in oxidative metabolism by skeletal muscle in intact man in the basal state. Measurements of oxygen and glucose uptake and carbon dioxide and lactate production in the forearm. *J. Clin. Invest.* 35:671.

Berger, M., Brechtold, P., Cüppers, H.J., Drost, H., Kley, H.K., Müller, W.A., Wiegelmann, W., Zimmermann-Telschow, H., Gries, F.A., Krüskemper, H.L., and Zimmermann, H., 1977, Metabolic and hormonal effects of muscular exercise in juvenile type diabetics. *Diabetologia* 13:355.

Björkman, O., Felig, P., Hagenfeldt, L., and Wahren, J., 1981, Influence of hypoglucagonemia on splanchnic glucose output during leg exercise in man. *Clin. Physiol.* 1:43.

Björkman, O., and Wahren, J., 1988, Glucose homeostasis during and after exercise, *in*: Exercise, Nutrition and Energy Metabolism, E.S. Horton, and R.L. Terjung, eds., Macmillan, New York.

Bondy, P.K., Bloom, W.L., Whitner, V.S., and Farrar, B.W., 1949, Studies of the role of the liver in human carbohydrate metabolism by the venous catheter technique. II. Patients with diabetic ketosis, before and after the administration of insulin. *J. Clin. Invest.* 28:1126.

Coderre, L., Srivastava, A.K., and Chiasson, J.-L., 1991, Role of glucocorticoids in the regulation of glycogen metabolism in skeletal muscle. *Am. J. Physiol.* 260:E927.

Coggan, A.R., 1991, Plasma glucose metabolism during exercise in humans. *Sports Med.* 11:102.

Cryer, P.E., 1989, Retraction. *Am. J. Physiol.* 256:E338.DeFronzo, R.A., Ferrannini, E., Sato, Y., and Wahren, J., 1981, Synergistic interaction between exercise and insulin on peripheral glucose uptake. *J. Clin. Invest.* 68:1468.

Devlin, J.T., Hirshman, M., Horton, E.D., and Horton, E.S., 1987, Enhanced peripheral and splanchnic insulin sensitivity in NIDDM men after a single bout of exercise. *Diabetes* 36:434.

Eriksson, K.-F., and Lindgärde, F., 1991, Prevention of type 2 (non-insulin-dependent) diabetes mellitus by diet and physical exercise. *Diabetologia* 34:891.

Galbo, H., 1986, The hormonal response to exercise. *Diabetes / Metabolism Rev.* 1:385.

Goodyear, L.J., Hirshman, M.F., and Horton, E.S., 1991, Exercise-induced translocation of skeletal muscle glucose transporters. *Am. J. Physiol.* 261:E795.

Hilstead, J., Galbo, H., and Christensen, N.J., 1980, Impaired responses of catecholamines, growth hormone, and cortisol to graded exercise in diabetic neuropathy. *Diabetes* 29:257.

Hirsch, I.B., Marker, J.C., Smith, L.J., Spina, R.J., Parvin, C.A., Holloszy, J.O., and Cryer, P.E., 1991, Insulin and glucagon in prevention of hypoglycemia during exercise in humans. *Am. J. Physiol.* 260:E695.

Hoelzer, D.R., Dalsky, G.P., Clutter, W.E., Shah, S.D., Holloszy, J.O., and Cryer, P.E., 1986a, Glucoregulation during exercise: hypoglycemia is prevented by redundant glucoregulatory systems, sympathochromaffin activation and changes in islet hormone secretion. *J. Clin. Invest.* 77:212.

Hoelzer, D.R., Dalsky, G.P., Schwartz, N.S., Clutter, W.E., Shah, S.D., Holloszy, J.O., and Cryer, P.E., 1986b, Epinephrine is not critical to prevention of hypoglycemia during exercise in humans. Am J Physiol 251:E104.

Holloszy, J.O., Schultz, J., Kusnierkiewicz, J., Hagberg, J.M., and Ehsani, A.A., 1986, Effects of exercise on glucose tolerance and insulin resistance: brief review and some preliminary results. *Acta Med. Scand. Suppl.* 711:55.

Horton, E.S., 1988, Exercise and diabetes mellitus. *Medical Clinics N. America* 72:1301.

Hübinger, A., Franzen, A., and Gries, F.R., 1987, Hormonal and metabolic response to exercise in hyperinsulinemic and non-hyperinsulinemic type 2 diabetics. *Diabetes Res.* 4:57.

Jenkins, A.B., Furler, S.M., Bruce, D.G., and Chisholm, D.J., 1986, Regulation of hepatic glucose output during exercise by circulating glucose and insulin in humans. *Am. J. Physiol.* 250:R411.

Jenkins, A.B., Furler, S.M., Bruce, D.G., and Chisholm, D.J., 1988, Regulation of hepatic glucose ouput during moderate exercise in non-insulin-dependent diabetes. *Metabolism* 37:966.

Jorfeldt, L., and Wahren, J., 1970, Human forearm muscle metabolism during exercise. V. Quantitative aspects of glucose uptake and lactate production during prolonged exercise. *Scand. J. Clin. Lab. Invest.* 26:73.

Karamanos, B.G., Christacopoulos, P.D., Andriotis, A.T., Tountas, C.P., and Komninos, Z.D., 1982, Metabolic and cardiac effect of mild exercise in non-insulin dependent diabetics. *in*: Diabetes and Exercise, M. Berger, P. Christacopoulos, and J. Wahren, eds., Huber Publishers, Bern.

Kelley, D.E., Mokan, M., and Mandarino, L.J., 1992, Intracellular defects in glucose metabolism in obese patients with NIDDM. *Diabetes* 41:698.

Kemmer, F.W., Tacken, M., and Berger, M., 1987, Mechanism of exercise-induced hypoglycemia during sulfonylurea treatment. *Diabetes* 36:1178.

Kjaer, M., Farrell, P.A., Christensen, N.J., and Galbo, H., 1986, Increased epinephrine response and inaccurate glucoregulation in exercising athletes. *J. Appl. Physiol.* 61:1693.

Kjaer, M., Hollenbeck, C.B., Frey-Hewitt, B., Galbo, H., Haskell, W., and Reaven, G.M., 1990, Glucoregulation and hormonal responses to maximal exercise in non-insulin-dependent diabetes. *J. Appl. Physiol.* 68:2067.

Koivisto, V.A., and DeFronzo, R.A., 1984, Exercise in the treatment of type II diabetes. *Acta Endocrinol. Suppl.* 262:107.

Krotkiewski, M., Lönnroth, P., Mandroukas, K., Wrobewski, Z., Rebuffé-Scrive, M., Holm, G., Smith, U., and Björntorp, P., 1985, The effects of physical training on insulin secretion and effectiveness and on glucose metabolism in obesity and Type 2 (non-insulin-dependent) diabetes mellitus. *Diabetologia* 28:881.

Lawrence, R.H., 1926, The effects of exercise on insulin action in diabetes. *Br. Med. J.* 1:648.

Lyngsjoe, J., Clausen, J.P., Trap-Jensen, J., Sestoft, L., Schaffalitzky de Muckadell, O., Holst, J.J., Nielsen, S.L., and Rehfeld, J.F., 1978, Exchange of metabolites in the leg of exercising juvenile diabetes subjects. *Clin. Sci. Mol. Med.* 55:73.

Marliss, E.B., Simantirakis, E., Miles, P.D.G., Purdon, C., Gougeon, R., Field, C.J., Halter, J.B., and Vranic, M., 1991, Glucoregulatory and hormonal responses to repeated bouts of intense exercise in normal male subjects. *J. Appl. Physiol.* 71:924.

Marker, J.C., Hirsch, I.B., Smith, L.J., Parvin, C.A., Holloszy, J.O., and Cryer, P.E., 1991, Catecholamines in prevention of hypoglycemia during exercise in humans. *Am. J. Physiol.* 260:E705.

Minuk, H.L., Vranic, M., Marliss, E.B., Hanna, A.K., Albisser, A.M., and Zinman, B., 1981, Glucoregulatory and metabolic response to exercise in obese noninsulin-dependent diabetes. *Am. J. Physiol.* 240:E458.

Paternostro-Bayles, M., Wing, R.R., and Robertson, R.J., 1989, Effect of life-style activity of varying duration on glycemic control in type II diabetic women. *Diabetes Care* 12:34.

Rogers, M.A., Yamamoto, C., King, D.S., Hagberg, J.M., Ehsani, A.A., and Holloszy, J.O., 1988, Improvement in glucose tolerance after 1 wk of exercise in patients with mild NIDDM. *Diabetes Care* 11:613.

Ruderman, N.B., Ganda, O.P., and Johansen, K., 1979, The effect of physical training on glucose tolerance and plasma lipids in maturity-onset diabetes. *Diabetes* 28(Suppl 1):89.

Schneider, S.H., Khachadurian. A.K., Amorosa, L.F., Gavras, H., Fineberg, S.E., and Ruderman, N.B., 1987, Abnormal glucoregulation during exercise in type II (non-insulin-dependent) diabetes. *Metabolism* 36:1161.

Schneider, S.H., Vitug, A., Ananthakrishnan, R., and Khachadurian, A.K., 1991, Impaired adrenergic response to prolonged exercise in Type I diabetes. *Metabolism* 40:1219.

Segal, K.R., Edano A., Abalos, A., Albu, J., Blando, L., Tomas, M.B., and Pi-Sunyer, F.X., 1991, Effect of exercise training on insulin sensitivity and glucose metabolism in lean, obese, and diabetic men. *J. Appl. Physiol.* 71:2402.

Vranic, M., 1992, Glucose turnover: a key to understanding the pathogenesis of diabetes (indirect effects of insulin). *Diabetes* 41:1188.

Vranic, M., Wasserman, D., and Bukowiecki, L., 1990, Metabolic implications of exercise and physical fitness in physiology and diabetes, *in:* Ellenberg and Rifkin's Diabetes Mellitus: Theory and Practice (4th ed), H. Rifkin, and D. Porte Jr, eds., Elsevier, New York.

Wahren, J., Felig, P., Ahlborg, G., and Jorfeldt, L., 1971, Glucose metabolism during leg exercise in man. *J. Clin. Invest.* 50:2715.

Wahren, J., Felig, P., Cerasi, E., and Luft, R., 1972, Splanchnic and peripheral glucose and amino acid metabolism in diabetes mellitus. *J. Clin. Invest.* 51:1870.

Wahren, J., Hagenfeldt, L., and Felig, P., 1975, Splanchnic and leg exchange of glucose, amino acids, and free fatty acids during exercise in diabetes mellitus. *J. Clin. Invest.* 55:1303.

Wallberg-Henriksson, H., Gunnarson, R., Henriksson, J., DeFronzo, R., Felig, P., Östman, J., and Wahren, J., 1982, Increased peripheral insulin sensitivity and muscle mitochondrial enzymes but unchanged blood glucose control in type I diabetics after physical training. *Diabetes* 31:1044.

Wasserman, D.H., and Abumrad, N.N., 1989, Physiological bases for the treatment of the physically active individual with diabetes. *Sports Med.* 7:376.

Wasserman, D.H., Greer, R.J., Rice, D.E., Bracy, D., Flakoll, P.J., Brown, L.L., Hill, J.O., and Abumrad, N.N., 1991, Interaction of exercise and insulin action in humans. *Am. J. Physiol.* 260:E37.

Wasserman, D.H., Mohr, T., Kelly, P., Lacy, D.B., and Bracy, D., 1992, Impact of insulin deficiency on glucose fluxes and muscle glucose metabolism during exercise. *Diabetes* 41:1229.

Wolfe, R.R., 1990, Isotopic measurement of glucose and lactate kinetics. *Annals Med.* 22:163.

Wolfe, R.R., Nadel, E.R., Shaw, J.H.F., Stephenson, L.A., and Wolfe, M.H., 1986a, Role of changes in insulin and glucagon in glucose homeostasis in exercise. *J. Clin. Invest.* 77:900.

Wolfe, R.R., Shaw, J.H.F., Jahoor, F., Herndon, D.N., and Wolfe, M.H., 1986b, Response to glucose infusion in humans: role of changes in insulin concentration. *Am. J. Physiol.* 250:E306.

Zinman, B., Marliss, E.B., Hanna, A.K., Minuk, H.L., and Vranic, M., 1981, Exercise in diabetic man: glucose turnover and free insulin responses after glycemic normalization with intravenous insulin. *Can. J. Physiol. Pharmacol.* 60:1236.

Zinman, B., Murray, F.T., Vranic, M., Albisser, M., Liebel, B.S., PA McClean, P.A., and Marliss, E.B., 1977, Glucoregulation during moderate exercise in insulin treated diabetics. *J. Clin. Endocrinol. Metab.* 45:641.

SUBSTRATES AND THE REGULATION OF HEPATIC GLYCOGEN METABOLISM

Jerry Radziuk, Susan Pye and Zi Zhang

Diabetes and Metabolism Research Unit
Ottawa Civic Hospital and the University of Ottawa
1053 Carling Ave., Ottawa, Canada, K1Y 4E9

INTRODUCTION

Hepatic glycogen has traditionally been considered as the storage form of glucose when an excess of carbohydrate is available, as after meals. In the post-absorptive period glucose is released from glycogen by the process of glycogenolysis with an increasing contribution from gluconeogenesis as glycogen stores are depleted (Duderman, 1975). The processes of glycogen formation and glycogenolysis are controlled by the activity of glycogen synthase and phosphorylase which, in turn, are subject to finely-tuned regulatory mechanisms (eg. Hers, 1976; van de Werve and Jeanrenaud, 1987; Nuttall et al., 1988).

It has long been evident that liver glycogen is synthesized, not just from glucose but also from gluconeogenic substrate (Wood et al., 1945; Topper and Hastings, 1949). More recent evidence demonstrated, initially in the rat (Shikama and Ui, 1978) as well as in humans (Radziuk, 1979; Radziuk, 1982), that even when a surfeit of glucose is available after meals, a large fraction of liver glycogen continues to be formed from substrates other than glucose itself, suggesting a diversion of the gluconeogenic process from the supply of glucose to the circulation to the formation of glycogen.

This may be important, not only for the maintenance of an adequate store of glycogen, but also from the point of view of the disposal of the products of obligatory anaerobic metabolism, primarily lactate as well as of amino acids which result from ongoing proteolysis. This role also is strongly suggested by the fact that, even when the glucose supply is adequate from exogenous sources, glucogenic substrates continue to provide a substantial portion of the precursors for glycogen. The potential dual role of glycogen as a sink for both glucose and the products of metabolism, suggests the possibility that not only glucose but also these metabolic products may play a role in the regulation of glycogen synthesis. In the following therefore we would like to examine glycogen metabolism in the liver from the following perspectives: (i) pathways, (ii) substrates and (iii) regulation, particularly by metabolites.

New Concepts in the Pathogenesis of NIDDM, Edited by
C. G. Östenson *et al.*, Plenum Press, New York, 1993

PATHWAYS OF HEPATIC GLYCOGEN SYNTHESIS

Glycogen Formation and the Hepatic Uptake of Glucose

The importance of glycogen as a storage form of glucose was at least partially predicated upon the observation that tolerance to glucose administered orally was much higher than to the same quantity of glucose given intravenously (eg. Scow and Cornfield, 1954). It was hypothesized that the augmented insulin secretion mediated by gut factor stimulated by the oral glucose (McIntyre et al., 1965; Unger and Eisentraut, 1975; Brown et al., 1975) as well as the increased portal availability of the newly absorbed glucose led to a hepatic uptake of 40-65% of the absorbed glucose (Felig et al., 1975). More recent estimates (Kelley et al., 1988; Marin et al., 1992) also attribute a large fractional first-pass uptake of glucose to the liver. An impairment of hepatic glucose uptake could therefore be a primary lesion in the development of glucose intolerance and fasting hyperglycemia in patients with NIDDM (Felig et al., 1978).

If the liver, in fact, was able to take up glucose at the rates suggested, its disposal would necessarily imply a role for glucose as the principal, if not the only, substrate for glycogen synthesis. In dogs, this has been implied to be the case, although some preliminary intrahepatic metabolism of glucose to 3-carbon products was included (Moore et al., 1991). The evidence against a large fractional hepatic uptake of glucose however is compelling. In humans, it was demonstrated using a double tracer method (one tracer infused intravenously to determine changing metabolic clearance rates of glucose and the other incorporated into an oral glucose load), that approximately 93% of the ingested glucose appeared in the peripheral circulation (Radziuk et al., 1978). These data were confirmed using splanchnic arterio-venous differences in humans which yielded estimates of 3% for splanchnic glucose uptake following glucose loading (Ferrannini et al., 1980). Direct observation using differential catheterization techniques of the liver in swine (Radziuk, 1987) also showed a 5% uptake of glucose, both during oral and intravenous glucose. These data therefore suggest that the centrally-mediated mechanism initiated by a porto-arterial glucose gradient, in stimulating hepatic glucose uptake, which is seen in dogs (Atkins et al., 1987) is, at least, greatly attenuated in the other species. Complementary evidence demonstrates convincingly that the primary site of glucose disposal and its modulation under different physiological circumstances (eg. oral vs intravenous) is the periphery, primarily muscle. As an example, the forearm uptake of glucose can be compared during oral and intravenous glucose administration (Fig. 1; Radziuk and Inculet, 1983). This uptake of glucose was increased by 50% during the absorption of a glucose load relative to that during the infusion of glucose at analogous rates. This increased uptake can be attributed to the higher insulin concentrations following oral glucose (Radziuk and Inculet, 1983). Again, in patients with NIDDM, there is a decrease in peripheral glucose uptake, reflected also in both decreased glucose oxidation and storage (DeFronzo, 1988) and in particular muscle glycogen synthesis (Young et al., 1988).

It can therefore be concluded that the site at which responsiveness to altered physiological and pathological circumstances, can be demonstrated at least from the viewpoint of glucose disposal to be the periphery rather than the liver. Modulation of liver glycogen formation would therefore be expected to occur by other means.

Glucogenic and Direct Pathways of Hepatic Glycogen Formation

It has long been known that liver glycogen can be formed from glucogenic substrates (Wood et al., 1945; Topper and Hastings, 1949; Hems et. al., 1972). More recently two important observations were added.

It was found that even when a surfeit of glucose is available, the gluconeogenic process predominates both in man (Radziuk, 1979, 1982, 1989a, 1989b; Magnusson et al., 1987) and in rats (Shikama and Ui, 1978; Baer and Radziuk, 1979). In man this was deduced from the fact that little glucose label (H^3-6-glucose) could be mobilized from the liver, using glucagon infusions, in the immediately post-absorptive period (Radziuk, 1982). Upon addition of C^{14}-bicarbonate to track the gluconeogenic process, much more glucose label was released from liver under the same circumstances (Radziuk, 1982). Quantitative evaluation of the relative importance of the direct incorporation of glucose into glycogen (the direct pathway) and its gluconeogenic formation (the gluconeogenic pathway) using substrate specific activities and accounting for the dilution of gluconeogenic label in the oxaloacetate pool (Hetenyi, 1979) yielded the following estimate: during loading with glucose by the oral route or during the administration of glucose intravenously at rates

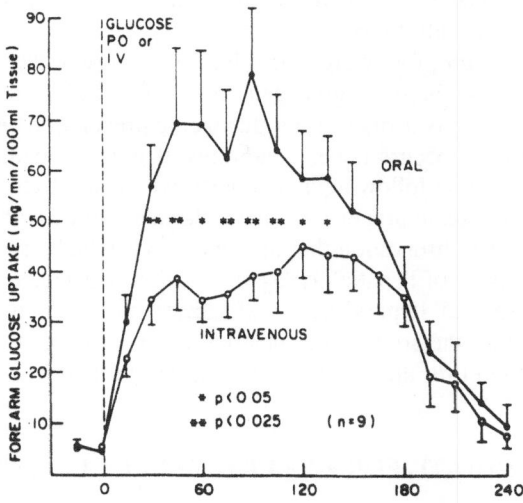

Figure 1. The forearm uptake of glucose following loading with oral and intravenous glucose in an equivalent fashion in man. (Reproduced with permission from Radziuk and Inculet, 1983.)

equivalent to those seen during glucose absorption , 40% of the glycogen formed is directly from glucose and 60% is by the gluconeogenic pathway. (Radziuk, 1982). These results were corroborated using the recycling of glucose label from the first carbon to the remaining carbons in the glycogen (Radziuk, 1989a, 1989b) and using label distributions in glucuronide derivatives of UDPG, the immediate precursor of liver glycogen (Magnusson et al., 1989b). Similar results were obtained in rats both qualitatively (Shikama and Ui, 1978) by comparing ratios of C^{14} (from C^{14}-bicarbonate) and H^3 (from H^3-glucose) in plasma glucose and glycogen as well as using the redistribution of recycling label from C^{14}-1-glucose into other positions (Baer and Radziuk, 1980; Newgard et al., 1983; Shulman et al., 1985.)

The second observation was that, not only is glycogen repleted gluconeogenetically, but this process is necessary to maintain net glycogen synthesis. This was demonstrated by blocking phosphoenolpyruvatecarboxykinase using 3-mercaptopicolinic acid during glucose loading with a resulting glycogen synthesis rate of near zero (Sugden et al., 1983; Newgard et al., 1984). The predominance of the gluconeogenic process in glycogen synthesis is also emphasized by the fact it appears to override the effect of fructose-2,6-bisphosphate (Fru-2,6-P_2) in this context.

It is now well established that fructose-2,6-bisphosphate exerts an important influence, in the regulation of hepatic glycogen and gluconeogenesis. When glucose levels are high, Fru-2,6-P_2 levels increase, favouring glycolysis and inhibiting gluconeogenesis. When glucose concentrations fall, the reverse situation occurs. (Claus et al., 1984) With glucose refeeding in starved rats Fru-2,6-P_2 levels remain low or are only slightly elevated (Kuwajima et al., 1984) but with sucrose refeeding those are 20-fold elevated (Kuwajima et al., 1984). Under both sets of conditions the glucogenic flux is primarily responsible for the new synthesis of glycogen. This divergent situation suggests the possibility that not only do the processes of glycolysis and gluconeogenesis take place in different metabolic zones of the liver (Jungermann et al., 1977; Jungermann and Katz, 1982) but that this zonation may extend to the concentration of Fru-2,6-P_2. In fact, hepatocytes enriched in periportal cells exhibit both increased glucose and glycogen synthesis (Chen and Katz, 1988) although in no population of parenchymal cells did significant glycolysis occur at physiological glucose concentrations.

In summary therefore, following a 100g load of glucose (either oral or intravenous), a maximum of 10g of hepatic glycogen is synthesized directly from glucose. Approximately another 15g is formed from glucogenic precursors. Although this portion between the two pathways occurs under conditions of glucose loading, it is particularly suitable to glycogen repletion following mixed meals where substrate enters the portal vein both as glucose and glucogenic precursors such as lactate, amino acids and fructose. This is also consistent with the observation that in diabetic rats which are necessarily catabolic the gluconeogenic pathway of hepatic glycogen repletion is enhanced under conditions of acute hyperinsulinemia and hyperglycemia (Giaccari and Rossetti, 1992). All these observations suggest that the sources of substrate are peripheral although this has been questioned. This is therefore the subject of the discussion below.

SOURCES OF SUBSTRATE FOR HEPATIC GLYCOGEN SYNTHESIS

The interesting observation has been made that, although muscle is a principal site of glycolysis, it contributes little, if any, lactate to the circulation following glucose loading (Jackson et al., 1973; Radziuk and Inculet, 1983). These data were obtained in man using forearm differences between arterialized venous and deep venous blood both after oral and intravenous glucose loading (Radziuk and Inculet, 1983). This appears to eliminate muscle (or at least muscle as typified by the forearm) as a source of lactate under these circumstances. Nevertheless, lactate (and pyruvate) is the only metabolite which increases significantly during the assimilation of a glucose load (et. Radziuk and Inculet, 1983). These two observations together have been taken to suggest the intrahepatic production from glucose of most of the glucogenic substrate which is subsequently incorporated into glycogen (Moore et al. , 1991). In fact, data in the dog appears to confirm such a hypothesis, since following glucose loading the dog liver is a net producer of lactate while taking up sufficient glucose to account for all glycogen formation. Gluconeogenesis contributes significant carbon to this formation (Moore et al., 1991). In humans it can also be estimated (Radziuk et al., 1978) that the liver takes up sufficient glucose (just) to account for all the glycogen formation (Radziuk, 1989a, 1989b). However, a body of evidence suggests that a somewhat more complex situation exists.

Incorporation of Label from Different Precursors or Markers into Liver Glycogen in Humans

Estimates of glycogen synthesis by the gluconeogenetic pathway were obtained in overnight-fasted humans using the incorporation of carbon-14 from three different markers or substrates: (i) C^{14}-bicarbonate (ii) C^{14}-U-lactate and (iii) C^{14}-1-glucose (Radziuk, 1989a, 1989b). These labels were infused concurrently with H^3-3-glucose at constant rates throughout each study. Following two hours of label infusion a 100g glucose load was administered orally (or infused intravenously at rates which are equivalent to the absorption rate of the oral glucose load). Since absorption takes place over approximately 4 hours (Radziuk et al., 1978), precursor label infusion was terminated and a graded infusion of glucagon was initiated after this time to mobilize glycogen newly formed during the absorption period. A third tracer (H^3-6-glucose) was infused during the last part of the study and used to calculate the rate of appearance of C^{14}-glucose in the circulation. In the case of C^{14}-1-glucose infusion, the mobilization of recycled C^{14}-glucose was estimated in this way by monitoring C^{14}-6-glucose appearance in the circulation. Exactly analogous experiments were performed without a glucose load to account for the incorporation of carbon-14 into glucose by new gluconeogenesis stimulated by the glucagon infusion. Results of these studies for C^{14}-U-lactate are demonstrated in Fig. 2 for the final, glycogen mobilization phase, of the study. The area between the two curves represents the minimum estimate of the C^{14} glucose mobilized from glycogen formed during the loading phase.

For each label then: (i) the total amount of label mobilized was estimated (ii) the substrate specific activity was determined: this was bicarbonate specific activity for this label and lactate specific activity for both the labelled lactate and glucose infusions (iii) an appropriate correction factor was estimated (Hetenyi, 1979; Radziuk, 1982, 1989b) to account for label dilution by Krebs cycle carbon in the oxaloacetate pool. This information was then used to estimate the total amount of glycogen found in the liver by the gluconeogenic pathway. Results of these calculations are summarized in Fig. 3.

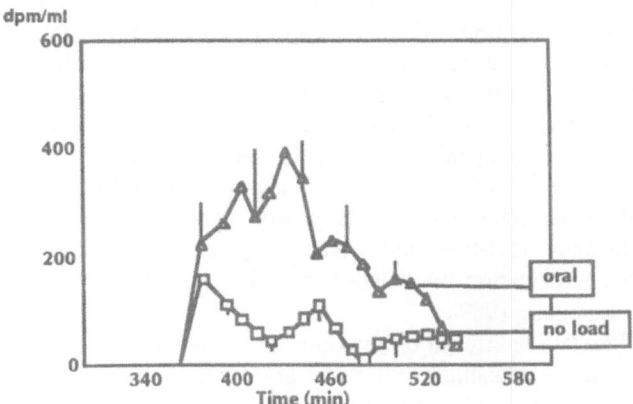

Figure 2. Rate of appearance of [C^{14}] glucose that is mobilized from glycogen postabsorptively and during a glucagon infusion (initiated at $t=410$). Calculation is performed using the concentrations of a third tracer ([6-H^3] glucose) infused intravenously after absorption has terminated. It can be seen that at least 3 times more label appears in circulation after a glucose load than without when the labelled precursor is lactate.

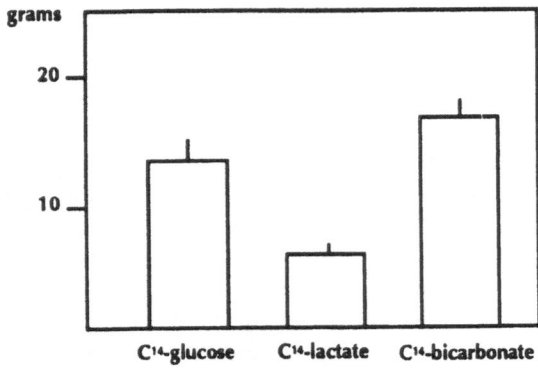

(dpm mobilized / substrate s.a.) x corr fact

Figure 3. Gluconeogenetic synthesis of liver glycogen in humans following a 100g oral glucose load when this process is followed using 3 different labels. (Results are corrected for dilution in the oxaloacetate pool.)

Approximately 15g of glycogen is estimated to be formed by this pathway when either C^{14}-bicarbonate or label randomization from the first position in glucose is used in the calculation. However this estimate is reduced to approximately 40% when C^{14}-label is used as label. The explanation of this observation is as follows:

The principal issue in question is the specific activity of the relevant substrate or marker in the hepatocyte. In humans, these precursors are however measured in the peripheral circulation. Any dilution of the label in the splanchnic bed will lead to an over estimate of the hepatocyte specific activity and hence underestimation of the amount of glycogen synthesized gluconeogenetically.

The ubiquitous nature of carbonic anhydrase (Marsolais et al., 1987; Radziuk, 1989b) ensures a rapid equilibration of $C^{14}O_2$ between intracellular and extracellular spaces. A small dilution may occur across the gut (about 5% in the rat - Ryan et al., 1993) from local unlabelled CO_2 production. When C^{14}-1-glucose is used as the initial label and C^{14}-6-glucose is measured as representative of label randomized to other positions during glycolysis and subsequent gluconeogenesis, C^{14}-lactate specific activity is used as substrate specific activity. It is likely that selective dilution of this specific activity will not occur to a great extent in the splanchnic bed since C^{14}-lactate production takes place in every tissue when C^{14}-glucose is metabolized - that is the gut and the liver will produce labelled rather than unlabelled lactate since the substrate, glucose, is labelled. Peripheral and hepatocyte specific activities of lactate are thus expected to be similar. In fact, the estimate of hepatic glycogen formation by gluconeogenesis using the two labels is remarkably similar (Fig. 3; Radziuk, 1989b).

When C^{14}-lactate is infused on the other hand and its incorporation into glycogen determined, less gluconeogenetic synthesis of glycogen appears to occur. This suggests that a significant production of lactate may take place in the splanchnic bed, diluting its specific activity. This could include both the gut and the liver.

Although, as discussed above, muscle does not appear to be a source of lactate during glucose loading, the above data suggests the possibility that other tissues could contribute significantly. Candidates for this role are discussed below:

The Liver. In dogs fasted for less than 36 hr, following glucose loading, the liver is a net producer of lactate. This suggested the possibility that all important sources of 3-carbon precursors of glycogen are within the liver (Moore et al., 1991). However, other data suggest a modification of this interpretation. Fig. 4 shows the results of studies in 18hr fasted dogs where C^{14}-lactate was infused concurrently with glucose. It is clear that net lactate output (a negative "fractional extraction") takes place during glucose loading (t > 150min). Simultaneously a 20% extraction of the labelled moiety continues at the same rate throughout the basal and loading periods. In the face of a net production, therefore a utilization of lactate occurs. This could take place at every point along the sinusoid. On the other hand, it is completely consistent with the concept of metabolic zonation which has now been well documented (Jungermann et al., 1977; Jungermann and Katz, 1982) for glycolysis and gluconeogenesis. Thus, in perivenous cells glycolysis predominates, whereas gluconeogenesis, and therefore substrate (lactate) uptake, occur primarily in the periportal cells. Since lactate concentrations increase during glucose administration, the data suggest an increase in lactate uptake consistent with the increased demand for substrate for glycogen synthesis. The perivenous cells simultaneously contribute lactate to the peripheral pool.

In contrast to the dog, the rat and the pig take up large amounts of lactate in the liver during glucose loading. 30-40% extractions have been demonstrated (Niewohner et al., 1984; Zhang and Radziuk, 1991; Radziuk, 1987). Little lactate output is therefore expected consistent with a liver which exhibits primarily "periportal" or gluconeogenetic characteristics under these experimental circumstances. Zonation of gluconeogenetic glycogen production would therefore not be expected in the glucose-loaded rat or pig. It has been hypothesized (Radziuk, 1988) that a functional "switching point" occurs in the liver sinusoid with respect to lactate uptake. On the portal side, there is a net lactate uptake which gradually diminishes until, at the switching point, a net production begins which is gradually accentuated as "perivenous" cells predominate. Finally, the uptake of lactate (or at least glucogenic substrate) appears to be necessary for the synthesis of glycogen (Zhang and Radziuk, 1991). This issue will be discussed further below.

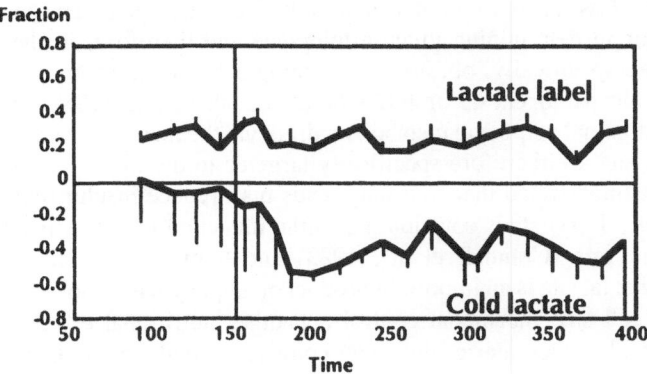

Figure 4. Fractional extraction of lactate across the liver in 18hr fasted dogs before and after the initiation of a glucose infusion. During glucose infusion the extraction of lactate label (C^{14}-U-Lactate) remains at 20% where the net production of lactate (negative "extraction") increases.

The Gut. When the splanchnic bed is considered as a unit as is most often the case in human studies, small amounts of net lactate production were detected as the glucose load increased (Brattusch-Marrain et al., 1980). The gut and, in particular, the epithelial and smooth muscle cells, exhibit significant glycolytic activity, with lactate the most important product of glucose metabolism. This can be clearly seen in the rat (Hanson and Parsons, 1976; Niewohner et al., 1984; Tormo et al., 1988). In fact, although this is not the case, it has been hypothesized that lactate is the principal form in which glucose is absorbed (Shapiro and Shapiro, 1979). In the pig, as illustrated in Fig. 5, 40% of the lactate which is extracted by the liver is produced by the gut during glucose loading (Radziuk, 1987). Limited human data is available but portal sampling (see Landau and Wahren, 1988) was demonstrated a net gut production of lactate by the gut. The gut is thus an important source of lactate and therefore substrate for glycogen synthesis during glucose loading.

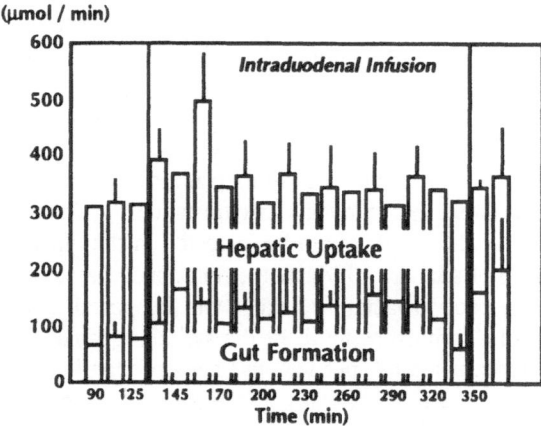

Figure 5. The hepatic uptake and gut production of lactate in a conscious pig during the intraduodenal infusion of glucose.

Adipose Tissue. Body fat distribution has recently been identified as a potentially important factor in determining glucose tolerance and therefore, in the development of NIDDM. More specifically, obesity representing a preponderance of abdominal fat is a risk factor in the pathogenesis of NIDDM (Kissebah et al., 1988). Interestingly, the products of abdominal adipocyte metabolism drain via the mesenteric vessels primarily into the portal vein and are therefore specifically targeted to the liver. As an example, it has recently been demonstrated that free fatty acids may reduce insulin removal by the liver (Svedberg et al., 1991) thus contributing to the induction of hyperinsulinemia.

Work in rats (Crandall et al., 1983) and humans (Jansson et al., 1990) has demonstrated that lactate is an important product of adipocyte metabolism. This production is sensitive to cell size, hormonal control and other nutritional effects (Crandall et al., 1983). In particular, mesenteric adipocytes convert more glucose to lactate (Newby et al., 1988) suggesting that part of the lactate which is deemed to arise from the gut, could be attributed to abdominal fat deposits. This establishes the possibility of a direct influence of lactate arising from abdominal fat on liver metabolism.

In diabetes where central obesity may predominate, abdominal fat may make an increasingly important contribution to portal hyperlactatemia enhancing the generalized elevation in lactate levels at the liver. Interestingly it has also been shown that there is an increase in gluconeogenetically formed glycogen in diabetic rats (Giaccari and Rossetti, 1992) suggesting that the availability of additional lactate in the portal vein may indeed affect hepatic glycogen metabolism in this way.

Abdominal fat, however is not the only source of lactate arising from adipocyte metabolism. Recent measurements using microdialysis probes have demonstrated that subcutaneous fat produces lactate and that this production increases following glucose loading (Hagstrom et al., 1990). The total contribution of adipose tissue to lactate formation following glucose administration or meals is thus highly significant.

The Skin. The skin is a large organ, constituting about 10% of the normal body weight. Its contribution to metabolism may have been underrated. Glucose metabolism has long been studied in the skin (for review, Johnson & Fusaro, 1972; Nguygen and Keast, 1991). In particular a high skin concentration of lactate has been found relative to that in the circulation suggesting a continuous and active synthesis of this metabolite in the skin with diffusion into the circulation (likely particularly rapid due to high blood flow to the skin) and participation in the Cori cycle and thus glycogen synthesis in the liver.

The CNS and Erythrocytes. Under normal physiological conditions the brain and other neural tissue metabolize glucose and are obligatory lactate producing organs (Siesjo, 1978). Similarly red blood cells produce lactate at a nearly consistent rate independently of glucose concentrations (Murphy, 1960). Although the quantitative contributions to the lactate pool may not be very large, this production is not consistent and subject to acute regulation. Gluconeogenesis leading to new glucose or glycogen formation is therefore likely an important mechanism for the elimination of this lactate.

THE REGULATION OF HEPATIC GLYCOGEN METABOLISM

Control of glycogen metabolism both at the level of synthesis (synthase) and breakdown (phosphorylase) has long been known to be mediated by enzyme cascades. Such "biochemical amplification" mechanisms lead to potentially exquisitely sensitive regulation - hormones circulating at extremely low concentrations can lead to substantial physiological responses. That hepatic glycogen is illustrative of such control is well established. The massive and rapid degree of glycogenolysis in response to increases in glucagon or catecholamine levels is generally considered to be an integral part of the "fight or flight" phenomenon. Vasopressin and angiotensin may modulate this response. Insulin can switch the liver from glycogenolysis to glycogen synthesis in a few minutes (Stalmans and van de Werve, 1981). Hormonal regulation of glycogen metabolism and its mechanism have been extensively reviewed (eg. Hers, 1976; Stalmans et al., 1987;van de Werve and Jeanrenaud, 1987; Nuttall et al., 1988).

The primary importance of hormones in the control of glycogen metabolism however may need to be tempered by observations such as those made by Terrettaz et al. (1986). These demonstrated that at euglycemia, hyperinsulinemia does not affect the glycogen content of the liver nor the activities of phosphorylase *a* or synthase *a*. These data suggest that concomitant changes in glucose levels may be necessary for enzymatic charges leading to an alteration in glycogen content. The potential dual role of substrates as precursors of glycogen and regulators of its metabolism is explored below.

Substrates and the Regulation of Glycogen Synthesis

As discussed above the principal substrates for glycogen synthesis are glucose (the direct pathway) and glucogenic substrate, primarily lactate and glucogenic amino acids. Clearly the presence of sufficient substrate is necessary for glycogen formation. It will also exert a mass effect as availability is altered. The seminal work on glucose regulation of glycogen synthesis was done by Hers (1976). In contrast to the mass effect or "push mechanism", he hypothesized a "pull mechanism" by which glucose stimulated its own incorporation into glycogen. Figure 6 outlines in simplified form the control of glycogen metabolism by synthase and phosphorylase (eg. Nuttall et al., 1988). Briefly, phosphorylase *a* acts as a glucose receptor for the liver. Glucose binds to this enzyme inhibiting its activity. The glucose phosphorylase complex is more amenable to conversion to the inactive *b* form. Synthase activation occurs when phosphorylase activity falls below the level which is inhibiting to synthase phosphatase yielding an effective "on-off" mechanism for glycogen synthesis. This has been largely confirmed in the fed situation. In the fasted state, after refeeding, synthase and phosphorylase are found to be both partially in the active form and further activation of synthase was not necessary to the net synthesis of glycogen (van de Werve and Jeanrenaud, 1984). The presence of hyperglycemia (as after glucose feeding) compared to lower glucose levels after mixed meals appears to determine the relative activation of synthase and phosphorylase (van de Werve and Jeanrenaud, 1987) suggesting that other factors may come into play during mixed meals.

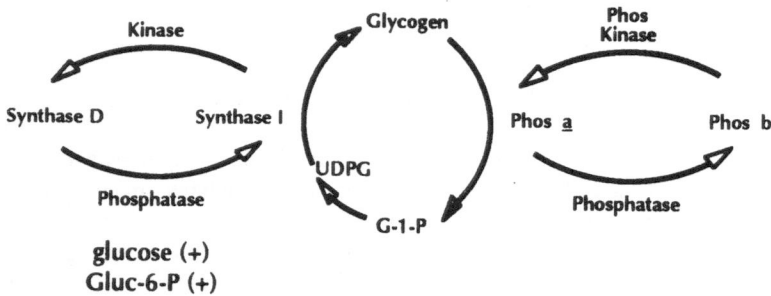

Figure 6. Simplified scheme of the control of glycogen synthesis and glycogenolysis.

The latter will increase the level of glucogenic substrates, particularly amino acids in addition to glucose and lactate. Addition of amino acids (glutamine, alanine) and lactate to glucose has been shown to stimulate glycogen synthesis synergistically in hepatocytes (gealen) or perfused liver (Whitton and Hems, 1975). Amino acids have also recently been shown to induce hepatocyte swelling (Meijer et al., 1992) which in itself can stimulate glycogen synthesis, likely mediated by intracellular changes in ion (Cl⁻) concentration. This suggests a mechanism for the synergy which occurs in the presence of both glucose and amino acids. Fructose has long been known to maximize glycogen synthesis in the presence of glucose (Hers, 1976; Geelen, 1977), likely due to an inhibition of phosphorylase by fructose-1-phosphate (Kaufmann and Froesch, 1973) an effect which has recently been confirmed (Youn et al., 1987).

It is lactate, however, that is an ubiquitous end-product of anaerobic metabolism of glucose. It is changed in response to most metabolic changes -- both meals and glucose loads as well as fasting, exercise etc. (Crandell et al., 1983; Hagstrom et al., 1990). Dual effects of glucose and lactate on glycogen synthesis have been shown in hepatocytes (Geelen, 1977). Our own work has therefore focussed on describing the interaction between glucose and the prototypic glucogenic substrate, lactate, since both concentrations are elevated following glucose administration, the situation in which liver glycogen synthesis has been studied to the greatest extent.

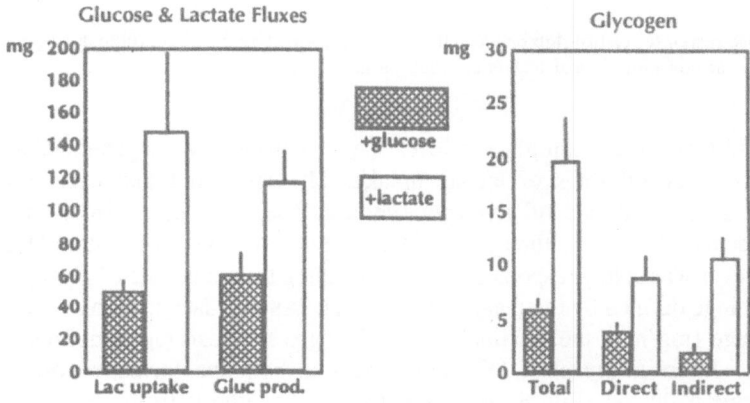

Figure 7. Uptake of lactate and glucose production in perfused rat liver when additional glucose and lactate are added to the perfusate. Glycogen formed by the direct and gluconeogenic pathways during these additions.

The Coordinated Control of Liver Glycogen Synthesis by Glucose and Lactate

This has recently been examined in a series of studies in rats (Zhang and Radziuk, 1991, 1993). In order to isolate, as much as possible the relationship between glucose, lactate and glycogen, extra-hepatic effects were removed by using a perfused liver preparation. Physiological circumstances were maintained as closely as possible by using blood from donor rats as perfusate. Blood gases were determined and were found to reflect closely those found in the portal and hepatic venous blood of the anaesthetized rat. The liver was initially perfused with physiological concentrations of glucose and lactate together with a glucose label (H^3 or C^{14}-glucose). Then either additional glucose or lactate was infused. When additional lactate was supplied, five times more glycogen was formed by the gluconeogenic pathway (predictably) but more than twice as much glycogen was also formed directly from glucose (Fig.7). Glucose production also increased proportionally to lactate uptake. These preliminary results prompted further studies to examine the relationship of glucose and lactate in promoting glycogen metabolism in the liver. A variety (nine) combinations of glucose and lactate were infused following appropriate initial

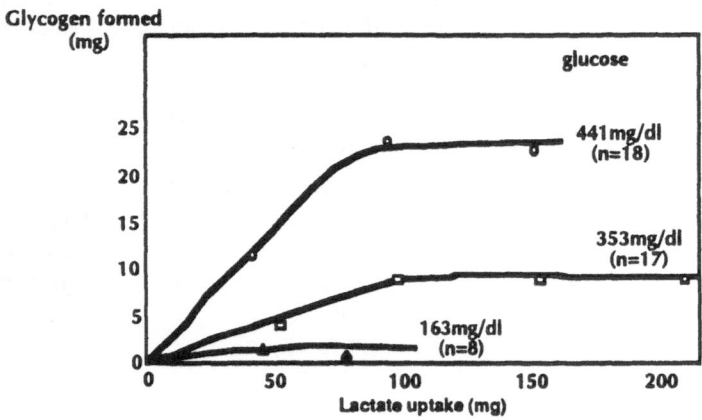

Figure 8. Dose-response relationship between the glycogen formed by the direct route during a 2hr perfusion of rat liver at varying glucose and lactate concentrations.

conditions so that three mean glucose levels (approximately) were generated and, at each glucose level, several rates of lactate uptake. It was found that each glucose level determined a maximal rate of glycogen synthesis (i.e. glycogen formed in 2hr). The uptake of lactate increased although nonlinearly, with the level of lactate. This uptake of lactate however was solely responsible for determining the actual rate of glycogen synthesis within the range defined by the mean glucose level. As total lactate uptake exceeds 100mg or, on average 1mg/min, the maximal rate of glycogen synthesis (again defined by glucose) is reached. Interestingly also, this rate applies to both the formation of glycogen by gluconeogenic pathways (Fig. 8) as well as by direct synthesis from glucose (Fig. 9). In contrast to the saturation in the rate of net glycogen synthesis with increased lactate uptake by the liver, within the range of uptakes used in those studies, and under the conditions when only one substrate (lactate) is present, glucose production exhibits a linear relationship with lactate uptake (Fig.10). This suggests that the control of the two processes may be different.

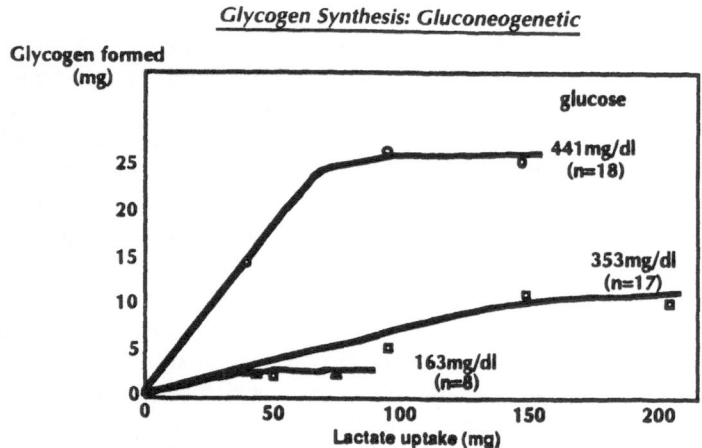

Figure 9. Dose response relationship between the glycogen formed by the gluconeogenetic pathway during a 2hr perfusion of rat liver at varying glucose and lactate concentrations.

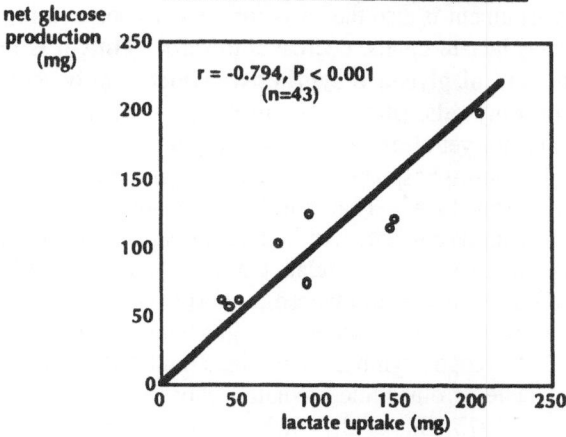

Figure 10. Relationship between net glucose production (glucose outflow-glucose inflow) and net lactate uptake over the 2hr period of perfusion of a rat liver at various concentrations of glucose and lactate.

The effect of lactate on glycogen synthesis is not simply permissive since its rate is determined (within limits) by the amount of lactate taken up. It is moreover, not only due to an increase of glucose-6-phosphate, the effective substrate for glycogen synthesis and a regulator of glycogen synthase (Nuttall et al., 1988), since the rate of glycogen synthesis saturates whereas the rate of formation of circulating glucose does not. In addition, glucose can provide glucose-6-phosphate but does not determine the actual rate of glycogen synthesis (only its upper limit).

In the context of these studies it has also been demonstrated that the hepatic uptake of glucose is directly proportional to the perfusate concentration whereas the uptake of lactate increases with lactate concentration but not in a linear fashion. Since glucose uptake is proportional to its concentration, we can also say that the maximal rate of glycogen synthesis is determined by the glucose uptake. Both parameters - the maximal glycogen synthetic rate and the actual rate - are therefore related to glucose and lactate which has been taken up into hepatocytes and therefore to their intracellular concentrations or that of their metabolites.

Net glycogen synthesis is dependent on the relative activities of glycogen synthetase and phosphorylase as well as the availability of substrate (glucose-6-phosphate). In the fed state glycogen synthase is stimulated (Hers, 1976; Hems et al., 1972) whereas under fasting conditions the principal factor may be phosphorylase inactivation (van de Werve and Jeanrenaud, 1984). Nevertheless, glucose-6-phosphate levels increase following glucose loading (Niewohner et al., 1984) in fasting rats. Both glucose and glucose-6-phosphate (Gilboe and Nuttall, 1983) have been found to stimulate synthase phosphatase activity. In the studies presented here, the net effect of this complex, and not yet completely understood, regulation is to set a maximal rate of glycogen synthesis.

This rate of synthesis prevails independently of lactate uptake when the latter is high - greater than about 1mg/min. Below this rate of lactate uptake the rate of net glycogen synthesis decreases. Previous work by Sugden et al (1983) and Newgard et al (1984) demonstrated that carbon flow through the gluconeogenetic pathway is essential to the continued synthesis of liver glycogen. In vivo, it has also been shown that during glucose loading, continued peripheral provision of lactate is necessary to this process (Radziuk, 1989b; Wajngot et al., 1989). Taken all together, the present observations imply that, in the system studied, lactate uptake by the liver determines the actual net rate at which glycogen is synthesized within the entire range from zero to the maximum set by the glucose levels. This argument is also the basis for the extrapolation in Fig. 8 and 9 in the glycogenic flux rates as lactate uptake decreases to zero. **Thus it is lactate availability that determines the rate of glycogen synthesis. Glucose only sets the range within which this regulation may take place.** The mechanisms of this regulation of glycogen synthesis by lactate are not yet clear. An increased gluconeogenetic flux would increase both the level of glucose-6-phosphate (and UDPG) and the stimulation of synthase phosphatase by this intermediate. That this is likely not the only mechanism can be inferred from the following argument. An increasing hepatic lactate uptake generates an increasing and approximately linearly related glucose output. This suggests that an increasing level of G6P is available for glucose-6-phosphatase. Yet, beyond a certain level of lactate uptake, no further net synthesis of glycogen occurs while glucose output continues to increase. Glycogen synthase could be stimulated and phosphorylase activity could be altered by lactate or other intermediates whose formation can be stimulated by lactate (Wimhurst et al., 1973; Hue et al., 1982). Thus lactate could function analogously to fructose which has been shown to stimulate synthase activity (Niewohner and Nuttall, 1986) and both to inhibit (via fructose-1-phosphate, Kaufmann and Froesch, 1973; Youn et al., 1987) or stimulate (van de Werve and Hers, 1979) phosphorylase activity. This regulatory activity of lactate would then be analogous to its hypothesized role in fat metabolism (Geelen et al., 1980).

SUMMARY

Glycogen metabolism is a complex process which depends on the metabolic circumstances and the hormonal milieu. In this overview an intriguing new possibility has been emphasized - the possible central role of lactate in coordinating, with glucose, the net synthesis of glycogen. Since lactate changes acutely under many physiological circumstances, it would be a logical candidate for a signal which communicates to the liver the metabolic states of the periphery. It would then acutely determine the synthetic rate of glycogen synthesis within the range determined by the glucose concentrations which in turn could be said to reflect the nutritional state of the system. Interestingly, after oral glucose loading, portal glucose levels would be about 25% higher (Radziuk et al., 1978) relative to arterial. As seen from Figs 8 and 9 however the glycogen synthetic rate appears very sensitive to glucose (at a given lactate uptake). Everything else being assumed equal therefore, more glycogen would be synthesized than during intravenous loading with an equivalent peripheral concentration. This is indeed the case (Shulman and Rossetti, 1989). On the other hand, during equivalent loads, peripheral glucose levels are higher and the same quantity of glycogen is synthesized (Radziuk, 1989a, 1989b).

If lactate is typical of other glucogenic substrates, then it is also logical that mixed meals with higher levels of portal substrate would maximize glycogen synthetic rates. Similarly, in diabetes where hyperglycemia and hyperlactatemia prevail, gluconeogenesis plays a predominant role in glycogen synthesis (Giaccari and Rossetti, 1992). This is consistent with the increased availability of glucogenic substrate and an increase in the proportion of gluconeogenic glycogen which was seen as lactate availability rose (Zhang and Radziuk, 1991). With the large effects exerted by metabolites, at least in the context of glycogen synthesis, it may be hypothesized that these are the primary regulatory factors and the role of hormones may be mainly modulatory.

References

Adkins, B.A., Myers, S.R., Hendrick, G.K., Stevenson, R.W., Williams, P.E., and Cherrington, A.D., 1987, Importance of the route of intravenous glucose delivery to hepatic glucose balance in the conscious dog, *J. Clin. Invest.*, **79**:557-65.

Baer, A. and Radziuk, J., 1980, Sources of hepatic glycogen formation in conscious rats during intraduodenal glucose loading, *Clin. Res.*,**28**:385A.

Brattusch-Marrain, P.R., Waldhausl, W.K., Gasic, S., Korn, A., and Nowotny, P., 1980, Oral glucose tolerance test: effect of different glucose loads on splanchnic carbohydrate and substrate metabolism in healthy man, *Metabolism*, **29**:289-95.

Brown, J.C., Dryburgh, J.R., Ross, S.A., and Dupre, J., 1975, Identification and actions of gastric inhibitory polypeptide, *Recent. Prog. Horm. Res.*, **31**:487-532.

Claus, T.H., El-Mahgrabi, M.R., Regen, D.M., Stewart, H.B., McGrane, M., Kountz, P.D., Nyfeler, F., Pilkis, J., Pilkis, S.J., 1984, The role of fructose-2,6-bisphosphate in the regulation of carbohydrate metabolism, *Curr. Top. Cell. Regul.*, **23**:57-86.

Chen, Kim S., and Katz, Joseph, 1988, Zonation of glycogen and glucose syntheses, but not glycolysis, iin rat liver, *Biochem. J.*, **255**:99-104.

Crandall, D.L., Fried, S.K., Francendese, A.A., Nickel, M., and DiGirolamo, M., 1983, Lactate release from isolated rat adipocytes: influence of cell size, glucose concentration, insulin and epinephrine, *Horm. Metabol. Res.*, **15**:326-9.

DeFronzo, R.A., 1988, The triumvirate: β-cell, muscle, liver: a collusion responsible for NIDDM, *Diabetes*, **37**:667-87.

Felig, P., Wahren, J., and Hendler, R., 1975, Influence of oral glucose ingestion on splanchnic glucose and gluconeogenic substrate metabolism in man, *Diabetes*, **24**:468-75.

Felig, P., Wahren, J., Hendler, R., 1978, Influence of maturity-onset diabetes on splanchnic glucose balance after oral glucose ingestion, *Diabetes*, **27**:121-26.

Ferrannini, E., Wahren, J., Felig, P., and DeFronzo, R.A., 1980, The role of fractional glucose extraction in the regulation of splanchnic glucose metabolism in normal and diabetic man, *Metabolism*, **29**:28-35.

Geelen, M.J.H., 1977, Restoration of glycogenesis in hepatocytes from starved rats, *Life Sciences*, **20**:1027-34.

Geelen, M.J.H., Harris, R.A., Beynen, A.C., and McCune, S.A., 1980, Short-term hormonal control of hepatic lipogenesis, *Diabetes*, **29(12)**:1006-22.

Giaccari, A., and Rossetti, L., 1992, Predominant role of gluconeogenesis in the hepatic glycogen repletion of diabetic rats, *J. Clin. Invest.*, **89**:36-45.

Gilboe, D.P. and Nuttall, F.Q., 1983, Direct glucose stimulation of glycogen synthase phosphatase activity in a liver glycogen particle preparation, *Arch. Biochem. Biophys.*, **228(2)**:587-91.

Hagstrom, E., Arner, P., Ungerstedt, U., and Bolinder, J., 1990, Sucutaneous adipose tissue: a source of lactate production after glucose ingestion in humans, *Am. J. Physiol.*, **258**:E888-93.

Hanson, P.J., Parsons, D.S., 1976, The utilization of glucose and production of lactate by in vivo preparations of rat small intestine: Effects of vascular perfusion, *J. Physiol.*, **255**:775-95.

Hems, D.A., Whitton, P.O., and Taylor, E.A., 1972, Glycogen synthesis in the perfused liver of the starved rat, *Biochem. J.*, **129**:529-38.

Hers, H.G., 1976, The control of glycogen metabolism in the liver, *Ann. Rev. Biochem.*, **45**:167-89.

Hetenyi, G. Jr., 1979, Correction factor for the estimation of plasma glucose synthesis from the transfer fo C^{14} atoms from labelled substrate in vivo: a preliminary report, *Can. J. Physiol. Pharmacol.*, **57**:767-70.

Hue, L., Blackmore, P.F., Shikama, H., Robinson-Steiner, A., Exton, J.H., 1982, Regulation of fructose-2,6,-bisphosphate content in rat hepatocytes, perfused hearts, and perfused hindlimbs, *J. Biol. Chem.*, **257**:4308-13.

Jackson, R.A., Peters, N., Advani, U., Perry, G., et al., 1973, Forearm glucose uptake during the oral glucose tolerance test in normal subjects, *Diabetes*, **22**:442-58.

Jansson, P.A., Smith, U., and Lonnroth, P., 1990, Evidence for lactate production by human adipose tissue in vivo, *Diabetologia*, **33**:253-6.

Johnson, J.A. and Fusaro, R.M., 1972, The role of skin in carbohydrate metabolism, *Advances in Metabolic Disorders*, **6**:1-55.

Jungermann, K., Katz, N., Teutsch, H., and Sasse, D., 1977, Possible metabolic zonation of liver parenchyma into glucogenic and glycolytic hepatocytes, *in*: "Alcohol and Aldehyde Metabolizing Systema", R.G. Thurman, J.R. Williamson, H.R. Drott, and B. Chance, eds., Academic, New York.

Jungermann, K. and Katz, N., 1982, Functional hepatocellular heterogeneity, *Hepatology*, **2**:385-95.

Kaufmann, U., and Froesch, E.R., 1973, Inhibition of phosphorylase-a by fructose-1-phosphate, α-glycerophosphate and fructose-1,6-diphosphate: Explanation for fructose-induced hypoglycemia in hereditary fructose intolerance and fructose-1,6-diphosphatase deficiency, *Europ. J. Clin. Invest.*, e:407-13.

Kelley, D., Mitrakou, A., Marsh, H., Schwenk, F., Benn, J., Sonnenberg, G.,Arcangeli, M., Aoki, T., Sorensen, J., Berger, M., Sonksen, P., and Gerich, J. , 1988, Skeletal muscle glycolysis, oxidation and storage of an oral glucose load, *J. Clin. Invest.*, **81**:1563-71.

Kissebah, A.H., Peiris, A.N., and Evans, D.J., 1988, Mechanisms associating body fat distribution to glucose intolerance and diabetes mellitus: window with a view, *Acta Med. Scand.*, **723**:79-89.

Kuwajima, M., Newgard, C.W., Foster, D.W. and McGarry, J.D., 1984, Time course and significance of changes in hepatic fructose-2,6-bisphosphate levels during refeeding of fasted rats, *J. Clin. Invest.*, **74**:1108-11.

Landau, B.R., Wahren, J., 1988, Quantification of the pathways followed in hepatic glycogen formation from glucose, *FASEB J.*, **2**:2368-75.

Magnusson, I., Chandramouli, V., Schumann, W.C., Kumaran, K., Wahren, J., and Landau, B.R., 1987, Quantitation of the pathways of hepatic glycogen formation on ingesting a glucose load, *J. Clin. Invest.*, **80**:1748-54.

Marin, P., Hogh-Kristiansen, I., Jansson, S., Krotkiewski, M., Holm, G., and Bjorntorp, P., 1992, Uptake of glucose carbon in muscle glycogen and adipose tissue triglycerides in vivo in humans, *Am. J. Physiol.*, **263**:E473-80.

Marsolais, C., Huot, S., David, F., Garneau, M. and Bruengraber, H., 1987, Compartmentation of $C^{14}O_2$ in the perfused rat liver, *J. Biol. Chem.*, **262**:2604-07.

McIntyre, N., Holdsworth, C.D., and Turner, D.S., 1965, Intestinal factors in the control of insulin secretion, *J. Clin. Endocrinol. Metab.*, **25**:1317-24.

Meijer, J., Baquet, A., Gustafson, L., van Woerkom, G. M., and Hue, L., 1992, Mechanism of activation of liver glycogen synthase by swelling, *J. Biol. Chem.* , **267(9)**:5823-8.

Moore, M.C., Cherrington, A.D., Cline, G., Pagliassoti, M.J., Jones, E.M., Neal, D.W., Badet, C., and Shulman, G.I., 1991, Sources of carbon for hepatic glycogen synthesis in the conscious dog, *J. Clin. Invest.*, **88**:578-87.

Murphy, J.R., 1960, Erythrocyte metabolism. II. Glucose metabolism and pathways, *J. Lab. Clin. Med.*, **55**:286.

Newby, F.D., Sykes, M., and DiGirolamo, M., 1988, Regional differences in adipocyte lactate production from glucose, *Am. J. Physiol.* , **255**:E716-22.

Newgard, C.B., Hirsch, L.J., Foster, D.W., McGarry, J.D., 1983, Studies on th mechanism by which exogenous glucose is converted into liver glycogen in the rat, *J. Biol. Chem.*, **258**:8046-52.

Newgard, C.B., Moore, S.V., Foster, D.W. and McGarry, J.D., 1984, Efficient hepatic glycogen synthesis in refeeding rats requires continued carbon flow through the gluconeogenic pathway, *J. Biol. Chem.*, **259**:6958-63.

Nguyen, D.T. and Keast, D., 1991, Energy metabolism and the skin, *Int. J. Biochem.*, **23**:1175-83.

Niewoehner, C.B. and Nuttall, F.Q., 1986, Mechanism of stimulation of liver glycogen synthesis by fructose in alloxan diabetic rats, *Diabetes*, **35**:705-11.

Niewohner, C.B., Gilboe, D.P., and Nuttal, F.Q., 1984, Metabolic effects of oral glucose in the liver of fasted rats, *Am. J. Physiol.*, **246**:E89-94.

Nuttall, F.Q., Gilboe, D.P., Gannon, M.C., Niewoehner, C.B., Tan, A.W.H., 1988, Regulation of Glycogen Synthesis in the Liver, *Am. J. Med.*, **85**:suppl 5A.

Radziuk, J., 1979, Hepatic glycogen formation by direct uptake of glucose following oral glucose loading in man, *Can. J. Physiol. Pharmacol.*, **57**:1196-99.

Radziuk, J., 1982, Carbon transfer in the measurement of glycogen synthesis from precursors during absorption of an ingested glucose load, *Fed. Proc.*, **41**:88-90.

Radziuk, J., 1987, Tracer methods and the metabolic disposal of a carbohydrate load in man, *Diabetes/Metabol. Rev.*, **3**:231-67.

Radziuk, J., 1988, The liver and glucose homeostasis, *in*: "Proceedings of the 13th Congress of the International Diabetes Federation", R.G. Larkins, P.Z. Zimmet, P.J. Chisholm, eds., Diabetes Excerpta Medica, Elsevier, Netherlands.

Radziuk, J., 1989a, Hepatic glycogen in humans. I. Direct formation after oral and intravenous glucose or after a 24-hr fast, *Am. J. Physiol.*, **257**:E145-57.

Radziuk, J., 1989b, Hepatic glycogen in humans. II. Gluconeogenentic formation after oral and intravenous glucose, *Am. J. Physiol.*, **257**:E158-69.

Radziuk, J. and Inculet, R., 1983, The effects of ingested and intravenous glucose on forearm uptake of glucose and glucogenic substrate in normal man, *Diabetes*, **32**:977-81.

Radziuk, J., McDonald, T.J., Rubenstein, D., and Dupre, J., 1978, Initial splanchnic extraction of ingested glucose in normal man, *Metabolism*, **27**:657-69.

Ruderman, N.B., 1975, Muscle amino acid metabolism and gluconeogensis, *Ann Rev Med*, **26**:245-58.

Ryan, C., Ferguson, K. and Radziuk, J., 1993, Glucose dynamics and gluconeogenesis during and following prolonged swimming in rats, *J. Appl. Physiol.*, in press.

Scow, R.O. and Cornfield, J., 1954, Quantitative relations between the oral and intravenous glucose tolerance curves, *Am. J. Physiol.*, **179**:435-38.

Shapiro, A. and Shapiro, B., 1979, Role of the liver in intestinal glucose absorption, *Biochem. Biophys. Acta.*, **586**:123-27.

Shikama, H. and Ui, M., 1978, Glucose load diverts hepatic glucogenic product from glucose to glycogen in vivo, *Am. J. Physiol.*, **235**:E354-60.

Shulman, G.I. and Rossetti, L., 1989, Influence of the route of glucose administration on hepatic glycogen repletion, *Am. J. Physiol.*, **257**:E681-5.

Shulman, G.I., Rothman, D.L., Smith, D., Johnson, C.M., Blair, J.B., Shulman, R.G., DeFronzo, R.A., 1985, Mechanism of liver glycogen repletion in vivo by nuclear magnetic resonance spectroscopy, *J. Clin. Invest.*, **76**:1229-36.

Siesjo, B.M., 1978, "Brain Energy Metabolism," John Wiley & Sons, New York.

Stalmans, W., Bollen, M., Mrumbi, L., 1987, Control of glycogen synthesis in health and disease, *Diabetes/Metab. Rev.*, **3**:126-61.

Stalmans, W. and van de Werve, G., 1981, Regulation of glycogen metabolism by insulin, *in*:"Short-Term Regulation of Liver Metabolism", L. Hue and G. van de Werve, eds., Eisevier, Netherlands.

Sugden, M.C., Watts, D.I., Palmer, T.N., and Myles, D.D., 1983, Direction of carbon flux in starvation and after refeeding: in vitro and in vivo effects of 3-mercaptopicolinate, *Biochem. Intern.*, **7**(3):329-37.

Svedberg, J., Stromblad, G., Wirth, A., Smith, U., and Bjorntorp, P., 1991, Fatty acids in the portal vein of the rat regulate hepatic insulin clearance, **88**:2054-8.

Terrettaz, J., Assimacopoulos-Jeannet, F. and Jeanrenaud, B., 1986, Inhibition of hepatic glucose production by insulin in vivo in the rat: contribution of glycolysis, *Am. J. Physiol.*, **250**:E346-51.

Topper, Y.T., and Hastings A.B., 1949, A study of the Chemical origin of glycogen by use of ^{14}C-labelled carbon dioxide, acetate and pyruvate, *J. Biol. Chem.*, **179**:1255-64.

Tormo, M.A., Zubeldia, M.A.G., Montero, J.L., and Campillo, J.E., 1988, In vitro study on the contribution of the rat intestine-pancreas to glucose homeostasis, *Diabetologia*, **31**:916-21.

Unger, R.H. and Eisentraut, A.M., 1969, Enteroinsular axis, *Arch. Intern. Med.*, **123**:261-6.

van de Werve, G. and Hers, H-G., 1979, Mechanism of activation of glycogen phosphorylase by fructose in the liver. Stimulation of phosphorylase kinase related to the consumption of adenosine triphosphate, *Biochem. J.*, **178**:119-26.

van de Werve, G. and Jeanrenaud, B., 1984, Synthase activation is not a prerequisite for glycogen synthesis in the starved liver, *Am. J. Physiol.*, **247**:E271-5.

van de Werve, G., and Jeanrenaud, B., 1987, Liver glycogen metabolism: an overview, *Diabetes/Metabol Rev*, **3**:47-8.

Wajngot, A., Chandramouli, V., Schumann, W.C., Kumaran, K., Efendic, S., and Landau, B.R., 1989, Testing of the assumptions made in estimating the extentof futile cycling, *Am. J. Physiol.*, **256**:E668-75.

Whitton, P.D. and Hems, D.A., 1975, Glycogen synthesis in the perfused liver of streptozotocin-diabetic rats, *Biochem. J.*, **150**:153-65.

Wimhurst, J.M., Manchester, K.L., 1973, Induction and suppression of the key enzymes of glycolysis and gluconeogenesis in isolated perfused rat liver in response to glucose, fructose and lactate, *Biochem. J.*, **134**:143-56.

Wood, H.G., Lifson, N. and Lorer, V, 1945, The position of fixed carbon in glucose from rat liver glycogen, *J Biol Chem*, **159**:475-89.

Youn, J.H., Kasloc, H.R., and Bergman, R.N., 1987, Fructose effect to suppress hepatic glycogen degradation, *J. Biol. Chem.*, **262**:11470-77.

Young, A.A., Bogardus, C., Wolfe-Lopez, D., Mott, D.M., 1988, Muscle glycogen synthesis and disposition of infused glucose in humans with reduced rates of insulin-mediated carbohydrate storage, *Diabetes*, **37**:303-8.

Zhang, Z. and Radziuk, J., 1991, Effects of lactate on pathways of glycogen formation in the perfused rat liver, *Biochem. J.*, **280**:415-9.

Zhang, Z. and Radziuk, J., 1993, The coordinated regulation of hepatic glycogen formatio in the perfused rat liver by glucose and lactate, *Am. J. Physiol.*, in press.

GLUCONEOGENESIS IN TYPE 2 DIABETES

John E. Gerich and Nurjahan Nurjhan

[1] The Whittier Institute for
Diabetes and Endocrinology
9894 Genesee Avenue
LaJolla, CA 92037

INTRODUCTION

There is now considerable evidence that increased hepatic glucose output rather than reduced peripheral glucose uptake is the primary factor responsible for both fasting and postprandial hyperglycemia in type 2 diabetes[1,2]. It is, therefore, appropriate to consider the mechanisms that may be involved in permitting and promoting this excessive hepatic output of glucose.

The liver releases glucose into the circulation via two processes: breakdown of its stored glycogen (glycogenolysis) and the formation of new glucose molecules (gluconeogenesis) from nonglucose precursors. Until recently, little was known regarding the relative contribution of these two processes to the increased hepatic output of glucose found in type 2 diabetes. Early studies using the hepatic catheter technique[3] or conventional isotope approaches[4-7] yielded inconclusive or conflicting assessment of whether gluconeogenesis was increased in type 2 diabetes. To a large extent, this was probably due to limitations of the techniques used and the inclusion of too few or poorly matched subjects.

Fasting Hyperglycemia: Evidence for Predominant Role of Gluconeogenesis

Recently, consistent results have been obtained which indicate that gluconeogenesis is increased in type 2 diabetes and suggest that it may be the predominant process. In 1988, Zawadski et al.[8] reported increased Cori Cycle activity in obese Pima Indians with type 2 diabetes. This supported the previously suggestive but inconclusive results of DeMeutter and Shreeve,[5] Reichard et al.[6] and Waterhouse and Keilson[7] whose use of small numbers of subjects prevented statistical inferences. Although the studies of Zawadski et al.[8] provided definite evidence that gluconeogenesis was increased in type 2 diabetes, they did not, however, define its relative importance.

Two major problems limiting the quantitative assessment of gluconeogenesis are the underestimation due to Krebs Cycle carbon exchange and the lack of a method to measure overall gluconeogenesis as opposed to incorporation of individual precursors into glucose. In 1985, Katz[9] proposed an approach using [2-^{14}C] acetate which theoretically could overcome these problems. Consoli et al.[10] used this approach to compare overall rates of gluconeogenesis in subjects with type 2 diabetes and age-weight matched nondiabetic volunteers. As shown in Figure 1, they found not only that was gluconeogenesis increased in type 2 diabetes but that it could essentially account for all of the increased hepatic glucose output. Unfortunately, the validity of this approach as a quantitative measure of overall gluconeogenesis has been questioned[11,12]. Nevertheless, other studies, not relying on this approach,[14] have clearly demonstrated that gluconeogenesis from lactate,[13,14] alanine,[13] and glycerol[15,16] is increased in type 2 diabetes.

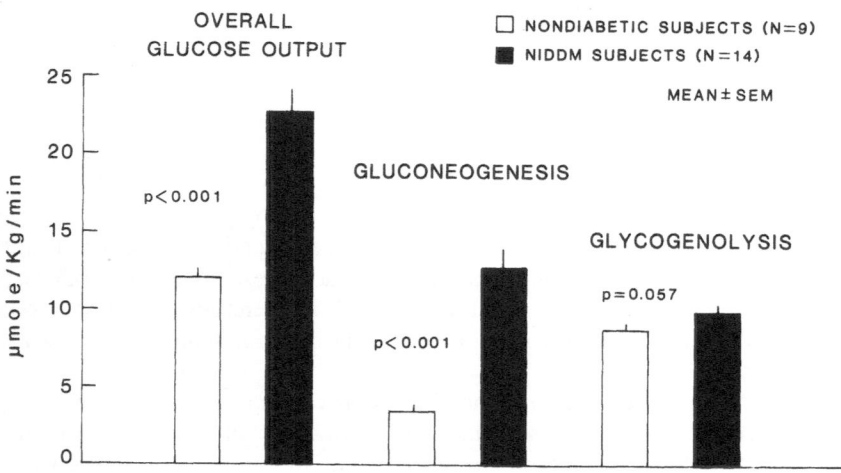

Figure 1. Rates of overall hepatic glucose output, gluconeogenesis, and glycogenolysis in normal volunteers and subjects with type 2 diabetes. Data from Consoli et al.[10] with permission.

Thus, it seems clear that gluconeogenesis is increased in type 2 diabetes but how much of the increased hepatic output can be accounted for by the increased gluconeogenesis remains to be firmly established. The recent preliminary results of Korytkowski et al.[14] support the original view of Consoli et al.[10] that gluconeogenesis is the predominant factor.

Korytkowski et al.[14] administered [6-^{3}H] glucose orally to subjects with type 2 diabetes whose hepatic glycogen stores had been depleted by fasting; after rendering the subjects euglycemic by an overnight infusion of insulin, the investigators infused [^{14}C] lactate along with [2-^{3}H] glucose and observed changes in the rates of appearance of [^{3}H] glucose and [^{14}C] glucose in plasma after stopping the insulin infusion and allowing the subjects to develop fasting hyperglycemia. They found that hepatic glucose output increased by almost 50 percent, and although the rate of appearance of [^{3}H] glucose (an index of glycogenolysis) did not increase, the rate of appearance of [^{14}C] glucose (an index of gluconeogenesis) increased nearly 3-fold. These observations provide evidence that increased hepatic glucose output in postabsorptive individuals with type 2 diabetes is largely if not exclusively due to increased gluconeogenesis.

Source of Gluconeogenic Precursors

Normally, the key gluconeogenic precursors in man are lactate,[17] amino acids, such as alanine[18] and glutamine,[19] and glycerol[15]. Approximate estimates of the contributions of these precursors to gluconeogenesis are: lactate 55-70%, alanine 10-15%, glutamine 15-20% and glycerol 5-10%. However, since a substantial proportion of lactate and alanine are derived from plasma glucose, glutamine and glycerol may be the key precursors for adding new carbons to the glucose pool[15]. Most of the circulating lactate and amino acids in the postabsorptive state are thought to originate from muscle while plasma glycerol is predominantly the result of lipolysis in adipose tissue[20]. In type 2 diabetes, the rates of

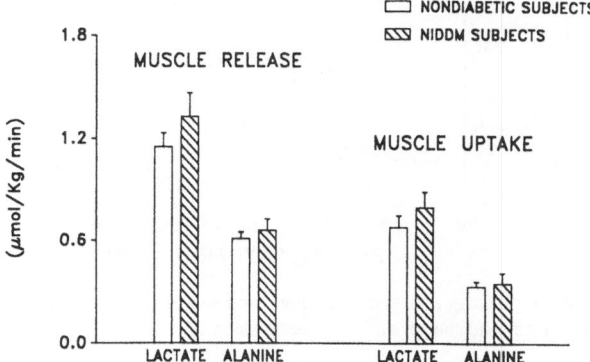

Figure 2. Muscle release and update of lactate and alanine in normal volunteers and subjects with type 2 diabetes estimated from forearm balance measurements. Data from Consoli et al.[13] with permission.

appearance in plasma of lactate, alanine and glycerol have been reported to be increased[13,15,16]. While the source of glycerol still appears to be adipose tissue, measurement of rates of release of lactate and alanine from forearm muscle indicate no increase in patients with type 2 diabetes (Figure 2) suggesting that tissues other than muscle may be responsible for the increased availability of these precursors. These tissues remain to be identified but adipose tissue, skin and the gastrointestinal are prime candidates.

Roles of Substrate Supply and Hepatic Efficiency

The increased gluconeogenesis from lactate, alanine, and glycerol found in type 2 diabetes could be due to either increased availability of precursors and increased hepatic efficiency in their uptake and conversion of glucose. Although the rates of delivery into plasma are increased in type 2 diabetes, so is the proportion of their turnover that is used for gluconeogenesis, at least as far as glycerol[15] and lactate[13] are concerned. Studies of gluconeogenesis from glycerol suggest that increased hepatic efficiency is the prominent mechanism[15]. As shown in Figure 3, although plasma glycerol concentrations are increased in type 2 diabetes, when normal volunteers are infused with glycerol to raise their plasma levels to those seen in type 2 diabetes, rates of gluconeogenesis are still far below those seen in type 2 diabetes. Thus for a given degree of substrate availability, there is greater incorporation of the substrate into glucose in individuals with type 2 diabetes. It is likely that accelerated intrahepatic conversion is also more important than increased substrate availability[13,15].

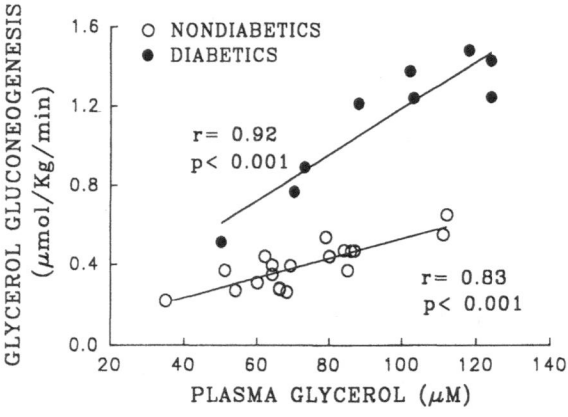

Figure 3. Relation between glycerol and glycerol conversion to glucose in normal volunteers and subjects with type 2 diabetes data from Nurjhan et al.[15] with permission.

Gluconeogenesis and Postprandial Hyperglycemia

So far, our discussion has centered on the postabsorptive state. After meal ingestion, the failure to suppress release of endogenous glucose is the prime factor responsible for postprandial hyperglycemia in type 2 diabetes[2]. Although it was originally thought that gluconeogenesis virtually ceased postprandially, there is now considerable evidence that this pathway remains active and is a major route for hepatic glycogen repletion[21]. Thus the failure to reduce appropriately endogenous hepatic glucose output after meal ingestion in type 2 diabetes could be due to failure to suppress glycogenolysis, failure to suppress gluconeogenesis or failure to direct the glucose produced via gluconeogenesis into glycogen appropriately. Recent preliminary studies from our laboratory using glycerol to trace postprandial gluconeogenesis[22] indicate that there is greater than normal gluconeogenesis from glycerol after meal ingestion in type 2 diabetes. Indeed with certain assumptions (e.g. glycerol gluconeogenesis represents 10% of overall gluconeogenesis), it could be estimated that gluconeogenesis might account for as much as 70% of the increased postprandial hepatic glucose output. Thus, it appears that abnormal gluconeogenesis plays a role in both postprandial hyperglycemia and postabsorptive hyperglycemia in type 2 diabetes.

Mechanisms for Increased Gluconeogenesis

There are several potential factors which may be responsible for or at least contribute to the increased hepatic conversions of gluconeogenic precursors into glucose. These include hepatic insulin resistance,[23] diminished insulin secretion, particularly in the postprandial state,[24] hyperglucogonemia,[24,25,26] increased hepatic fatty acid oxidation[27]). All of these factors are present in type 2 diabetes and have been shown either in vivo or in vitro to augment hepatic glucose output.

Summary and Future Directions

At the present time, it appears that increased hepatic glucose output is the primary factor responsible for both fasting and postprandial hyperglycemia in type 2 diabetes. Moreover, it seems equally well established that gluconeogenesis is increased in type 2 diabetes. It is likely though not yet firmly established that this increased gluconeogenesis is largely responsible for the excessive hepatic glucose output. Thus one critical area for future studies is to unequivocally determine whether or not gluconeogenesis is the primary process responsible for the increased hepatic glucose output since this could have important implications regarding development of new therapeutic modalities.

Whether or not gluconeogenesis is the major process, it must contribute to the increased hepatic glucose output found in type 2 diabetes. Therefore, it is of interest for future studies to determine the relative importance of the various factors responsible for the increased gluconeogenesis. Such an effort will require development of improved ways to quantitate gluconeogenesis in vivo and new clinical tools to specifically manipulate factors which influence gluconeogenesis such as glucagon antagonists, hepatic fatty acid and oxidation inhibitors, and blockers of metabolic pathways in liver.

ACKNOWLEDGMENT

We thank Laura Brinker for superb editorial assistance. The studies reported on from our laboratory were supported in part by National Institutes of Health, DK20411 and University of Pittsburgh General Clinical Research Center, 5M01 RR00056.

REFERENCES

1. J.E. Gerich, Is muscle the major site of insulin resistance in type 2 (noninsulin-dependent) diabetes mellitus?, *Diabetologia*. 34:607-610 (1991).
2. A. Mitrakou, D. Kelley, T. Veneman, T. Jenssen, T. Pangburn, J. Reilly, and J. Gerich, Contribution of abnormal muscle and liver glucose metabolism to postprandial hyperglycemia in noninsulin-dependent diabetes mellitus, *Diabetes*. 39:1381-1390 (1990).
3. J. Wahren, P. Felig, E. Cerasi, and R. Luft, Splanchnic and peripheral glucose and amino acid metabolism in diabetes mellitus, *J Clin Invest*, 51:1870-1878 (1972).
4. R.H. Chochinov, H.F. Bowen, and J.A. Moorhouse, Circulating alanine disposal in diabetes mellitus, *Diabetes*. 27:420-26 (1978).

5. R.C. DeMeutter and W.W. eeve, Conversion of $_{DL}$-lactate-2-^{14}C or pyruvate-2-^{14}C to blood glucose in humans: effect of diabetes, insulin, tolbutamide and glucose load, *J Clin Invest.* 42:523-33 (1963).

6. G.A. Reichard, F.N. Moury, N.J. Hochella, A.L. Patterson and S. Weinhouse, Quantitative estimation of the Cori Cycle in humans, *J Biochem.* 238:495-501 (1963).

7. C. Waterhouse and J. Keilson, The contribution of glucose to alanine metabolism in man, *J Lab Clin Med.* 92:803-812 (1978).

8. J. Zawadski, R. Wolfe, S. Mott, S. Lillioja, B. Howard and C. Bogardus, Increased rate of Cori Cycle in obese subjects with NIDDM and effects of weight reduction, *Diabetes.* 37:154-159 (1988).

9. J. Katz, Determination of gluconeogenesis in vivo with [^{14}C]-labelled substrates, *Am J Physiol.* 248:R331-R339 (1985).

10. A. Consoli, N. Nurjhan, F. Capani, and J.E. Gerich, Predominant role of gluconeogenesis in increased hepatic glucose production in NIDDM, *Diabetes.* 38(5)550-557 (1989).

11. C. Desrosiers, F. David, M. Garneau and H. Brunengraber, Nonhomogeneous labeling of liver mitochondrial acetyle CoA, *J Biol Chem.* 266:1574-1578 (1991).

12. W. Schumann, I. Magnusson, V. Chandramouli, K. Kumaran, J. Wahren and B. Landau, Metabolism of [2-^{14}C] acetate and its use in assessing hepatic Krebs cycle activity and gluconeogenesis, *J Biol Chem.* 266:6985-6990 (1991).

13. A. Consoli, N. Nurjhan, J. Reilly, D. Bier, and J. Gerich, Mechanism of increased gluconeogenesis in noninsulin-dependent diabetes mellitus, *J Clin Invest.* 86:2038-2045 (1990).

14. M. Korytkowski, A. Consoli, W. Pimenta, and J. Gerich, Pathogenesis of fasting hyperglycemia in NIDDM, *Diabetes.* 41(1): 10(A)#40 (1992) (Abstract).

15. N. Nurjhan, A. Consoli, and J. Gerich, Increased lipolysis and its consequences on gluconeogenesis in noninsulin-dependent diabetes mellitus, *J Clin Invest.* 89:169-175 (1992).

16. A. Virkamaki, I. Puhakainen, N. Nurjhan, J. Gerich and H. Yki-Jarvinen, Measurement of lactate formation from glucose using [6-^{3}H] and [6-^{14}C] glucose in humans, *Am J Physiol.* 259:397-404 (1990).

17. R. Kreisberg, Glucose-lactate interrelations in man, *N Engl J Med.* 287:132-137 (1972).

18. P. Felig, The glucose-alanine cycle, *Metabolism.* 22:179-207 (1973).

19. A. Bucci, I. Toft, T. Jenssen, D. Bier, and N. Nurjhan, Glutamine metabolism an its contribution to glucose and alanine production in man, *Diabetes.* 41(1):68A;#249 (1992) (Abstract).

20. N. Nurjhan, F. Kennedy, A. Consoli, C. Martin, J. Miles and J. Gerich, Quantification of the glycolytic origin in plasma glycerol as an index of lipolysis in vivo, *Metabolism.* 37:371-377 (1988).

21. J. Katz and J. McGarry, The glucose paradox: is glucose a substrate for liver metabolism, *J Clin Invest.* 74:1901-1909 (1984).

22. N. Nurjahan, *Diabetologia.* 33:A1-A4 (1990).

23. P. Campbell, G. Bolli, and J. Gerich, Quantification of the relative impairment in actions of insulin on hepatic glucose production and peripheral glucose uptake in noninsulin-dependent diabetes mellitus, *Metabolism.* 37:15-22 (1988).

24. A. Mitrakou, D. Kelley, T. Veneman, T. Pangburn, J. Reilly, and J. Gerich, Role of reduced suppression of hepatic glucose output and diminished early insulin release in impaired glucose tolerance, *N Engl J Med.* 326:22-29 (1992).

25. A.R. Baron, L. Schaeffer, P. Shragg and O.G. Kolterman, Role of hyperglucogenesis in maintenance of increased rates of hepatic glucose output in type 2 diabetes, *Diabetes.* 36:274-83 (1987).

26. R. Unger and L. Orci, Physiology and pathophysiology of glucagon, *Physiol Rev.* 56:779-826 (1976).

27. J. Williamson, R. Kreisberg, and P. Felta, Mechanism for the stimulation of gluconeogenesis by fatty acids in perfused rat liver, *Proc Nat'l Acad Sci.* 56:247-254 (1966).

REGULATION OF ADIPOSE TISSUE LIPOLYSIS, IMPORTANCE FOR THE METABOLIC SYNDROME

Peter Arner

Karolinska Institute
Department of Medicine
Huddinge Hospital
Huddinge, S-141 86
Sweden

INTRODUCTION

Adipose tissue plays a key role in the regulation of the energy balance. Energy rich free fatty acids are continuously stored as triglycerides in fat cells through esterification and released from adipose tissue through hydrolysis (lipolysis) of triglycerides in adipocytes. Triglycerides in fat cells and blood can be exchanged through lipoprotein lipase in fat cells, which breaks down triglyceride rich lipoproteins (mainly of the very low density type) into free fatty acids and glycerol. Free fatty acids can then be taken up by the fat cells and be esterified to triglycerides, then they are released again from the fat cells through lipolysis. Fatty acids leaving adipose tissue are bound to albumin and thereafter transported in blood to the liver, where they are used as substrate for esterification to triglycerides and incorporation into lipoproteins. Small changes in the turnover rate of free fatty acids in adipose tissue may in the long-term cause marked alterations of triglycerides in plasma and in adipose tissue leading to hypertriglyceridemia and/or obesity.

Free fatty acids may also cause disturbances in carbohydrate metabolism through the so-called Randle's cycle. There seems to be a competition between fatty acids and glucose as energy substrates in skeletal muscle, favouring the substrate that is in excess. Thus, high levels of circulating free fatty acid may impair peripheral glucose uptake leading to glucose intolerance.

New Concepts in the Pathogenesis of NIDDM, Edited by
C. G. Östenson *et al.*, Plenum Press, New York, 1993

It is increasingly evident that some metabolic and cardiovascular diseases form a syndrome, which is unified by hyperinsulinemia and insulin resistance[1]. Hypertension, type II diabetes, hypertriglyceridemia or other lipid abnormalities, atherosclerotic cardiovascular diseases and abdominal obesity are parts of this syndrome. It is obvious from the discussion above that disturbances in adipose tissue metabolism may play an important role in the so-called metabolic (or insulin resistance) syndrome.

Adipose tissue metabolism has been investigated intensely for almost 40 years. Many modern sophisticated techniques in analytical chemistry, molecular biology, physiology and radiology can be directly applied to adipose tissue metabolism. However, it is still rather difficult to study the synthesis of triglycerides in adipose tissue. This is mainly due to that most of the enzymes involved in the synthesis pathway are not well defined. In addition, it is not possible to directly quantitate free fatty acid esterification to triglycerides in fat cells. Instead, indirect radioactive methods have to be used. In contrast, the lipolytic cascade starting with hormone receptor binding and ending with acceleration or retardation of triglyceride hydrolysis is known in some detail. The processes is easily quantified by measuring the end products (glycerol or free fatty acids) and many of the regulatory enzymes and receptors are purified, sequenced and cloned.

This review will focus on the regulation of lipolysis in human fat cells. The possible role of adipose tissue lipolysis for the metabolic syndrome is discussed.

REGULATION OF LIPOLYSIS IN MAN

The human white fat cell is unique in many ways as regards lipolysis regulation. The only hormones with a pronounced lipolytic activity are catecholamines in adult man and TSH during the infancy period[2]. Parathyroid hormones, cholecystokinin and cortisol have only a weak lipolytic action[3-5]. Other hormones such as glucagon, secretine, vasopressine and ACTH, which are potent lipolytic agents in other species, are uneffective in human fat cells. On the other hand, several antilipolytic hormones and parahormones can inhibit lipolysis in human fat cells through distinct receptors. This includes insulin[6], adenosine[7], prostaglandin[8] and insulin-like growth factors[9].

As reviewed[10] catecholamines can stimulate lipolysis in human fat cells through $beta_1$- and $beta_2$-adrenoceptors and inhibit lipolysis through $alpha_2$-receptors. The latter receptor is frequently lacking or poorly expressed in other species. The recently cloned $beta_3$-adrenoceptor is also expressed in human fat cells[11] but its importance in relation to the other beta-adrenoceptor subtypes is unclear at present.

REGIONAL DIFFERENCES IN LIPOLYSIS

It has been known since long from in vitro studies that the lipolytic activity varies between the adipose depots in man. This potentially important observation has been subjected to intensive research and has recently been reviewed in detail[12].

The lipolytic activity in different fat depots is visceral > abdominal subcutaneous>gluteal-femoral subcutaneous adipose tissue. The mechanism behind these variations have been partly elucidated and involves the two major lipolysis regulating hormones insulin and catecholamines. The antilipolytic action of insulin is omental<subcutaneous adipose tissue which can be explained by differences in receptor affinity and variations in the ill-defined post-receptor pathways of insulin action. The lipolytic action of catecholamines is visceral>abdominal subcutaneous>gluteal - femoral subcutaneous adipose tissue. These differences are largely due to variations in beta-receptor number following the same rank order in the different depots as the lipolytic effect of catecholamines, although variations in alpha$_2$-receptor expression and affinity may also play a role.

Thus, regional variations in the antilipolytic effect of insulin and in the lipolytic effect of catecholamines may explain why the lipolytic activity is higher in visceral than in subcutaneous adipose tissue. In this respect it is of importance to note that regional variations in lipolysis have also been demonstrated in vivo with independent techniques such as microdialysis[13] and radioisotope turnover[14]. Visceral adipose tissue is believed to play a major role for metabolic abnormalities associated with the metabolic syndrome[15]. Free fatty acids from visceral adipose tissue are in direct contact with the liver through the portal system. A high influx of free fatty acids to the liver may, first, cause glucose intolerance through the Randle's cycle and, second, cause increased very low density lipoprotein production by the liver since free fatty acids are major substrates for triglycerides. Furthermore, free fatty acids may directly alter hepatic insulin action. They can inhibit clearance of insulin by the liver[16] and also cause gluteal insulin resistance due to interactions with insulin receptors in the hepatocytes[17].

It is well documented that the body distribution differs between the sexes. The high rate of lipolysis in central as compared to peripheral adipose tissue may participate in the normal female fat distribution and can play a role for development of the less harmful female (i.e. peripheral) obesity. In addition, regional variations in lipolysis are more marked in women than in men. It is less easy to explain the more dangerous male type of central (abdominal) obesity on the basis of the regional variations in lipolysis. This is of importance for the metabolic syndrome, since abdominal adipose tissue may be a cornerstone for this condition. It has been demonstrated[18] in elderly men with type II diabetes that hyperinsulinemia and insulin resistance only occur in patients who have concomitant abdominal obesity. In elderly diabetic male subjects with normal body-weight insulin sensitivity is normal and the diabetes is caused by a marked insulin secretory defect. It is, however, possible that in certain individuals the rate of lipolysis in the central adipose depots is decreased favouring abdominal obesity. Evidence for this assumption has recently been presented showing blunted catecholamine-induced lipolysis in the abdominal region but normal hormone induced lipolysis in the gluteal-femoral areas of men with abdominal obesity[19]. The underlying mechanisms may be enhanced alpha$_2$-adrenoceptor responsiveness in the abdominal adipocytes.

In summary, a lot of evidence favouring a role of lipolysis for the development of regional obesity is present. In particular, the data can explain normal female fat distribution and peripheral obesity. Mobilization of free fatty acids from the visceral fat cells may play a direct pathophysiological role in the metabolic syndrome.

LIPOLYSIS IN OBESITY

It is an attractive hypothesis that a lipolysis defect could be involved in the development of obesity in certain individuals. As reviewed in detail[20] the regulation of lipolysis in obesity is not well defined. This may in part be due to difficulties in finding a reliable denominator for the lipolytic rate and to establish optimal conditions to study this rate.

When the rate of lipid mobilization from adipose tissue is measured in vivo, it is necessary to, first, consider the regional variations in lipolysis discussed above and, second, to relate the rate to the total fat mass (which is difficult to measure). Lipolysis in vitro, on the other hand, is largely dependent upon fat cell size, making it necessary to distinguish variations in lipolysis rate that merely are related to fat cells size from changes that are independent of cell size in obesity studies.

The most important methodological question is probably to define the optimal conditions for measuring the lipolytic rate in vivo or in vitro, because lipolysis in man is a very complex process largely dependent upon interactions between hormones and parahormones[10,20]. This is of particular importance for insulin, since the lipolytic rate induced by this hormone in human fat cells is controlled by the preventing catecholamine concentration[21]. Therefore, in order to determine the true antilipolytic action of insulin it is necessary to perform detailed dose-response experiments varying the concentrations of insulin as well as catecholamines. Such type of studies have not yet been published in the field of obesity. Consequently, increased, normal or decreased antilipolytic action of insulin has been observed in obesity as reviewed[20]. When catecholamine-induced lipolysis is concerned, it is probably less important to consider other hormones, since catecholamines are more powerful as lipolytic hormones than insulin is as an antilipolytic hormone in vivo and in vitro[21,22]. The findings with catecholamine-induced lipolysis in obesity are more uniform than those with insulin, showing a blunted catecholamine effect in vivo[23-25]. The mechanisms behind this resistance remain to be established. The findings with basal (resting) lipolysis are also consistent as reviewed[20]. The basal rate of lipolysis is increased in obesity in vivo as well as in vitro even if the increment in total fat mass, regional lipolysis and increased fat cell size are taken into consideration.

In conclusion, abnormalities in the lipolytic rate are described in obesity which can be of pathophysiological importance for the metabolic syndrome. The increase in basal lipolysis rate may elevate the circulating free fatty acid level and cause glucose intolerance, insulin resistance, impaired insulin clearance as well as hypertriglyceridemia according to mechanisms discussed above. The blunted catecholamine action can be a contributing factor to the development of obesity in certain individuals and may lead to resistance to slimming therapies. Whether the antilipolytic action of insulin is altered or not in obesity is still an open question.

LIPOLYSIS IN TYPE II DIABETES

It is well established that all forms of untreated or uncompensated diabetes are associated with acceleration of the lipolysis rate, probably reflecting the absolute or relative insulin deficiency. Whether or not lipolysis is altered in treated type II diabetics is a matter of question. The antilipolytic action of insulin has been found to be normal or even increased in vivo and in vitro after therapy with diet + sulphonylurea[26] and in vivo studies[27] suggest increased catecholamine-induced lipolysis in treated type II diabetics; the rate in vitro may, however, be dependent upon the degree of metabolic control[28]. It should be noted that, the insulin studies suffer from the same methodological problems as the obesity studies discussed above and in the catecholamines studies only the effects of maximum effective pharmacological doses have been investigated. It is necessary to perform dose-response experiments with physiological catecholamine concentrations in order to evaluate hormone action in a proper way. However, it appears that the lipolysis rate is accelerated in vitro in type II diabetics who are receiving antidiabetic therapy.

On the other hand, numerous of studies have shown increased circulating free fatty acid levels in treated type II diabetics as reviewed[29]. In addition, there is a direct correlation between degree of plasma free fatty acid concentrations and degree of fasting hyperglycemia[30]. These data suggest further that there is an increased lipolytic activity in vivo in treated type II diabetics, but this may be dependent upon the degree of metabolic control. Alternatively, the increase in circulating free fatty acids can be due to other factors besides lipolysis such as decreased re-esterification in adipose tissue or inhibited utilization by in peripheral tissue. The relative in vivo contributions of lipolysis, re-esterification and peripheral utilization of free fatty acids in treated type II diabetes are not well defined.

Anyhow, the increment in the circulating level of free fatty acid in type II diabetes is most likely of importance for the metabolic syndrome. A high substrate availability may be of importance for the development of hypertriglyceridemia[31,32] The increased free fatty acid levels may further deteriorate the glucose intolerance through the Randle's cycle and further impair insulin resistance, since treatment of type II diabetes with antilipolytic drugs improves insulin resistance and peripheral glucose utilization[33,34].

LIPOLYSIS IN HYPERTRIGLYCERIDEMIA

Surprisingly little is known about lipolysis regulation in hyperlipidemia and the studies are published almost 20 years ago, when there was a lack of sophisticated in vitro and in vivo methods. Insulin action on lipolysis is reported to be blunted and catecholamine action increased in endogenous hypertriglyceridemia in vitro[35]. The mechanisms behind the changes are not known. The basal lipolysis rate seems, however, to be normal in vitro in this condition[36]. However, the ability of human adipose tissue to esterify free fatty acids to triglycerides is impaired in hypertriglyceridemia[36-37].

CONCLUDING REMARKS

Adipose tissue lipolysis is a central process in lipid metabolism and seems to be disturbed in some conditions associated with the metabolic syndrome. The basal rate, occuring between meals, at rest and at night is increased in obesity; this condition is a corner stone in the metabolic syndrome. An acceleration of the lipolytic rate elevates the circulating free fatty acid levels, which in its turn may cause glucose intolerance, insulin resistance, decreased insulin clearance and hypertriglyceridemia. Catecholamine-induced lipolysis, operating in stressful situations such as physical work, mental awareness or distress and during cold is blunted in obesity, which may play a role for weight gain or resistance to slimming. The existence of regional variations in lipolytic activity can play a protective role in women, making them more prone to develop peripheral obesity, which is less frequently associated with the metabolic syndrome than abdominal obesity, the latter usually occuring in men. The link between visceral adipose tissue and the liver through the portal system makes the latter organ very sensitive to regional variations in lipolysis. The increased lipolytic activity in visceral fat cells may cause impaired function and clearance of insulin by the liver and enhance the hepatic esterification of fatty acids to triglycerides through increased substrate availability. The circulating free fatty acid level is increased in type II diabetic who are subjected to antidiabetic which at least partly is due to accelerated lipolytic rate. The increased availability of free fatty acids in type II diabetics may aggravate glucose intolerance and insulin resistance and also cause hypertriglyceridemia according to the same mechanisms that are operating in obesity.

REFERENCES

1. R.A. DeFronzo. Insulin resistance. A multifaceted syndrome responsible for   NIDDM, obesity, hypertension, dyslipidemia, and atherosclerotic cardiovascular   disease. Diabetes Care 14: 173 (1991).

2. C. Marcus, H. Ehrén, P. Bolme, P, Arner. Regulation of lipolysis during the neonatal period. J Clin Invest 82: 1793 (1988)

3. A. Taniguchi, K. Kataoka, T. Kono, F. Oseko, H. Okuda, I. Nagata, H. Imura. Parathyroid hormone-induced lipolysis in human adipose tissue. J. Lip. Res. 28: 490 (1987).

4. W.O. Richter, P. Schwandt. Cholecystokinin 1-21 stimulates lipolysis in human adipose tissue. Horm. Metabol. Res. 21: 216 (1989).

5. G.D. Divertie, M.D. Jensen, J.M. Miles. Stimulation of lipolysis in humans by physiological hypercortisolemia. Diabetes 40: 1228 (1991).

6. M. Amatruda, J. Livingston. D. Lockwood. Insulin receptor: role in the resistance of human obesity to insulin. Science 188: 264 (1975).

7. A. Green, S. Swenson, J.L. Johnson, M. Partin. Characterization of human adipocyte adenosine receptors. Biochem. Biophys. Res. Commun. 163: 137 (1989).

8. R. Richelsen, E.F. Eriksen, H. Beck-Nielsen, O. Pedersen. Prostaglandin E_2 receptor binding and action in human fat cells. J. Clin. Endocrinol. Metab. 59: 7 (1983).

9. P.A. Kern, M.E, Svoboda, R.H. Eckel, J.J. van Wyk. Insulinlike growth factor action and production in adipocytes and endothelial cells from human adipose tissue. Diabetes 38: 710 (1989).

10. P. Arner. Adrenergic receptor function in fat cells. Am. J. Clin. Nutr. 55: 228S (1992).

11. S. Krief, F. Lönnqvist, S. Raimbault, B. Baude, P. Arner, D. Strosberg, D. Ricquier, L.J. Emorine. Tissue distribution of beta3-adrenergic receptor mRNA in man. J. Clin. Invest. in press (1993).

12. R.L. Leibel, N.K. Edens, S.K. Fried. Physiologic basis for the control of body fat distribution in humans. Annu. Rev. Nutr. 9: 417 (1989).

13. P. Arner, E. Kriegholm, P. Engfeldt, J. Bolinder. Adrenergic regulation of lipolysis in situ at rest an during exercise. J. Clin. Invest. 85: 893 (1990).

14. M.D. Jensen. Regulation of forearm lipolysis in different types of obesity. J. Clin. Invest. 87: 187 (1991).

15. P. Björntorp. Metabolic implications of body fat distribution. Diabetes Care 14: 1132 (1991).

16. J. Svedberg, G, Strömblad, A. Wirth, U. Smith, P. Björntorp. Fatty acids in the partal vein of the rat regulate hepatic insulin clearance. J. Clin. Invest. 88: 2054 (1991).

17. J. Svedberg, P. Björntorp, U. Smith, P. Lönnroth. Free-fatty acid inhibition of insulin binding, degradation and action in isolated rat hepatocytes. Diabetes 39: 570 (1990).

18. P. Arner, T. Pollare, H. Lithell. Different aetiologies of type 2 (non-insulin- dependent) diabetes mellitus in obese and non-obese subjects. Diabetologia 34: 483 (1991).

19. P. Maurigège, J.P. Després, D. Prud'homme, M.C. Pouliot, M. Marcotte, A. Tremblay, C. Bouchard. Regional variation in adipose tissue lipolysis in lean and obese men. J. Lipid Rés. 32: 1625 (1991).

20. P. Arner. Control of lipolysis and its relevance to development of obesity in man. Diabetes/Metabolism Rev 4: 507 (1988).

21. P. Engfeldt, J. Hellmér, H. Wahrenberg, P. Arner. Effects of insulin on adreno- ceptor binding and the rate of catecholamine-induced lipolysis in isolated human fat cells. J. Biol. Chem. 263: 15553 (1988).

22. E. Hagström-Toft, P. Arner, U. Johansson, L.S. Eriksson, U. Ungerstedt, J. Bolinder. Effect of insulin on human adipose tissue metabolism in situ. Interactions with beta-adrenoceptors. Diabetologia 35: 664 (1992).

23. R.R. Wolfe, E.J. Peters, S. Klein, O.B. Holland, J. Rosenblatt, J.r.H. Gary. Effect of short-term fasting on lipolytic responsiveness in normal and obese human subjects. Am. J. Physiol. 252: E189 (1987).

24. M.D. Jensen, M.W. Haymond, R.A Rizza, P.E. Cryer, J.M. Miles. Influence of body fat distribution on free fatty acid metabolism in obesity. J. Clin. Invest. 83: 1168 (1989).

25. A.A. Connacher, W.M. Bennet, R.T. et al. Effect of adrenaline infusion on fatty acid and glucose turnover in lean and obese human subjects in the post-absorptive and fed state. Clin. Sci. 81: 635 (1991.

26. J. Bolinder, P. Arner. Antilipolytic effect of insulin in non-insulin-dependent

diabetes mellitus after conventional treatment with diet and sulfonylurea. Acta Med. Scand. 224: 451 (1988).

27. S. Nordlander, J. Östman, E. Cerasi, R. Luft, L.G. Ekelund. Occurrence of diabetic type of plasma FFA and glycerol response to physical exercise in prediabetic subjects. Acta Med. Scand. 193: 9 (1973).

28. P. Arner, P. Engfeldt, J. Östman. Blood glucose control and lipolysis in diabetes mellitus. Acta Med. Scand. 208: 297 (1980).

29. G.M. Reaven. Role of insulin resistance in human disease. Diabetes 37: 1595 (1988).

30. C. Bogardus, S. Lillioja, B.V. Howard, G.M. Reaven, D. Mott. Relationship between insulin secretion, insulin action and fasting plasma glucose concentration in nondiabetic and non-insulin-dependent diabetic subjects. J. Clin. Invest. 74: 1238 (1984).

31. M. Greenfield, O. Kolterman, J. Olefsky, G.M. Reaven. Mechanism of hyper-triglyceridaemia in patients with fasting hyperglycaemia. Diabetologia 18: 441 (1980).

32. G.M. Reaven, M.S. Greenfield. Diabetic hypertriglyceridaemia: evidence for three clinical syndromes. Diabetes 30: 66 (1981).

33. G.M. Reaven, H. Chang. H. Ho, C.Y. Jeng, B. Hoffman. Lowering of plasma glucose in diabetic rats by antilipolytic agents. Am. J. Physiol. 254: E23 (1988).

34. A. Vaag, P. Skött, P. Damsbo, M-A. Gall, E.A. Richter, H. Beck-Nielsen. Effect of the antilipolytic nicotinic acid analogue acipimox on whole-body and skeletal muscle glucose metabolism in patients with non-insulin-dependent diabetes mellitus. J. Clin. Invest. 88: 1282 (1991).

35. B. Larsson, P. Björntorp. J. Holm. T. Schersten, L. Sjöström, U. Smith. Adipocyte metabolism in endogenous hypertriglyceridemia. Metabolism 24: 1375 (1975).

36. L.A. Carlson, G. Walldius. Fatty acid incorporation into human adipose tissue in hypertriglyceridaemia. Eur. J. Clin. Invest. 6: 195 (1976).

37. P. Rubba. Fractional fatty acid incorporation into human adipose tissue (FIAT) in hypertriglyceridemia. Atherosclerosis 29: 39 (1978).

CELLULAR AND MOLECULAR FACTORS IN ADIPOSE TISSUE GROWTH AND OBESITY

Daniel A.K. Roncari and Bradford S. Hamilton

Department of Medicine
Sunnybrook Health Science Centre - University of Toronto
Toronto, Ontario M4N 3M5

INTRODUCTION

One of the major challenges in biology and medicine is the unravelling of the fundamental abnormality imparting vulnerability to the development of obesity. The experimental approaches adopted to date have simply not provided the solution. It is pertinent, particularly in the context of this meeting, that obesity frequently triggers or aggravates non-insulin-dependent diabetes. Thus, prevention or effective treatment of obesity would also have a prophylactic or ameliorative effect on type II diabetes. This paper will review and provide new data about the proposed functions of heparin-binding (fibroblast) growth factors in relation to adipose cell dynamics and processes opposing adipose differentiation.

PARACRINE/AUTOCRINE FACTORS AND EXCESSIVE PROLIFERATION OF PREADIPOCYTES

Pursuant to our original finding [1] indicating that omental preadipocytes from massively obese persons (body mass index >37 $kg \cdot m^{-2}$) replicated to a significantly greater degree than cells from lean subjects, a search has been carried out for the responsible mitogenic principles. Using a cDNA probe for basic fibroblast growth factor (bFGF, heparin-binding growth factor 2), we have discovered by Northern blot analysis that omental preadipocytes express the gene related to this protein.[2] These results were confirmed by reverse transcription-polymerase chain reaction. While there was considerable variation between cell strains from different individuals, on the average, there was significantly greater expression by preadipocytes from appreciably obese persons (correlation with body mass index at r = 0.71). As reported, preadipocytes from the massively obese release into the medium significant quantities of compounds mitogenic on preadipocytes. With antibodies directed against different segments of bFGF, Western blot analysis indicated the presence of two reactive proteins, M_r 66,000 and 32,000. These were not related to each other or to bFGF (M_r 18,000) by association through non-covalent bonds or disulfide bonds, since the same results were obtained in the presence or absence of sodium dodecyl sulfate (SDS) or a sulfhydryl reducing agent. The mitogenic activity of these proteins was retained after binding to and elution from heparin-agarose columns, gel filtration chromatography, and SDS polyacrylamide gel electrophoresis (Fig. 1). In view of the known structure of "classic" bFGF, the adipose-derived mitogenic proteins must be different, but related. As was the case for gene expression, superimposed on a background of prominent interindividual variation, there was good correlation between the production of the adipose-derived mitogenic proteins and body mass index (e.g. r = 0.72 for M_r 66,000 protein).

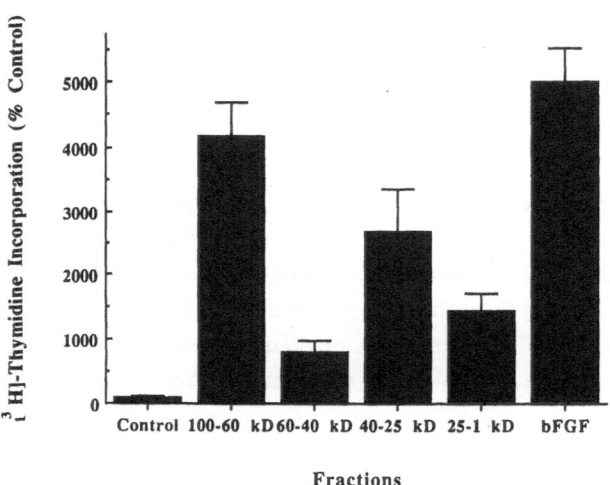

Figure 1. Mitogenic activity of conditioned medium proteins and recombinant bFGF electroeluted from SDS-PAGE. Conditioned medium and bFGF (100 ng) were run on 12.5% slab gels and the proteins were eluted and concentrated from slices of the gel corresponding to different molecular masses. The eluted proteins were tested for mitogenic activity. The results are expressed as percent above control (mean ± S.E.M.), and are representative of 2 experiments.

We propose that the heparin-binding growth factors related to bFGF, produced by preadipocytes, are major determinants of the clonal adipose cell composition, in different fat depots, and thus the normal regional variation. Further, the exaggerated production of these proteins in substantive obesity plays a major role in establishing the adipose cell hyper-cellularity characteristic of massive obesity. As well, inordinate regional production of these factors might contribute to abdominal obesity, as well as other localized forms of adiposity.

INHIBITION OF ADIPOSE DIFFERENTIATION

Basic fibroblast growth factor, and probably the larger related proteins produced by adipose cells, not only stimulate the replication of preadipocytes, but also potently inhibit their differentiation.[3] These dual functions act in concert to maintain complements of undifferentiated or partially differentiated preadipocytes, conforming with our view that the heparin-binding growth factors dictate the pools of these cells. Then, in the case of adipose tissue, specific hormones, notably glucocorticoid and insulin, would trigger and promote the specific process of adipose differentiation.[3]

ADIPOSE DIFFERENTIATION AND TRIGLYCERIDE ACCRETION

No matter how extensive their pools, preadipocytes cannot contribute significantly to adipose expansion in obesity since these cells do not contain appreciable quantities of triglyceride. Mass can only be affected by preadipocyte differentiation and triglyceride accretion. For massive obesity, we have corroborative evidence not only for excessive preadipocyte proliferation, but also augmented differentiation. Indeed, in contrast to preadipocytes from lean persons, some clones from the massively obese differentiate spontaneously (without the addition of specific inducers and promoters), and when fused to certain epithelial cells reveal prominent differentiation which persists in successive subcultures.[4] We also investigated, as an example, the promoter/enhancer regions for the gene encoding the human adipocyte lipid-binding protein (aP2).[5] In contrast to fibroblasts and HeLa cells, human preadipocytes had specific positive and negative regulators. In addition, our early evidence indicates greater activity of these regulators of aP2 promoter/enhancer regions in preadipocytes derived from massively obese persons.

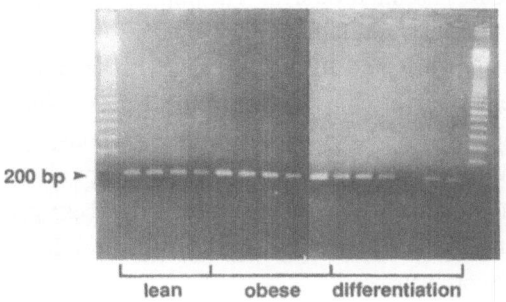

Figure 2. Relative levels of basic fibroblast growth factor mRNA in human omental preadipocytes and differentiated preadipocytes. Total RNA isolated from the preadipocytes was reverse-transcribed and amplified using polymerase chain reaction (30 cycles; 97°C 1 min, 63°C 1 min, 72°C 3 min). The product was electrophoresed in a 1.5% agarose gel and stained with ethidium bromide.

During adipose differentiation induced in primary culture of human omental preadipocytes with dexamethasone and insulin, the expression of the gene for the heparin-binding growth factors related to bFGF decreases progressively, as we have established by Northern blot hybridization and reverse transcription-polymerase chain reaction (Fig. 2). Mitogenesis and inhibition of differentiation,[3] two properties of bFGF and its related proteins, are processes which prevent or retard the differentiation program. Thus, progressive decrease in the expression of the gene(s) for these factors which antagonize differentiation through two complementary influences, is at least synchronous with promotion of differentiation. We actually propose that the decreased expression of this gene facilitates or enables the progress of differentiation. We are currently exploring the possibility that decreased expression of the gene for the proteins related to bFGF occurs earlier or to a greater extent in preadipocytes from massively obese persons.

ADIPOSE DE-DIFFERENTIATION

We have confirmed definitively that mature human omental fat cells, isolated directly from fat tissue, undergo de-differentiation, as originally reported.[6] Omental adipose tissue obtained, after informed consent, from individuals undergoing elective abdominal surgery was gently digested with collagenase and the floating adipocytes were attached to inverted flasks to begin "de-differentiation". During reversion, the triglyceride-replete, rather immobile adipocytes lose the bulk of their triglycerides and become motile fusiform cells, increasingly similar to preadipocytes. Such reversion is reflected not only by the morphological changes, but also by congruent molecular genetic and biochemical alterations. For example, gene expression and enzyme activity of glycerophosphate dehydrogenase and lipoprotein lipase are decreased significantly within 7 days of de-differentiation. In contrast, and as might have been expected from the previous section, gene expression and production of heparin-binding growth factors rise during reversion, consonant with the regained capability of replication. Significantly, reverted adipose cells retain the "memory of their origin"; de-differentiated cells from massively obese subjects proliferate to a significantly greater extent than cells from lean individuals.[7] As illustrated in Fig. 3, moreover, gene expression for β-actin increases during reversion. The immense potential significance of the possibly slower renewed (during de-differentiation) expression in the case of cells from massively obese persons, will be discussed in the context of our proposed hypothalamic-neural efferent-cytoskeletal pathway, which we have coined the "N" pathway.

We have demonstrated that bFGF accelerates the process of adipose de-differentiation; it is most probable that the related heparin-binding growth factors produced by preadipocytes have the same effect. Notably, preadipocytes from massively obese persons are relatively resistant to this augmenting influence of bFGF, another (newly) discovered abnormality characteristic of appreciable corpulence.

SYNTHESIS

As summarized in Table 1, heparin-binding growth factors have a central role in adipose cell dynamics. Through three effects, they channel molecular pathways and processes toward the preadipocyte state. Indeed, these proteins stimulate the replication of preadipocytes, inhibit their differentiation, and augment their de-differentiation.

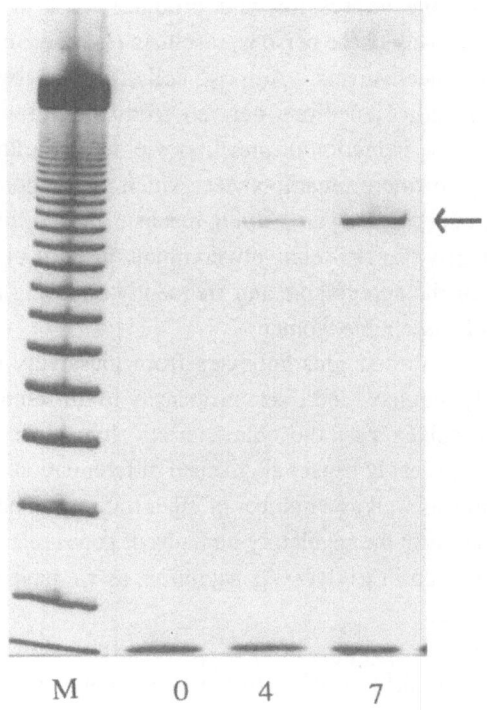

Figure 3. Relative levels of β-actin mRNA in reverting human omental adipocytes. Total RNA from "de-differentiating" adipocytes was reverse-transcribed and amplified using polymerase chain reaction (25 cycles; 94°C 1 min, 60°C 1 min, 72°C 1 min). The products were electrophoresed in a 6% polyacrylamide gel and silver stained. Predicted size of the β-actin product was 1125 bp. M: 100 bp DNA ladder, 0: initial conditions, 4: 4th day of culture, 7: 7th day of culture.

Table 1. Actions of Heparin-Binding Growth Factors in Adipose Cell Dynamics

Process	Action
Replication	Stimulation of preadipocyte mitogenesis
Differentiation	Inhibition of adipose differentiation
De-differentiation	Acceleration of mature adipocyte de-differentiation
Cell Motility	Stimulation of adipose cell motility

The exaggerated gene expression and production of bFGF-related proteins by preadipocytes from massively obese persons, results in excessive proliferation of these cells by paracrine/autocrine mechanisms. Adipose cells form estrogens, mainly estrone by aromatization of 3,17-androstenedione derived from the adrenal gland and the ovary. Estrogens contribute to the induction of preadipocyte-derived mitogenic proteins.[8] These stimulate the formation of more preadipocytes, which make more estrogens, leading to a greater number of preadipocytes, an amplifying mechanism. This effect of estrogens may be responsible, at least partly, for the relatively common onset of obesity in susceptible girls during puberty, and for the appreciable and frequently enduring aggravation of adiposity with each pregnancy of susceptible women.

As already stated, cultured preadipocytes from massively obese subjects not only proliferate inordinately, but have increased propensity to differentiation. Especially under conditions of nutritional excess, the characteristic hyperinsulinemia and accelerated glucocorticoid turnover probably trigger augmented differentiation and triglyceride accretion. Glucocorticoid receptors as well as a number of other transcriptional factors activate specific promoter/enhancer regions in the regulatory domains of genes related to receptors, enzymes, and other proteins involved in triglyceride accretion, as we have shown for the adipocyte lipid-binding protein, aP2.

Opposite events, i.e. suppression of gene expression, occur during adipose differentiation, even earlier than the described promotion of expression. Indeed, suppression of genes encoding cytoskeletal elements, e.g. β-actin and α-tubulin precedes the even relatively early expression related to lipid metabolism, e.g. lipoprotein lipase.[9] This temporal primacy of expression of genes related to cytoskeletal proteins may have profound significance in relation to the new concept which will be presented.

Mature fat cells can revert or "de-differentiate" to earlier forms, which increasingly assume the properties of preadipocytes. bFGF accelerates reversion, but mature adipocytes from massively obese persons reveal resistance to this influence of bFGF. We envision that under conditions of prolonged nutrient energy control and regular exercise, adipocytes de-differentiate. However, if derived from the the appreciably corpulent, they are genetically susceptible to renewed excessive proliferation and differentiation when triggered by nutrient energy overload and sedentariness. Thus, with each "relapse", a new wave of enlarged fat cells is formed. Consequently, cycles of compliance and then relapse result, like a staircase, in an ever increasing number of enlarged fat cells. In this context, it is better to have stable, mild obesity than the described undulating pattern, which also has harmful metabolic effects. As described, heparin-binding growth factors related to bFGF are probably the major determinants of the clonal cellular composition of adipose depots, and of the adipocyte hyperplasia characteristic of massive obesity. Thus, these growth factors participate in the molding of the varying sizes of different depots, both normal and expanded. During such molding, these factors work in concert with hormone receptors (e.g. insulin and adrenergic receptors), transducing machineries, lipid-related enzymes and their modifiers, as well as lipid carrier proteins to confer topologic specialization to adipose depots, and to mediate disproportionate expansion in certain forms of obesity. However, we do not believe that these local events in fat tissue, including excessive preadipocyte proliferation, differentiation, lipid assimilation, deposition and storage, constitute the fundamental, primary abnormality imparting vulnerability to the development of obesity.

CONCEPT EXPLAINING ENERGY OVERLOAD IN OBESITY

The concept which we developed a few years ago,[10] for which we have increasing experimental support, is entirely different from any previously proposed, in that it does not invoke a primary abnormality of lipid metabolism, in either white or brown fat cells. Rather, this concept targets the primary problem to the cytoskeleton. According to this proposal, individuals vary on a genetic basis in the degree of cytoskeletal activity of motile cells such as preadipocytes and fibroblasts. Further, for each subject, there is a particular reciprocal relationship between the energy consumed for cytoskeletal functions and the remaining energy available for chemical storage, mainly as triglyceride in adipocytes. At one extreme, at the level of "supermassive" obesity, a probable mutation in a cytoskeletal element, possibly in a gene encoding for a chemomechanical ATPase motor such as kinesin or dynein, dampens cytoskeletal activity, leaving available the highest quantity of energy for triglyceride storage.

In addition, the complete concept invokes a new pathway, the hypothalamic-efferent neural-cytoskeletal pathway, which we have named the "N" pathway.[11] The efferent neural outflow might be related to the efferent β-sympathetic pathway, but would innervate receptors specific for the cytoskeleton, and would thus be distinct, physiologically and pharmacologically.

By fortuitous coincidence, but of extreme relevance and significance, fibroblast growth factors turn out to be involved in the new concept. Recent studies indicate that the membrane transducing signalling mechanisms for a number of growth factors, probably including heparin-binding growth factors, is mediated by the *Rho* family of GTP-binding proteins, which stimulate GTPase activity.[12, 13, 14] Through this action, *rac* 1, a member of the *Rho* family, results in membrane ruffling, a process mediated by actin and its polymerization.[14] Similarly, *Rho* stimulates the formation of stress fibres (based on actin and its polymerization), which project from specific regions of the plasma membrane called focal adhesions, where clusters of integrin receptors bind to such extracellular matrix proteins as collagen and fibronectin.[13] It is interesting that bFGF and probably other heparin-binding growth factors are stored in the extracellular matrix, from which they are recruited in certain situations.[15]

Fibroblast growth factors stimulate a variety of motile functions including changes in cell shape. Studies with mutant fibroblast growth factor receptors have revealed that some of these effects, including modification of cell shape, are mediated through stimulation of phospholipase C_γ activity which catalyzes the hydrolysis of phosphatidylinositol-(4,5)-bisphosphate, resulting in diacylglycerol and inositol trisphosphate.[16, 17] The latter brings about calcium ion mobilization, which is involved in cellular motility. Diacylglycerol, along with other intermediates in triglyceride and phospholipid synthesis, notably monoacylglycerophosphate (lysophosphatidate) and diacylglycerophosphate (phosphatidate), which can actually be formed from diacylglycerol in certain cells, also have various stimulating effects on the cytoskeleton. The mutant fibroblast growth factor receptors have also disclosed that abrogation of phosphoinositide hydrolysis, does not influence the mitogenic effect, which is mediated by a phosphorylation cascade initiated by receptor tyrosine autophosphorylation.

Thus, through their influence on cell motility, bFGF and other heparin-binding growth factors probably modulate the bioenergetic relationships described by the new concept. Mutations affecting receptor function or signal transduction would channel energy toward triglyceride storage. Is this why preadipocytes from certain massively obese persons are relatively resistant to the acceleration of de-differentiation by bFGF? The acceleration which normally occurs might indeed be secondary to the stimulatory effects of heparin-binding growth factors on cell motility.

SUMMARY

Heparin-binding growth factors related to basic fibroblast growth factor are major determinants of the cellular clonal composition of adipose tissue. By providing and maintaining varying complements of preadipocytes in different fat depots, these factors contribute to the varying sizes and functions of different regions, including the hypercellularity in appreciable obesity. Thus, differing levels and activities of the heparin-binding growth factors contribute to variations in depots within the same individual and between individuals, in lean and obese states.

In contrast to regional differences in adiposity, which are accounted by factors resident in adipose tissue, we believe that obesity results from a generalized energy overload. According to our concept, there are genetic variations in cytoskeletal activity and thus differing quantities of energy are utilized for biomechanical processes. In a reciprocal relationship, the higher the cytoskeletal activity, the lesser the energy available for chemical energy storage, mainly in the form of triglyceride in adipocytes. At the extreme of "supermassive" obesity, a mutation in a gene related to a cytoskeletal protein would lead to appreciable dampening of cytoskeletal activity, with consequently the greatest quantity of energy remaining available for eventual triglyceride storage. Moreover, the new concept, for which we have have increasing experimental evidence, invokes a hypothalamic-efferent neural-cytoskeletal pathway, which would modulate the activity of the cytoskeleton.

REFERENCES

1. D. A. K. Roncari, D. C. W. Lau and S. Kindler, Exaggerated replication in culture of adipocyte precursors from massively obese persons, *Metabolism.* 30:425 (1981).
2. K. Teichert-Kuliszewska, B. S. Hamilton, M. Deitel and D. A. K. Roncari, Augmented production of heparin-binding mitogenic proteins by preadipocytes from massively obese persons, *J. Clin. Invest.* 90:1226 (1992).
3. D. A. K. Roncari and P. E. Le Blanc, Inhibition of rat perirenal preadipocyte differentiation, *Biochem. Cell Biol.* 68:238 (1990).
4. P. E. Le Blanc, D. A. K. Roncari, D. I. Hoar and A. Adachi, Exaggerated triglyceride accretion in human preadipocyte-murine renal line hybrids composed of cells from massively obese subjects, *J. Clin. Invest.* 81:1639 (1988).
5. H.-S. Ro and D. A. K. Roncari, The C/EBP-binding region and adjacent sites regulate

expression of the adipose P2 gene in human preadipocytes, *Mol. Cell. Biol.* 11:2303 (1991).

6. R. L. R. Van, C. E. Bayliss and D. A. K. Roncari, Cytological and enzymological characterization of adult human adipocyte precursors in culture, *J. Clin. Invest.* 58:699 (1976).

7. D. A. K. Roncari, S. Kindler and C. H. Hollenberg, Excessive proliferation in culture of reverted adipocytes from massively obese persons, *Metabolism.* 35:1 (1986).

8. S. C. Cooper and D. A. K. Roncari, 17-Beta-Estradiol increases mitogenic activity of medium from cultured preadipocytes of massively obese persons, *J. Clin. Invest.* 83:1925 (1989).

9. B. M. Spiegelman and H. Green, Decrease in tubulin and actin gene expression prior to morphological differentiation of 3T3-adipocytes, *Cell.* 29:53 (1982).

10. D. A. K. Roncari, Individual variations in energy utilized for biomechanical processes and molecular mobility account for diverse susceptibility to obesity, *Medical Hypotheses.* 23:11 (1987).

11. D. A. K. Roncari, Relationships between the hypothalamus and adipose tissue mass, *Advances Exp. Med. Biol.* 291:99 (1991)

12. J. Downward, Rac and Rho in tune, *Nature.* 359:273 (1992).

13. A. J. Ridley and A. Hall, The small GTP-binding protein rho regulates the assembly of focal adhesions and actin stress fibers in response to growth factors, *Cell.* 70:389 (1992).

14. A. J. Ridley, H. F. Paterson, C. L. Johnston, D. Diekmann and A. Hall, The small GTP-binding protein rac regulates growth factor-induced membrane ruffling, *Cell.* 70:401 (1992).

15. A. Yayon and M. Klagsbrun, Autocrine regulation of cell growth and transformation by basic fibroblast growth factor, *Cancer and Metastasis Reviews.* 9:191 (1990).

16. M. Mohammadi, C. A. Dionne, W. Li, N. Li, T. Spivak, A. M. Honeggerr, M. Jaye and J. Schlessinger, Point mutation in FGF receptor eliminates phosphatidylinositol hydrolysis without affecting mitogenesis, *Nature* 358:681 (1992).

17. K. G. Peters, J. Marie, E. Wilson, H. E. Ives, J. Escobedo, M. Del Rosario, D. Mirda and L. T. Williams, Point mutation of an FGF receptor abolishes phosphatidylinositol turnover and Ca^{2+} flux but not mitogenesis, *Nature.* 358:678 (1992).

REGIONAL OBESITY AND NIDDM

Per Björntorp

Department of Heart and Lung Diseases
Sahlgrenska Hospital
University of Göteborg
Sweden

ABSTRACT

Obesity and NIDDM are clearly linked. The subgroup of abdominal, visceral obesity has been shown to have a particularly close link to the development of diabetes. This is probably due to the marked insulin resistance of that condition. Epidemiological data show a predictive power for the development of NIDDM in both sexes, in signs of insulin resistance, visceral obesity and, in women, hyperandrogenicity. In men a relative hypogonadism may be of importance.

Experimental evidence suggests cause-effect relationships between these factors. In both sexes visceral fat may contribute to insulin resistance in the liver and the periphery by excess production of FFA. Hyperandrogenicity in women may also cause insulin resistance, although the reverse sequence of events cannot be excluded. The relative hypogonadism may well contribute to insulin resistance in men, as well as to the accumulation of visceral fat.

There are observations of additional endocrine aberrations in visceral obesity, suggesting a central, neuroendocrine disturbance, which might be a primary factor for the pathogenesis of the syndrome.

Key words: Visceral obesity, insulin, androgens, FFA, cortisol

There is an established statistical relationship between obesity and NIDDM. However, all obese subjects do not develop diabetes. Two components seem to be needed, both a strain on the ß-cell apparatus, and an inherent weakness of the ß-cells to resist the extra load of obesity. The damage obesity is causing to the insulin-producing machinery is generally believed to be mediated via a peripheral insulin resistance, which causes a compensatory overproduction of insulin, requiring a healthy ß-cell apparatus to cope with this extra load. Development of NIDDM then seems to require two components, peripheral insulin resistance, combined with an insufficient capacity for compensatory production of insulin.

Recent research, continuing a previous development by particularly Jean Vague in Marseille, has clearly revealed that insulin resistance is not equally distributed in the obese population. When adipose tissue excess is localizedmainly to central, abdominal, particularly visceral regions, then insulin resistance is more pronounced. This area is currently under intensive development and has been reviewed repeatedly recently, and the reader is directed to these overviews for detailed references (1-4). This review will briefly summarize the current developments, attempting to interpret the informations, and add recent, pertinent observations.

EPIDEMIOLOGY

Abdominal obesity, measured conveniently as the waist/hip circumference ratio (WHR), has been shown to be a risk factor for the development of NIDDM in both men and women. There is also a report available indicating that visceral fat seems to be the crucial factor. Obesity without regional distinction probably also contributes, in other words at equally elevated WHR, addition of fat mass with the same distribution seems to amplify the risk. Whether this is due to a concomitant enlargement of visceral fat masses only, or is a consequence of additional risk associated with large subcutaneous fat depots, is not definitely resolved.

Risk Factors for the Development of NIDDM in Women

There are several endocrine abnormalities following visceral obesity which by themselves may cause, or at least, contribute to insulin resistance. In women hyperandrogenicity is closely statistically associated with visceral obesity, and a low concentration of sex hormone binding globulin (SHBG), indicating hyperandrogenicity, is also a strong predictor for the development of

NIDDM. Hyperandrogenicity in turn is tightly statistically coupled to insulin resistance. We then have a triangle of insulin resistance, hyperandrogenicity and visceral obesity, which all are statistical predictors for the development of NIDDM in women. This set-up of factors seems to explain a majority of risk to develop NIDDM in women.

The question is of course if these factors are causally coupled, and in that case, how this is occurring. Several possibilities are apparent as indicated by experimental work. First, it is known from work in women that introduction of a hyperandrogenic condition by exogenous hormones is followed by insulin resistance. This is clearly also the case in female rats which become severely insulin resistant after the administration of small to moderate doses of testosterone. The localization of the lesion seems to be mainly at the level of the muscle, particularly the glycogen synthase system. This is of interest because there are now several reports suggesting that this seems to be the main point of explanation for insulin resistance in NIDDM in women, discoverable already in the prediabetic state.

It thus seems reasonably clear that introduction of hyperandrogenicity is followed by insulin resistance in women and in female rats. It can, however, not be excluded that the reverse chain of events is the correct one, hyperinsulinemia causing hyperandrogenicity as reviewed in (5).

The next question is how visceral obesity may be linked into a potential cause-effect chain. Let us first look at the potential role of hyperandrogenicity. This connection is currently not easy to visualize. Testosterone in men is lipolytic, and inhibits lipid uptake in adipose tissue, particularly in visceral depots. Although data on this are missing in women, there is no apparent reason to believe that hyperandrogenicity would contribute to depot fat accumulation in visceral depots in women. There is a remaining possibility, however, that other hormonal aberrations in women would be more powerful, overriding the androgen effect. Cortisol secretion is also elevated in women with visceral obesity (6), and cortisol clearly directs excess triglycerides to visceral fat depots, as seen dramatically in Cushing's syndrome. To sum up, it is thus difficult with the current status of information to see how visceral obesity and the hyperandrogenicity in women might be causally connected.

The remaining link then is to try to understand a potential cause-effect relationship between visceral fat and insulin resistance. Here there is information available which suggests a linkage. Visceral depot triglycerides have a rapid turnover, in the order of at least 50 % higher than abdominal subcutaneous fat, and still higher than femoral depot triglycerides. The visceral depot is therefore pouring out an excess of free fatty acids (FFA) into the portal vein, particularly under conditions of stimulation of fat mobilization. This probably has several consequences. First, previous and recent studies suggest that this will trigger gluconeogenesis in the liver, causing a tendency to increased blood glucose concentration, a step towards the clinical diagnosis of NIDDM. Furthermore, there is considerable evidence to suggest that the hepatic secretion of very low density lipoproteins will follow, which might explain the common phenomenon of hypertriglyceridemia in the pre-diabetic and diabetic conditions.

Recent studies also suggest the possibility that FFA may inhibit hepatic clearance of insulin, contributing to the peripheral hyperinsulinemia. This in turn might cause peripheral insulin resistance. It should, however, not be overlooked that massive secretion of FFA into the portal circulation might be followed also by elevated FFA concentrations in the peripheral circulation. This is particularly the case when visceral adipose tissue is enlarged, and can be illustrated by the following example. If a woman with visceral obesity has a visceral fat depot of 30-40 % of total body fat, and this depot has twice the lipolytic potential as other fat depots then 60-80 % of total FFA production will originate from the visceral depots. However, about 50 % of FFA from visceral fat is normally cleared by the liver, leaving 30-40 % of peripheral FFA from the visceral depots. With a contribution from, particularly, subcutaneous abdominal fat (which is considerably easier to mobilize than femoral subcutaneous fat) peripheral FFA concentrations will be considerably elevated.

There are two additional factors in this context which might be of importance. First, observations suggest that not only FFA concentrations in the periphery are elevated in visceral obesity in women, also the fractional turnover rate is high. In other words, the flux out from circulation into tissues is elevated. Both the concentrations and the peripheral uptake of FFA are thus elevated above normal, exposing peripheral tissues to an excess of FFA. Furthermore, since the visceral depots are less sensitive to the inhibitory effect of insulin on FFA mobilization, the associated hyperinsulinemia might be comparably insufficient to check the FFA-outflow from visceral fat depots, exaggerating peripheral FFA fluxes further.

The end result of peripheral elevation of FFA concentrations may well be that particularly muscle becomes insulin resistant via post-receptor mechanisms, the so-called Randle effect. A considerable part of such an effect might thus be the consequence of enlarged visceral fat depots. This area has also been reviewed in more detail recently (7).

There is thus considerable evidence to suggest, in women, a cause-effect relationship between hyperandrogenicity and insulin resistance, via muscle effects, and between visceral fat depot enlargement, and insulin resistance via FFA effects on both muscle and liver. However, for the moment there seem to be very weak, if any, suggestions of a causal relationship between androgens and visceral obesity. The chain of events then seems to be, hypothetically, that visceral obesity and hyperandrogenicity may explain, at least partly, insulin resistance in women developing NIDDM. We have also recent evidence that both these factors remain in the condition of clincal NIDDM (8).

Risk Factors for the Development of NIDDM in Men

The situation is clearly different in men, particularly, of course, as far as androgens are concerned. In men excess androgen exposure also seems to be followed by insulin resistance, but, more pertinent to the question discussed, too low levels of circulating testosterone are also associated with insulin resistance. Men with visceral obesity frequently have lower than normal testosterone in the circulation. Substitution is improving insulin resistance, suggesting a cause-effect relationship. This can be seen clearly in rats where castration is followed by marked insulin resistance, normalized after testosterone substitution. Again, like hyperandrogenicity in females, the lesion seems to be localized mainly to the glycogen synthase system in muscle.

The FFA mechnism discussed for women, is clearly a possibility also for men. As a matter of fact, this may even be more effective in men, because of their larger proportion of visceral fat mass.

In men there is also a possibility that their relative hypogonadism is actually contributing to visceral fat accumulation. With the lack of sufficient amounts of testosterone, lipid will tend to accumulate in these fat depots due to a lack of stimulation of lipid mobilization, and a check of lipid uptake. These events are mediated via a specific androgen receptor, the density of which is positively regulated by testosterone itself. Low testosterone levels will consequently also be followed by a low density of androgen receptors, further weakening the testosterone effects on the fat depots.

The prospective, epidemiological evidence in men is less complete than in women. However, visceral obesity has been established as an independent risk factor for the development of NIDDM in men. There is, however, no conclusive evidence for a potential, prospective risk for NIDDM with a relative hypogonadism. However, insulinresistance is also a risk factor in men as it is in women. Attempting to explain, again hypothetically, the available risk factor information in terms of cause-effect relationships may thus leave us with the following picture.

Insulin resistance is clearly a corner-stone for NIDDM-development. The relative hypogonadism of men with visceral obesity, as well as the visceral depots themselves, via FFA, may thus explain, at least partly, the insulin resistance in both muscle and liver. In the case of men it is thus also possible to see a link between the relative hypogonadism and visceral fat accumulation, which would then tend to form a vicious circle, amplifying potential FFA effects on insulin resistance.

Other Factors of Importance for Insulin Resistance

This overview has so far focused mainly on androgens and FFA as potential factors for the creation of insulin resistance in women and men. There may, however, well be other factors involved. Recently, we have found that there is also an increased secretion of cortisol in visceral obesity (6). This of course would most likely be followed by insulin resistance, a well-known effect of cortisol. In addition, cortisol tends to increase hepatic gluconeogenesis, an additional pre-diabetic feature. There are no data available of elevated cortisol secretion as a potential predictor for the development of NIDDM, perhaps due to the relatively complicated task to determine cortisol secretion in quantitative and qualitative terms in large materials.

Neuroendocrine Abnormalities

The summary of available information, and the attempts to a synthesis, has indicated that both androgen and corticosteroid secretions are disturbed in visceral obesity. There are also other endocrine aberrations. First, female sex hormone secretion is also abnormal, with anovulation and decreased progesterone secretion as a consequence. Furthermore, we have recently observed low IGF-I levels in abdominal obesity, suggesting a blunted growth hormone secretion, particularly with visceral obesity (9). Finally there are observations of dysregulation of hemodynamics, which may have an origin in the central nervous system (10).

Taken together, all these aberrations suggest central neuroendocrine disturbances which might be of primary importance for the syndrome, characterized among a number of phenomena, by visceral fat accumulation. Clearly, attempts to disclose the more detailed nature and the origin and cause to such aberrations is a research area of high priority. Observations and interpretations pertinent to this question have been summarized in (11).

REFERENCES

1.   P. Björntorp, Abdominal obesity and the development of non-

insulin-dependent diabetes mellitus, *Diabetes/Metabolism Rev.* 4:615-622 (1988).

2. A.H. Kissebah and A.N. Peiris, Biology of regional body fat distribution: Relationship to non-insulin dependent diabetes mellitus, *Diabetes/Metabolism Rev.* 4:622-632 (1988).

3. P. Björntorp, Obesity, insulin resistance and diabetes, *in*: "The Diabetes Annual" K.G.M.M. Alberti and L.P. Krall, eds., Elsevier, Amsterdam, 6:347-370 (1991).

4. P. Björntorp, Metabolic implications of body fat distribution, *Diabetes Care* 14:1132-1143 (1991).

5. L. Poretsky and M.L. Karlin, The gonadotropic function of insulin, *Endocrin Rev.* 8:132-141 (1987).

6. P. Mårin, N. Darin, T. Amemiya, B. Andersson, and P. Björntorp, Cortisol secretion in relation to body fat distribution in obese premenopausal women, *Metabolism*, in print, (1992).

7. P. Björntorp, "Portal" adipose tissue as a generator of risk factors for cardiovascular disease on diabetes, *Arteriosclerosis* 10:493-496 (1990).

8. B. Andersson, L. Lissner, A. Vermeulen, M. Krotkiewski, and P. Björntorp, Steroid hormones and muscle morphology in women with non-insulin diabetes mellitus, submitted for publication (1991).

9. P. Mårin, H Kvist, L. Sjöström, and P. Björntorp, Low concentrations of insulin-like growth factor I in abdominal obesity, *Int. J. Obesity*, in print (1992).

10. S. Jern, A. Bergbrant, P. Björntorp, and L. Hansson, Relation of central hemodynamics to obesity and body fat distribution, *Hypertension*, in print (1991).

11. P. Björntorp, Psychosocial factors and fat distribution, *in*: "Obesity in Europe", Proc 3rd Eur. Congress of Obesity, G. Ailhaud, B. Guy-Grand, M. Lafontan, and D. Ricquier, eds., Libbey, London (1992), pp 377-388.

HYPERINSULINEMIA AND VLDL KINETICS

George Steiner

The Toronto Hospital (Toronto General Division)
Toronto, Ontario, Canada. M5G 2C4

LIPOPROTEIN CONCENTRATIONS IN NIDDM

The most frequent lipoprotein abnormality in diabetes mellitus is hypertriglyceridemia. The frequency of hypercholesterolemia is not very different from that found in the nondiabetic population.[1,2] In those diabetic individuals who are grossly insulin deficient, the hypertriglyceridemia is due to an accumulation of chylomicrons, the primary transporters of dietary lipids. However, most diabetics, particularly those with NIDDM, do not have such severe insulin deficiency. In fact, as will be indicated below, many are hyperinsulinemic.[3] In them the hypertriglyceridemia is due to an increased concentration of those triglyceride-rich lipoproteins that transport endogenous (i.e. nondietary) triglyceride. Such lipoproteins are generally called VLDL. However, they are in fact a population of lipoproteins that cover a wide spectrum of size and density. We have found, in nondiabetic individuals, that 75% of this population of lipoproteins is in the smaller, S_f 12-60 subpopulation of lipoproteins, a population also referred to as IDL.[4] Based on kinetics considerations, this subpopulation in humans conforms to the characteristics of VLDL remnants.[5] The remnant lipoproteins have been shown by many to be associated with an increased incidence of coronary artery disease.[6-8]

PLASMA INSULIN LEVELS IN NIDDM

One feature common to most patients with NIDDM, particularly those who are also obese, is resistance to insulin. The pancreas may attempt to compensate for this by secreting more insulin than would be the case in an individual who is normally sensitive to insulin. This will result in hyperinsulinemia, particularly in the hepatic portal circulation. In those who develop NIDDM, the pancreas is unable to secrete sufficient insulin at the right time to maintain normal glucose homeostasis. However, in general, they too will have hyperinsulinemia, even if this is too little to maintain normoglycemia.

New Concepts in the Pathogenesis of NIDDM, Edited by
C. G. Östenson *et al.*, Plenum Press, New York, 1993

RATIONALE FOR STUDYING THE EFFECTS OF HYPERINSULINEMIA ON VLDL METABOLISM

Some years ago we began to investigate the kinetics of the triglyceride-rich lipoproteins in diabetes mellitus. There were several reasons for this. Atherosclerosis is the most frequent chronic problem facing those with diabetes.[9] As noted above, elevated levels of the small triglyceride-rich lipoproteins are associated with an increased incidence of coronary artery disease.[6-8] An increase in the triglyceride-rich lipoprotein population is the most frequent form of hyperlipoproteinemia in diabetic individuals.[1,2] Insulin has long been recognized to play a major role in regulating triglyceride metabolism. Hyperinsulinemia has been shown to be associated with coronary artery disease originally in nondiabetic populations, and more recently in diabetics.[10-15] Therefore, since hyperinsulinemia exists chronically in many patients with NIDDM, we started by examining the effects of chronic hyperinsulinemia on triglyceride-rich lipoprotein kinetics in humans.

EFFECT OF CHRONIC HYPERINSULINEMIA ON VLDL-TG PRODUCTION IN HUMANS

There is no perfect model of chronic hyperinsulinemia in humans. However, the most common circumstance in which this is seen is obesity. To study the effects of the chronically hyperinsulinemic state morbidly obese humans were studied before and at the end of several weeks of total fasting, and then during hypocaloric refeeding. Thus within the same individual triglyceride kinetics could be determined at several different plasma insulin levels. Using the radioactive glycerol method it was found that the rate of VLDL-triglyceride production was directly proportional to the serum level of insulin at the time of the kinetic study.[16] A similar relationship between serum insulin and VLDL-triglyceride production was observed in patients who were being treated with alternate day glucocorticoids and were studied both on the day of treatment with glucocorticoid and on the day on which glucocorticoid was omitted.[17] Although such studies were consistent with the possibility that the chronically hyperinsulinemic state was accompanied by an increase in VLDL-triglyceride production, each of these models had too many other variables to permit a more definitive conclusion.

EFFECT OF CHRONIC HYPERINSULINEMIA IN VIVO ON VLDL-TG PRODUCTION IN RATS

In order to study a more controllable model of the chronically hyperinsulinemic state in vivo, we turned to a model in which rats were given gradually increasing doses of NPH insulin over a two week period, ultimately receiving six units in divided doses twice daily. The insulin was give by subcutaneous injection. In order to avoid profound hypoglycemia, the animals were supplied with 10% sucrose solutions rather than water to drink. The behaviour of the insulin treated group was compared to that of two other groups, one receiving the same 10% sugar solution to drink but no insulin injections (sugar supplemented group) and the other receiving water to drink and no insulin injections (chow control group). All groups had free access to standard rat chow.[18]

Initially the insulin treated group and the sugar supplemented group gained weight more rapidly than did the chow controls. However, by the end of the two weeks all groups had attained the same average weight. Furthermore, the weights of their epididymal fats pads did not differ and the mean size of the adipocytes in these pads was the same.[19] The insulin treated rats were found to be resistant to insulin. Based on studies of insulin binding to adipocytes and on the responses of glucose oxidation by adipocytes to increasing concentrations of insulin, the insulin resistance was characterized as being at both the binding and post-binding level.[19] The rats were normoglycemic and did not have any change in their glucagon levels.[20] Thus, presumably any effects of the chronically hyperinsulinemic state were not due to changes in the levels of "counterregulatory" hormones.

Initially the effects of the chronically hyperinsulinemic state on VLDL-triglyceride production was determined. It was found that the rate of VLDL-triglyceride production was increased in the chronically hyperinsulinemic rats.[18] This was inspite of a low level of free fatty acids and raised the possibility that the fatty acids of the VLDL-triglyceride originated from a source other than plasma free fatty acids. The source was suggested by experiments in which the effects of substituting glucose or fructose for sucrose in the rats' drinking solution was examined. In those studies it became apparent that the chronic hyperinsulinemia induced increase in VLDL-triglyceride production was greatest in the rats drinking 10% fructose. We speculated that this may be related to the liver taking up a greater proportion of orally ingested fructose than it does of orally ingested glucose. The increased availability of substrate for lipogenesis in the fructose supplemented rats was combined with an increase that was found in the activity of hepatic lipogenic enzymes.[21] Such a combination certainly could account for the postulated increase in the rate of production of triglyceride fatty acids from a source other than plasma free fatty acids.

Inspite of the increase in VLDL-triglyceride production observed in chronically hyperinsulinemic rats, their plasma levels of triglyceride declined. This suggested that, in the chronically hyperinsulinemic state, the removal of VLDL-triglyceride was accelerated even more than was VLDL-triglyceride production. An increase in the activity of lipoprotein lipase could accomplish this and was in fact found to occur.[21] Thus, the final level of plasma triglyceride reflected the balance between VLDL-triglyceride production and removal.

This balance could be altered by changing the route by which insulin was delivered to the body. Rats were given insulin at the same rate by either the subcutaneous route or intraperitoneally. In rats receiving their insulin subcutaneously, the serum levels of insulin in the peripheral and hepatic portal blood were the same. In rats receiving their insulin intraabdominally the peripheral insulin levels were lower than when the insulin was given subcutaneously, and were also lower than the hepatic portal insulin levels. This presumably reflected the hepatic extraction of the portally delivered insulin. In the rats receiving subcutaneous insulin, the activity of lipoprotein lipase was higher than in rats receiving intraperitoneal insulin. This parallelled the higher peripheral insulin levels in the former group. Both groups of rats had similar VLDL production rates and this reflected their similar hepatic portal insulin levels. Thus, both groups had similar production rates, but the subcutaneous group had higher levels of lipoprotein lipase and rates of VLDL removal. Hence, the subcutaneously treated group had lower plasma triglyceride levels.[21]

Thus, the chronically hyperinsulinemic state in vivo is associated with an increase in VLDL-triglyceride production that is not related to changes in adiposity, and presumably not to changes in "counterregulatory" hormones. The increase in production reflects, at least in part, an increase in de novo fatty acid synthesis. Finally, the balance between production and removal of triglyceride reflected the relative levels of insulin in the hepatic portal and the peripheral circulation.

EFFECT OF INSULIN IN VITRO ON VLDL PRODUCTION BY ISOLATED HEPATOCYTES

The mechanisms underlying the increased rate of VLDL-triglyceride synthesis in vivo in the chronically hyperinsulinemic state are made more difficult to explain by virtue of the observations made with cultures of hepatocytes. In that in vitro model, the direct addition of insulin for short periods of time reduces the secretion of VLDL. This has led some to suggest that insulin's effect on VLDL secretion is inhibitory.[22-24] However, other studies have shown that if hepatocytes are exposed to insulin for long periods, such as 48hrs, insulin no longer inhibits, but in fact stimulates VLDL secretion.[24] Some have suggested that such longer exposure produces insulin resistance in the hepatocytes and therefore insulin is no longer inhibitory with respect to VLDL secretion. If this were the case, it could explain a restoration of VLDL secretion to a rate similar to that seen in the absence of any added insulin. However, it would not explain a stimulation of insulin secretion. Thus, even in the in vitro model the response of VLDL secretion to insulin is complex.

Many of the studies of insulin's effect on cultures of hepatocytes were conducted in the absence of insulin in the medium. In fact, when fatty acids were added to the medium they were able to overcome the insulin-induced inhibition of VLDL secretion. Therefore, it is possible that some of the differences between the observations made in the chronically hyperinsulinemic state in vivo and in the hepatocytes to which insulin was added in vitro could result from either, or both differences in the duration of exposure to insulin and the presence of free fatty acids.

ACUTE EFFECTS OF INSULIN ON VLDL PRODUCTION IN VIVO IN HUMANS AND RATS

To examine whether the duration of exposure to insulin affected its impact on VLDL secretion we conducted experiments in both humans and rats. Humans were studied before and during a 6 hr euglycemic-hyperinsulinemic clamp.[25,26] During the clamp period, compared to baseline, their rate of both VLDL-triglyceride and of VLDL-apoB production was reduced. While this would be consistent with acutely hyperinsulinemic state inhibiting VLDL production and the chronically hyperinsulinemic state enhancing it, other changes could also account for the acute effects of insulin.

Acute exposure to insulin also reduced plasma levels of free fatty acids.[25,26] Therefore, ti became necessary to examine whether acute exposure to insulin inhibited VLDL-triglyceride production if free fatty acid levels did not fall. To examine this we used rat models. In rats, as in humans, immediately after injecting insulin the plasma levels of free fatty acids fell and the rate of VLDL-triglyceride secretion fell. However, when the plasma levels of free fatty acids were maintained by infusing an albumin-oleate complex, VLDL-triglyceride production actually increased in response to an acute injection of insulin.[27]

These in vivo observations closely matched those made with perfused livers. In that in vitro model, perfusate levels of free fatty acid were maintained constant, and we found that the immediate response to insulin was also an increase in VLDL-triglyceride production.[28]

MULTIPLE INFLUENCES AFFECTING THE IN VIVO RESPONSE OF VLDL TO CHRONIC HYPERINSULINEMIA

These studies lead one to conclude that the effects of the chronically hyperinsulinemic state on the triglyceride-rich lipoproteins are complex. First, they represent a balance between the rates of production and removal.

The production side of this balance reflects the impact of a number of influences. The first among these is the availability of substrate for VLDL-triglyceride. The response to acute exposure to insulin is greatly influenced by the concentration of circulating free fatty acids. However, in the chronically hyperinsulinemic state, although plasma free fatty acid levels are reduced VLDL-triglyceride production is still increased. This appears to reflect an increase in the rate of de novo synthesis of triglyceride fatty acids. In turn, this increase in de novo lipogenesis is probably the consequence of an increase both in the supply of substrate from fructose and in the activity of hepatic lipogenic enzymes. Thus, the increase in de novo lipogenesis is not an immediate response to hyperinsulinemia, but takes some time to develop. The second group of influences on VLDL production is the entire hormonal milieu of the liver. Even though the hyperinsulinemic rats were normoglycemic and did not have a change in their levels of glucagon, it is probable that glucocorticoids and other hormones, even in basal concentrations will influence hepatic lipoprotein production. A third group of influences relate to insulin itself, its dose, route of delivery, and timing (i.e. acute or chronic). Finally, all of these influences can act at a number of hepatic metabolic sites, the balance between which can influence VLDL synthesis and secretion. These include not only de novo lipogenesis as discussed above, but also the mobilization of stored intracellular triglyceride fatty acids and the "partition" of fatty acids between oxidation and esterification, the synthesis and intracellular hydrolysis of apoB, the assembly of VLDL's apolipoproteins and lipids, and ultimately the secretion of the lipoprotein particle.

The removal side of the balance that determines plasma levels of VLDL, reflects primarily the activity of the enzyme lipoprotein lipase. As is well known, this enzyme's activity is increased by insulin. The present studies show that the route of insulin delivery can influence the level of insulin to which peripheral tissues are exposed. This, in turn, can influence the degree to which lipoprotein lipase activity and VLDL removal will be enhanced. The rate of lipolysis will also be determined by the composition of the substrate lipoprotein. Insulin, diabetes and even fructose feeding are among the many factors that can influence the lipid and apolipoprotein composition of the triglyceride-rich lipoproteins in a way that can affect their ability to serve as substrates for lipoprotein lipase. Finally, lipoprotein lipase regulation by nutritional and hormonal factors differs in adipose tissue and muscle. The interaction of all of these factors will determine the rate of VLDL degradation.

The balance between that rate and the rate of production will then ultimately determine the level of VLDL in the circulation. As well, it will influence the rate at which the smaller atherogenic catabolic remnants of the triglyceride-rich lipoproteins are made. This, then, may be one of the several ways in which the chronically hyperinsulinemic state can be linked to atherogenesis.

REFERENCES

1. Zimmerman BR, Palumbo PJ, O'Fallon WA, et al. A prospective study of peripheral arterila occlusive disease in diabetics, III. Initial lipid and lipoprotein findings. Mayo Clin Proc. 56:223(1981)

2. Barrett-Connor E, Grundy SM, Holdbrook MJ. Plasma lipids and diabetes mellitus in an adult community. Am J Epidemiol. 115:657(1982)

3. Steiner G. Insulin regulation of triglyceride metabolism. Atherosclerosis Revs. 22:2(1991)

4. Poapst M, Reardon M, Steiner G. Relative contribution of triglyceride-rich lipoprotein particle size and number to plasma triglyceride concentration. Arteriosclerosis 5:381(1985)

5. Reardon MF, Steiner G. The use of kinetics in investigating metabolism of very low density and intermediate density lipoproteins in "Lipoprotein Kinetics and Modeling" Berman M, Grundy SM, Howard BV, eds. Academic Press, New York (1982)

6. Steiner G, Schwartz L, Shumak S, Poapst M. The association of increased levels of intermediate-density lipoproteins with smoking and with coronary artery disease. Circulation 75:124(1987)

7. Reardon MF, Nestel PJ, Craig IH, HarperRW. Lipoprotein predictors of the severity of coronary artery disease in men and women. Circulation. 71:881(1985)

8. Krauss RM, Lindgren FT, Williams PT, et al. Intermediate density lipoproteins and progression of coronary artery disease in hypercholesterolemic men. Lancet II:62(1987)

9.   Steiner G. Atherosclerosis, the major complication of diabetes. *in* "Comparison of Type I and Type II Diabetes." Vranic M, Hollenberg CH, Steiner G, eds. Plenum Publishing Corp., Toronto, (1985)

10. Welborn TA, Wearn K. Coronary heart disease incidence and cardiovascular mortality in Busselton with reference to glucose and insulin concentrations. Diabetes Care 2:131(1979)

11. Fontbonne A, Charles MA, Thibult N, et al. Hyperinsulinemia as a predictor of coronary heart disease mortality in a healthy population: the Paris Prospective Study, 15 year follow-up. Diabetologia 34:356(1991)

12. Pyorala K, Savolainen E, Kaukola S, et al. Plasma insulin as a coronary heart disease risk factor: Relationship to other risk factors and predictive value during 9 year follow-up of the Helsinki Policeman Study population. Acta Med Scand. 701(Suppl):38(1985)

13. Ronnemaa T, Laakso M, Pyorala K, et al. High fasting plasma insulin is an indicator of coronary heart disease in noninsulin-dependent diabetic patients and nondiabetic subjects. Arterioscl. and Thromb. 11:80(1991)

14. Ronnemaa T, Laakso M, Puukla P, et al. Atherosclerotic cardiovascular disease in middle aged insulin treated diabetic patients. Association with endogenous insulin secretion capacity. Arteriosclerosis. 8:23(1988)

15. Nagi D, Hendra TJ, Ryle AJ, et al. The relationships of concentrations of insulin, intact proinsulin and 23-33 split proinsulin with cardiovascular risk factors in type 2 (non-insulin-dependent) diabetic subjects. Diabetologia 33:532(1990)

16. Streja DA, Marliss EB, Steiner G. The effects of prolonged fasting on plasma triglyceride kinetics in man. Metabolism 26:505(1977)

17. Cattran DC, Steiner G, Wilson DR, Fenton SSA. Hyperlipidemia after renal transplantation: natural history and pathophysiology. Ann Int Med. 91:554(1979)

18. Steiner G, Haynes FJ, Yoshino G, Vranic M. Hyperinsulinemia and in vivo very-low-density lipoprotein-triglyceride kinetics. Am J Physiol 246:E187(1984)

19. Martin C, Desai KS, Steiner G. Receptor and post-receptor insulin resistance induced by in vivo hyperinsulinemia. Can J Physiol Pharmacol. 61:802(1983)

20. Brubaker LP, Kazumi T, Hirano T, Vranic M, Steiner G. Failure of chronic hyperinsulinemia to supress pancreatic glucagon in vivo in the rat. Can J Physiol Pharmacol. 69:437(1991)

21. Kazumi T, Vranic M, Bar-On H, Steiner G Portal v peripheral hyperinsulinemia and very low density lipoprotein triglyceride kinetics. Metabolism 35:1024(1986)

22. Durrington PN, Newton RS, Weinstein DB, Steinberg D. Effects of insulin and glucose on very-low-density lipoprotein triglyceride secretion by cultured rat hepatocytes. J Clin Invest. 70:63(1982)

23. Patsch W, Franz S, Schonfeld G. Role of insulin in lipoprotein secretion by cultured rat hepatocytes. J Clin Invest. 71:1161(1983)

24. Bartlett SM, Gibbons GF. Short- and long-term regulation of very-low density lipoprotein secretion by insulin, dexamethasone and lipogenic substrates in cultured hepatocytes. A biphasic effect of insulin. Biochem J. 249:37(1988)

25. Shumak SL, Zinman B, Zunig-Guarjardo S, Poapst M,Steiner G. Triglyceride-rich lipoprotein metabolism during acute hyperinsulinemia in hypertriglyceridemic humans. Metabolism. 37:461(1988)

26. Lewis GF, Uffelman KD, Szeto LW, Steiner G. Acute hyperinsulinemia decreases very low density lipoprotein (VLDL) triglyceride and VLDL apolipoprotein (Apo) B production in normal weight but not in obese men. Diabetes. in press (1993)

27. Ferguson K, Mamo J, Steiner G. (in preparation)

28. Raman M, Steiner G. Effect of insulin on VLDL-triglyceride secretion and glucose production in the perfused rat liver. Diabetes 39(suppl 1):45A(1990)

HYPERINSULINISM AND DYSLIPIDEMIAS AS CORONARY HEART DISEASE RISK FACTORS IN NIDDM

Marja–Riitta Taskinen

Third Department of Medicine
University of Helsinki
Haartmaninkatu 4
00290 Helsinki, Finland

Since abnormalities of serum lipids and lipoproteins are highly prevalent in NIDDM they are considered as potential factors contributing to the increased CHD risk. On the other hand the link between hyperinsulinism and dyslipidemias is well–established although the causal relationship is not clear despite extensive studies. There is ample evidence that both conditions are associated with various metabolic abnormalities like hypertension, upper body obesity and NIDDM which are part of disease cluster named Syndrome X by Reaven[1] or the so–called "deadly quartet"[2]. Growing evidence indicate that hypertriglyceridemia and low HDL cholesterol, which characterize insulin resistance, play a role in atherogenesis and contribute to CHD risk.

It is well documented that hyperinsulinism may preceed for years the development of impaired glucose tolerance (IGT) and the onset of NIDDM[3]. Serum insulin concentrations increase parallelly to the plasma glucose from normal through the IGT range[4]. The measurement of plasma insulin reflects well the degree of insulin resistance in non–diabetic population. The question how hyperinsulinism is linked to dyslipidemias and coronary heart disease and if hyperinsulinemia destinates a person to later development of cardiovascular diseases has important implications for prevention.

IS HYPERINSULINISM A CARDIOVASCULAR RISK FACTOR?

A possible association between hyperinsulinism and coronary artery disease (CAD) was discovered already in 1965. Plasma insulin and its response to glucose were reported to be abnormally high in survivors of myocardial infarction[5,6]. Today, more than 25 years later, strong evidence has accumulated to support that hyperinsulinism is an independent risk factor of coronary heart disease (CHD). Three prospective studies, where plasma insulin levels were measured at the baseline, consistently found that high insulin values were independent predictors of CAD events[7,8,9]. The Helsinki Policemen Study and the Paris Prospective Study both based on cohorts of middle–aged men[8,9]. In the Helsinki Policemen Study the age adjusted 9 1/2 year incidence of CHD events among 982 men was significantly higher in the highest decile of 2 h insulin (Fig 1). In multivariate regression analyses the predictive value of 2 h insulin level turned out to be independent of other risk factors. The Paris Prospective Study included 6903 healthy men at the entry[9]. During 15 year follow–up 174 deaths could be attributed to CHD. Annual mortality rates for CHD were highest in the last quintiles of both fasting and 2 h post–load insulin values[9]. When the Cox regression model was used to test the predictive value of different variables 2 h post load insulin values, when entered as a categorial variable, appeared as an independent predictor in addition to systolic blood pressure, number of cigarettes per day and plasma cholesterol level. If we accept hyperinsulinism to be a reliable marker of insulin resistance the data from the three prospective studies suggest that insulin resistance is associated with increased CHD risk. Recently Laakso et al.[10] demonstrated that asymptomatic atherosclerosis, documented using ultrasonography investigations of femoral or carotied arteries, was associated with insulin resistance measured by using euglycemic clamp technique.

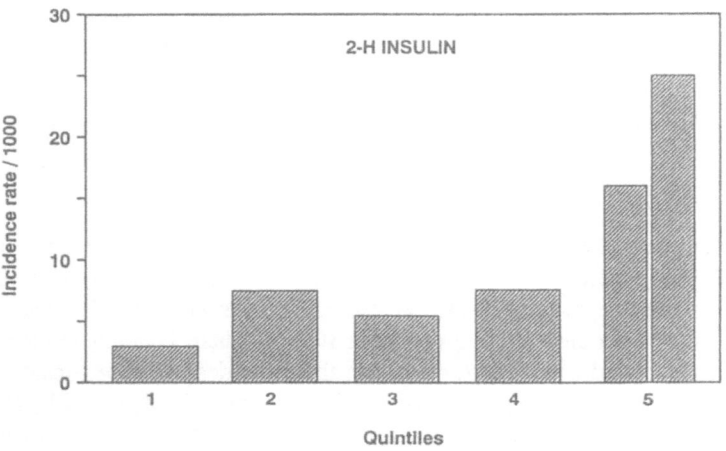

Figure 1. The age–adjusted 9½–year incidence of CHD events among 982 men by quintiles of 2 h plasma insulin. Note that the top quintile has been divided into two deciles. p<0.016 demotes the overall statistical significance of CHD risk over quintiles of plasma insulin (Pyörälä K. et al., ref. 8).

Several studies have also demonstrated that subjects with impaired glucose tolerance (IGT) exhibit increased risk of CHD and increased prevalence of CHD risk factors[3]. Hyperinsulinism prevails in IGT as demonstrated by the inverted horse–shoe curve of insulin response, and it is a predictor of CHD events in IGT[3,4]. Recent data from a cross sectional study in NIDDM population showed that hyperinsulinism also was a significant indicator of CHD among NIDDM men and women[11]. In this study a significant clustering of dyslipidemias, hypertension and BMI occurred in those NIDDM subjects belonging to the highest insulin quintiles. Similar data has also emerged from few previous studies which are based on smaller cohorts[12,13].

A question rises if hyperinsulinism is directly atherogenic or if it is only a marker of other CHD risk factors which play a causative role in atherogenesis. The concept that hyperinsulinism is incriminated in the pathogenesis of atherogenesis was suggested by Stout[14]. As highlighted twenty years later several line of experiental evidence indicate that insulin has multiple direct effects on arterial tissues which may promote the atherosclerotic process[15]. These actions of insulin on arterial wall include stimulation of cholesterol synthesis and accumulation, enhancement of LDL receptor activity and proliferation of smooth muscle cells. All these actions seem to occur at physiological concentrations of insulin.

INSULIN RESISTANCE AND DYSLIPIDEMIAS

Elevation of triglycerides and lowering of HDL cholesterol are charateristic features in NIDDM. The pattern of lipid abnormalities among hyperinsulinemic patients has been reported to be similar to that in NIDDM[16,17]. Notably fasting hyperinsulinemia was the best discriminator for lipid abnormalities also in asymptomatic hyperglycemia ie among subjects with IGT or newly diagnosed NIDDM[18]. Under the assumption that hyperinsulinemia is a reliable marker of insulin resistance the data strongly suggest that insulin resistance is associated with elevation of serum triglycerides and lowering of HDL cholesterol. Recent studies have confirmed that the rate of glucose disposal indeed correlates inversely to fasting serum triglycerides but positively to HDL cholesterol not only in non–diabetic subjects but also in subjects with IGT and NIDDM (Fig. 2)[19,20]. Interestingly these associations seem to be independent of other possible confounding factors like BMI or the waist/hip ratio[20].

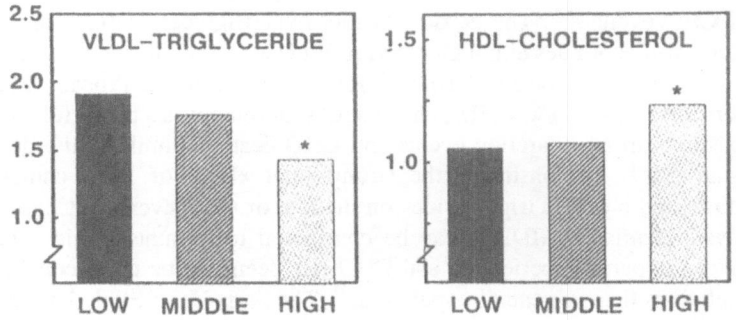

Figure 2. VLDL–triglycerides and HDL cholesterol levels by glucose disposal rate tertiles in a cohort including subjects with normal glucose tolerance (n=50), IGT (n=28) and NIDDM (n=54) (Laakso M. et al., ref. 20).

Is insulin resistance a cause rather than a consequence of lipoprotein abnormalities or just a coincidence? Since metabolic abnormalities in NIDDM represent a long term outcome, where a vicious cycle is continuously operating, the end–stage of metabolic abnormalities does not allow to reveal the causal connections or to define the initial mechanisms behind lipoprotein abnormalities. Consequently it has been questioned if hyperinsulinism preceeds or predicts the development of lipoprotein abnormalities. Family members of NIDDM patients represent a target group which commonly expresses hyperinsulinism as an initial hallmark of insulin resistance even in the presence of a complete normal OGTT[21]. In the Botnia–study, which is an ongoing family study of NIDDM patients in the western part of Finland, we have examined the interrelationship between fasting insulin concentrations and cardiovascular risk factors[22]. Our preliminary data, based on a cohort of 1013 first degree relatives of NIDDM probands, demonstrate that there is a stepwise elevation of serum triglycerides but lowering of HDL cholesterol over quintiles of fasting insulin. The data is consistent with the concept that hyperinsulinemia may predict the development of dyslipidemias as demonstrated in the San Antonio Heart

Study which is a population based study of Mexican Americans[23]. Haffner et al.[23] reported that baseline fasting insulin levels were significantly higher in those individuals who developed hypertriglyceridemia and/or lowering of HDL cholesterol during follow-up of 8 years. The data suggest that hyperinsulinemia may preceed the appearance of lipid abnormalities and also suggest that hyperinsulinemia is the underlying disorder. The clinical implication of these observations is that adverse CHD risk profile may antecede for years the onset of NIDDM. Therefore it is not surprising that in overt NIDDM the concentrations of serum lipids and lipoproteins are not as good predictors for future CHD events as expected[24]. The lack of predictive power probably reflects the enhanced atherogenesis during several preceeding years. The practical conclusion is that a family history of NIDDM is an indication to screen for serum lipids and lipoproteins and that hypertriglyceridemia and/or lowering of HDL cholesterol in such a subject indicates insulin resistance and increased CHD risk.

ATHEROGENEITY OF DYSLIPIDEMIAS IN NIDDM

Although the causal connection between low HDL cholesterol and CHD risk is not fully understood numerous studies have consistently demonstrated that low HDL cholesterol is a strong risk factor of CHD[25,26]. The data from recent intervention trials suggest that in addition to the lowering of LDL cholesterol also raising HDL cholesterol contributes to the reduction of CHD events[27,28]. The debate whether hypertriglyceridemia is an independent CHD risk factor still continues although growing evidence has confirmed that hypertriglyceridemia is a significant marker of CHD[29,30]. Recent data from the Procam Study suggest that the hypertriglyceridemia/low HDL cholesterol syndrome was a powerful risk factor for both nonfatal myocardial infarction events and CHD death[31]. Similarly the data from the Helsinki Heart Study demonstrated the strong joint effect of LDL cholesterol/HDL cholesterol ratio and elevated triglycerides on the rate of CHD events[32]. Consequently the pattern of dyslipidemia in NIDDM can be considered to be atherogenic. The potential relation between hypertriglyceridemia and CHD risk seems to be more consistent among diabetic populations than in general populations[33]. The Paris Heart Study has demonstrated that hypertriglyceridemia was a significant predictor of CHD death in IGT and NIDDM patients[33]. Recently Grundy and Vega has focused attention on the fact that hypertriglyceridemia has several metabolic consequences which are potentially atherogenic (Table 1). First elevation of VLDL triglycerides is accompanied by exaggerated postprandial lipemia and consequently elevated concentration of circulating remnant particles which are considered to be particularly atherogenic. Notably VLDL particles of NIDDM subjects show structural alterations which may increase their atherogeneity[34]. Recently it has been reported that diabetic VLDL is enriched in apo E which may enhance the uptake of particles into macrophages[35]. Elevation of VLDL triglycerides also results in changes of LDL and HDL subclasses and particle structure. Growing evidence suggest that hypertriglyceridemia is a significant determinant for the concentration of small dense LDL[36,37,38] which is more readily filtered into arterial cells than large LDL particles. In addition substantial evidence indicates that small dense LDL is more prone to oxidation than large LDL. Consequently small dense LDL is considered to be highly atherogenic and indeed small dense LDL associates closely with CHD risk in non-diabetic populations[39]. Hypertriglyceridemia is accompanied also with compositional changes of HDL particles which increase their removal rate and thus reduce HDL concentration[40]. Thus elevation of triglycerides, lowering of HDL and small dense LDL are a network of causally interrelated abnormalities which frequently prevail among NIDDM individuals. Finally it should be recognized that hypertriglyceridemia is associated with multiple changes in clotting system which promote trombogenesis. Recently elevation of plasminogen activator inhibitor (PAI-

1) has been included in the cluster of metabolic syndrome[41]. Available data suggest that impaired fibrinolysis coexists with hypertriglyceridemia and insulin resistance[41].

SUMMARY

In conclusion hypertriglyceridemia is accompanied by multiple metabolic disturbances which are potentially atherogenic. Atherogenic pattern of risk factors in insulin resistance syndrome request early intervention to prevent the development of CHD. The practical implication is that the presence of cardiovascular risk factors like hyperinsulinism and dyslipidemias particularly in a person with a family history of NIDDM or hypertension, deserves attention before overt diseases develope.

Table 1. Metabolic consequences of hypertriglyceridemia

1.	Exaggerated postprandial lipemia and delayed removal of remnant particles
2.	Alterations of VLDL subclass distribution and structure
3.	Changes of LDL and HDL subclass distribution and structure
4.	Alterations of haemostatic system

REFERENCES

1. G.M. Reaven, Role of insulin resistance in human disease, *Diabetes 37:1595–1607* (1988).

2. N.M. Kaplan, The deadly quartet: upper–body obesity, glucose intolerance, hypertriglyceridemia, and hypertension, *Arch Intern Med* 149:1514–1520 (1989).

3. P.Z. Zimmet, Challenges in diabetes epidemiology – from west to the rest, *Diabetes Care* 15:232–252 (1992).

4. R.A. DeFronzo and E. Ferrannini, Insulin resistance: a multifaceted syndrome responsible for Nobesity, hypertension, dyslipidemia, and atherosclerotic cardiovascular disease, *Diabetes Care* 14:173–194 (1991).

5. N. Peters and C.N. Hales, Plasma insulin concentrations after myocardial infarction, *Lancet* I:1144–1145 (1965).

6. E.A. Nikkilä, T.A. Miettinen, M.–R. Vesenne and R. Pelkonen, Plasma–insulin in coronary heart–disease, *Lancet* II:508–511 (1965).

7. T.W. Welborn and K. Wearne, Coronary heart disease incidence and cardiovascular mortality in Busselton with reference to glucose and insulin concentration, *Diabetes Care* 2:154–160 (1979).

8. K. Pyörälä, E. Savolainen, S. Kaukola and J. Haapakoski, Plasma insulin as coronary heart disease risk factor: relationship to other risk factors and predictive value during 9½–year follow–up of the Helsinki Policemen Study population, *Acta Med Scand* 701(Suppl):38–52 (1985).

9. A. Fontbonne, M.A. Charles, N. Thibult, J.L. Richard, J.R. Claude, J.M. Warnet, G.E. Rosselin and E. Eschwége, Hyperinsulinaemia as a predictor of coronary heart disease mortality in a healthy population: the Paris Prospective Study, 15–year follow–up, *Diabetologia* 34:356–361 (1991).

10. M. Laakso, H. Sarlund, R. Salonen, M. Suhonen, K. Pyörälä, J.T. Salonen and P. Karhapää, Asymptomatic atherosclerosis and insulin resistance, *Arterioscler Thromb* 11:1068–1076 (1991).

11. T. Rönnemaa, M. Laakso, K. Pyörälä, V. Kallio and P. Puukka, High fasting plasma insulin is an indicator of coronary heart disease in non–insulin–dependent diabetic patients and nondiabetic subjects, *Arterioscler Thromb* 11:80–90 (1991).

12. R.M. Hillson, T.D.R. Hockaday, J.I. Mann and D.J. Newton, Hyperinsulinaemia is associated with development of electrocardiographic abnormalities in diabetics, *Diabetes Research* 1:143–149 (1984).

13. W.Y. Fujimoto, Y. Akanuma, Y. Kanazawa, S. Mashiko, D. Leonetti and P. Wahl, Plasma insulin levels in Japanese and Japanese–American men with type 2 diabetes may be related to the occurrence of cardiovascular disease, *Diab Res Clin Pract* 6:121–127 (1989).

14. M.A. Menser and S.G. Purvis–Smith, Insulin and atheroma, *Lancet* I:1078–1080 (1969).

15. R.W. Stout, Insulin and atheroma, *Diabetes Care* 13:631–654 (1990).

16. M. Modan, H. Halkin, A. Lusky, P. Segal, Z. Fuchs and A. Chetrit, Hyperinsulinemia is characterized by jointly disturbed plasma VLDL, LDL, and HDL levels, *Arterioscler* 8:227–236 (1988).

17. I. Zavaroni, E. Bonora, M. Pagliara, E. Dall'Aglio, L. Luchetti, G. Buonanno, P.A. Bonati, M. Bergonzani, L. Gnudi, M. Passeri and G. Reaven, Risk factors for coronary artery disease in healthy persons with hyperinsulinemia and normal glucose tolerance, *N Engl J Med* 320:702–706 (1989).

18. M. Laakso and E. Barrett–Connor, Asymptomatic hyperglycemia is associated with lipid and lipoprotein changes favoring atherosclerosis, *Arterioscler* 9:665–672 (1989).

19. W.G.H. Abbott, S. Lillioja, A.A. Young, J.K. Zawadzki, H. Yki–Järvinen, L. Christin and B.V. Howard, Relationships between plasma lipoprotein concentrations and insulin action in an obese hyperinsulinemic population, *Diabetes* 36:897–904 (1987).

20. M. Laakso, H. Sarlund and L. Mykkänen, Insulin resistance is associated with lipid and lipoprotein abnormalities in subjects with varying degrees of glucose tolerance, *Arterioscler* 10:223–231 (1990).

21. J. Eriksson, A. Franssila–Kallunki, A. Ekstrand, C. Saloranta, E. Widén, C. Schalin and L. Groop, *N Engl J Med* 321:337–343 (1989).

22. J. Eriksson, M.–R. Taskinen, M. Nissén, B.–O. Ehrnström, B. Forsén, B. Snickars and L. Groop, The Botnia Study: concomitants of abdominal obesity in persons predisposed to type 2 diabetes, *Diabetologia* 35(Suppl 1):A68 (1992).

23. S.M. Haffner, R.A. Valdez, H.P. Hazuda, B.D. Mitchell, P.A. Morales and M.P. Stern, Prospective analysis of the insulin–resistance syndrome (syndrome X), *Diabetes* 41:715–722 (1992).

24. M.I.J. Uusitupa, L.K. Niskanen, O. Siitonen, E. Voutilainen and K. Pyörälä, 5–year incidence of atherosclerotic vascular disease in relation to general risk factors, insulin level, and abnormalities in lipoprotein composition in non–insulin–dependent diabetic and nondiabetic subjects, *Circulation* 82:27–36 (1990).

25. D.J. Gordon, J.L. Probstfield, R.J. Garrison, J.D. Neaton, W.P. Castelli, J.D. Knoke, D.R. Jacobs, S. Bangdiwala and A. Tyroler, High–density lipoprotein cholesterol and cardiovascular disease, *Circulation* 79:8–15 (1989).

26. D.J. Gordon, Role of circulating high–density lipoprotein and triglycerides in coronary artery disease: risk and prevention, *Endocrin Metab Clin North Am* 19:299–309 (1990).

27. M.H. Frick, O. Elo, K. Haapa, O.P. Heinonen, P. Heinsalmi, P. Helo, J.K. Huttunen, P. Kaitaniemi, P. Koskinen, V. Manninen et al. Helsinki Heart Study: primary– prevention trial with gemfibrozil in middle–aged men with dyslipidemia, *N Engl J Med* 317:1237-1245 (1987).

28. G. Brown, J.J. Albers, L.D. Fisher, S.M. Schaefer, J.-T. Lin, C. Kaplan, X.-Q. Zhao, B.D. Bisson, V.F. Fitzpatrick and H.T. Dodge, Regression of coronary artery disease as a result of intensive lipid–lowering therapy in men with high levels of apolipoprotein B, *N Engl J Med* 323:1289–1298 (1990).

29. M.A. Austin, Plasma triglyceride as a risk factor for coronary heart disease, *Am J Epidemiol* 129:249–259 (1989).

30. S.M. Grundy and G.L. Vega, Two different views of the relationship of hypertriglyceridemia to coronary heart disease, *Arch Intern Med* 152:28–34 (1992).

31. G. Assmann and H. Schulte, Relation of high–density lipoprotein cholesterol and triglycerides to incidence of atherosclerotic coronary artery disease (the PROCAM experience), *Am J Cardiol* 70:733–737 (1992).

32. V. Manninen, L. Tenkanen, P. Koskinen, J.K. Huttunen, M. Mänttäri, O.P. Heinonen and M.H. Frick, Joint effects of serum triglyceride and LDL cholesterol and HDL cholesterol concentrations on coronary heart disease risk in the Helsinki Heart Study, *Circulation* 85:37–45 (1992).

33. A. Fontbonne, E. Eschwége, F. Cambien, J.-L. Richard, P. Ducimetiére, N. Thibult, J.-M. Warnet, J. R. Claude and G.-E. Rosselin, Hypertriglyceridaemia as a risk factor of coronary heart disease mortality in subjects with impaired glucose tolerance or diabetes, *Diabetologia* 32:300–304 (1989).

34. M.-R. Taskinen, Hyperlipidaemia in diabetes, *Baill Clin Endocrin Metab* 4:743–775 (1990).

35. E.L. Bierman, Atherogenesis in diabetes, *Arterioscler Thromb* 12:647–656 (1992).

36. E.G. Richards, S.M. Grundy and K. Cooper, Influence of plasma triglycerides on lipoprotein patterns in normal subjects and in patients with coronary artery disease, *Am J Cardiol* 63:1214–1220 (1989).

37. M. Tilly–Kiesi, T. Kuusi, S. Lahdenperä and M.-R. Taskinen, Abnormalities of low density lipoproteins in normolipidemic type II diabetic and nondiabetic patients with coronary artery disease, *J Lipid Res* 33:333–342 (1992).

38. J.R. McNamara, J.L. Jenner, Z. Li, P.W.F. Wilson and E.J. Schaefer, Change in LDL particle size is associated with change in plasma triglyceride concentration, *Arterioscler Thromb* 12:1284–1290 (1992).

39. M.A. Austin, J.L. Breslow, C.H. Hennekens, J.E. Buring, W.C. Willett and R.M. Krauss, Low–density lipoprotein subclass patterns and risk of myocardial infarction, *JAMA* 260:1917–1921 (1988).

40. E.A. Brinton, S. Eisenberg and J.L. Breslow, Increased apo A–I and apo A–II fractional catabolic rate in patients with low high density lipoprotein–cholesterol levels with or without hypertriglyceridemia, *J Clin Invest* 87:536–544 (1991).

41. P. Vague and I. Juhan-Vague, Insulin and the fibrinolytic system. A link between metabolism and thrombogenesis, in: "Hypertension as an insulin–resistant disorder". U. Smith, N.E. Bruun, T. Hedner and B. Hökfelt, ed., Excerpta Medica, Amsterdam (1991).

A PARADIGM TO LINK CLINICAL RESEARCH TO CLINICAL PRACTICE: THE CHALLENGE IN NON-INSULIN DEPENDENT DIABETES MELLITUS

Phillip Gorden, Maureen I. Harris,
Robert Silverman, and Richard Eastman

National Institute of Diabetes and Digestive
and Kidney Diseases
National Institutes of Health
Bethesda, Maryland 20892 USA

INTRODUCTION

At the outset we would like to present several generalizations. For the most part these generalizations are supported by a strong body of scientific data, but we will not attempt to rigorously document each point. Instead, we will consider this document a framework for discussion. Further, this discussion is derived from the work of many investigators and appropriate and deserved attribution will not always be given.

We start with the generalization that the term Non-Insulin Dependent Diabetes (NIDDM) describes a collection of genetic diseases that ultimately result in hyperglycemia.[1] The earliest phase of the pathogenetic sequence that has thus far been identified is insulin resistance.[2,3] From an etiologic point of view, however, these diseases must be based in genetic aberrations. At present, in a few forms of NIDDM, genetic loci have been identified and in others mutations have been identified in candidate genes. While on theoretical grounds it is now clear that NIDDM is a polygenic disease, this has early experimental verification. In sum, a genetic background is necessary for the clinical expression of NIDDM.

One of the great enigmas of a genetic disease such as NIDDM is that it is clinically expressed in middle and older age.[4] There are monogenic diseases in which there is a long latent phase from birth to clinical expression. These include examples such as Huntington's disease where the neurologic deterioration is a late clinical manifestation. In other conditions such as phenylketonuria the genetic defect is complete at birth or before, but the "metabolic load" modifies the clinical expression of the disease. This is an especially interesting example for NIDDM. Thus, in NIDDM the genetic defect is probably established at conception and can not be specifically modified. However, the "metabolic load" can be environmentally modified. Changes in "metabolic load" can, in turn, modify two key elements, insulin secretion and insulin action.

GOALS OF THERAPY

We will first focus on pathogenetic relationships that are best understood. Diabetic retinopathy and nephropathy constitute microvascular complications of diabetes. Hyperglycemia appears to be necessary for these complications to develop and this process is primarily influenced by the duration or intensity of exposure to hyperglycemia. These points are supported by numerous epidemiologic studies.[4-12] Evidence that the rate of progression of these complications can be modified is provided by short-term clinical trials and by meta analysis.[13] However, major evidence for the efficacy of intervention will come from the Diabetes Control and Complications Trial (DCCT). If this trial, in the next 1 to 2 years, confirms the efficacy of intensive insulin therapy in diminishing or delaying retinopathy in insulin-dependent diabetes mellitus (IDDM), this has direct application to NIDDM. Not only will we know that there is benefit from intensive insulin treatment, but we will know the quantitative level of glycosylated Hgb where we can begin to see this effect. This clinical data will be further bolstered by a better understanding of how advanced glycosylation products lead to retinal and renal damage. Thus, while other factors may influence retinopathy and nephropathy, such as hypertension or possibly genetic defects, hyperglycemia remains a prime therapeutic target.

In IDDM the onset of hyperglycemia is dramatic and can be precisely documented. Thus, there is no retinopathy at the onset of hyperglycemia, and the risk of retinopathy is cumulative as a function of duration and intensity of hyperglycemia.[14,15] In NIDDM the situation is somewhat different.[16] The onset of hyperglycemia is slow and not dramatic. However, at the time of diagnosis of NIDDM, up to 21 percent of patients have retinopathy.[17] Their further risk, as in IDDM, is then a function of the intensity and duration of the hyperglycemia. When the time of onset of NIDDM, which is the starting point of their risk for retinopathy, is extrapolated it would appear that diabetes has been present for 7-12 years before the clinical diagnosis of diabetes is made.[17] Thus, one of the first goals of prevention is to find these patients prior to the onset of hyperglycemia and retinopathy.

TREATMENT OF MICROVASCULAR DISEASE

The First Step

If hyperglycemia is a prime therapeutic target, then prevention or more realistically delay in hyperglycemia becomes a major therapeutic goal. Our therapeutic strategy is limited by available technology and again some assumptions are necessary. If we are dealing with a progressive genetic disease, it is possible that a given therapeutic strategy will only work at one particular stage of the disease. For instance, diet and exercise are almost always recommended and almost always fail. While there are many reasons for failure based on patient compliance, it is also possible that this form of therapy works best in the pre-hyperglycemic phase of the disease. Diet and exercise influence insulin resistance, the earliest pathogenetic manifestation of diabetes.[18] In addition, diet and exercise decrease the metabolic load on the islet. Further, there are epidemiologic data to suggest that the prevalence of NIDDM may be reduced by diet and exercise.[19-25] Using this argument, therefore, it is important to determine whether diet and exercise can delay the conversion of IGT to glucose tolerance diagnosed diabetes (2-hr glucose $\geq$ 200 mg/dl). It is further possible that diet and exercise will delay the conversion of this post-glucose load hyperglycemia to overt fasting hyperglycemia ($\geq$ 140 mg/dl). This again must be demonstrated by clinical trial. The major point is that diet and exercise would be directed at the pre-hyperglycemic phase of the disease in an attempt to delay fasting and post-challenge hyperglycemia.

The Second Step

It seems clear from experience and numerous trials that during the early fasting hyperglycemic phase, diet and exercise are no longer effective in restoring normoglycemia. This phase of the disease is characterized by progressive beta cell failure, and while this may be ameliorated by diet and exercise in the short-term, the effect is usually not sustained. This may be due in part to lack of patient compliance, but it may also be due to progressive deterioration of the beta cell or "metabolic atrophy." If the goal of therapy is to maintain near euglycemia, then pharmacologic treatment must be added to diet and exercise. What is key at this point is that the DCCT may provide a quantitative goal for glycemia, as measured by glycosylated hemoglobin. Appropriate therapy here is both conventional non-insulin pharmacologic therapy or possibly one or more of the emerging non-insulin pharmacologic therapies.[26] The important point is that the therapeutic goal is clear, i.e., a targeted glycosylated hemoglobin. Many of the non-insulin pharmacologic therapies have only a limited period of effectiveness. This could be due to some form of refractoriness or to the progressive beta cell failure.

The Third Step

Since the goal of therapy with respect to microvascular disease is clearly defined, then insulin becomes the only available means of treatment once non-insulin pharmacologic therapy fails. Conventional insulin therapy may be used and hopefully some of the experimental modes of insulin delivery will be clinically useful.

MACROVASCULAR DISEASE

Macrovascular disease has always and continues to be a central problem in NIDDM. The issues are complex, but it is important to try to isolate the individual variables and to determine what are the goals of clinical practice and what are the fundamental experimental questions. The experimental questions must then be prioritized in terms of their importance and of the feasibility of reaching an answer.

Until this point, we have suggested that the pathogenesis of microvascular disease in IDDM and NIDDM are the same and that strategies to prevent, delay, or control hyperglycemia are appropriate. To some extent, this logic can be followed in thinking about macrovascular disease. First, from epidemiologic studies we know that the prevalence of macrovascular disease increases from normal to IGT to overt diabetes.[27-29] Further, the rate of defined cardiovascular events such as myocardial infarction is 2-3 fold greater in newly diagnosed patients with NIDDM than in the non-diabetic population. Further, over 60 percent of newly diagnosed diabetics are hypertensive and up to 75 percent have elevated serum cholesterol ($\geq$ 240 mg/dl).[17] Even though macrovascular disease does not have the same degree of dependency on hyperglycemia as microvascular disease, macrovascular disease is likely to benefit from the same diet and exercise strategies that we have advocated for microvascular disease.

Obesity, hypertension, and lipid disorders are all common in NIDDM. They are all independent risk factors for macrovascular disease. Treatment of these disorders, therefore, is not experimental but constitutes good clinical practice. Several caveats remain, however. These must be stratified according to research priority and feasibility and, like most things of this nature, these decisions are controversial. For example, weight reduction is desirable but how can it be achieved? Are very low calorie diets or pharmacologic intervention beneficial? Normal blood pressure should be a therapeutic goal, but is a stepped reduction below the conventional normal beneficial or would this impose an even greater risk? While the standard recommendations for controlling blood lipids should be clinically practiced, should all diabetics be treated as patients with known vascular disease? This might imply the use of lipid lowering drugs in all diabetics if the goal is to achieve an LDL cholesterol of 100 mg/dl. These are a group of complex questions that relate to the patient with NIDDM, but perhaps one of the most difficult is the question of whether intense glycemic control will decrease events such as myocardial infarction. This must be put into perspective when it is remembered that all of the cholesterol trials, which included a total of about 40,000 subjects,[30] have not shown that intervention decreases overall mortality, and it required almost 4,000 high risk subjects to demonstrate a 19 percent decrease in myocardial infarction.[31] This gives some indication of the enormous size and duration of a study that would be required to show that glycemic control decreased mortality in NIDDM. A study that would demonstrate a decrease in specific cardiovascular events such as myocardial infarction and stroke would likewise be large, of long duration, and extremely expensive.

Treatment of Advanced Complications

Treatment of diabetic retinopathy is the best example of successful therapy in the management of diabetes complications. In the paradigm that we have presented, the prevention or delay in hyperglycemia could have an enormous effect in ameliorating the manifestations of retinopathy. This, coupled with successful treatment, would essentially compress the problem from both ends.

Does the same argument apply to diabetic nephropathy? This is the most frequent cause of end-stage renal disease (ESRD) and accounts for one-third of new cases of ESRD in the United States.[32] Approximately half of these patients have IDDM and the other half have NIDDM.[33] Again, delay or amelioration of hyperglycemia could have an enormous effect on this complication. Early renal disease is more difficult to detect than early retinal disease and, therefore, is treated at a later stage. In both the eye and the kidney, adequate control of blood glucose is likely to have a much greater effect on preventing or delaying the problem than on affecting it once it is fully developed.

Animal studies have suggested that agents that modify renal blood flow or control blood pressure may be of benefit in diabetic nephropathy. Small scale human trials have shown that angiotensin converting enzyme inhibitors (ACE) can modify proteinuria and may protect against a decline in renal function. Larger trials will be completed in the near future with a more clear answer to this question. Thus, the experimental questions remain. Are ACE inhibitors the drug of choice in treating hypertension in diabetes and do these drugs have a beneficial effect in the normotensive patient when there is no further reduction in blood pressure? Is a targeted lowering of blood pressure of benefit? We now have the opportunity to test the question of whether the reduction of advanced glycosylation products will modify the course of renal disease. Studies with aminoguanidine, a drug that inhibits the formation of advanced glycosylation products, are soon to begin.

CONCLUSION

The challenge for the future is to pursue goals that are important yet are also achievable. Obviously new ideas and new technologies will lead us in new directions, but we must have a comprehensive vision of where we are and where we are going. A few examples are relevant.

We must maintain a strong basic research agenda. This should include a major emphasis on elucidating the genetic etiologies of the diseases we call NIDDM. Islet function and insulin action must be understood at a more fundamental level so that the critical rate-limiting steps of each can be clearly defined. The biochemical events leading to microvascular disease must be delineated. Further, we must understand better the specific risk of the diabetic patient for macrovascular disease and separate this from the general risk factors that are prevalent in diabetics. With this new and more complete picture of the disease, our therapeutic interventions will become more imaginative and also more focused.

Our clinical research agenda should include studies that will determine whether strategies can be employed to prevent or delay hyperglycemia in patients at high risk of developing diabetes. We must determine whether it will be useful to screen and find these patients at the earliest possible stage so that we can determine whether delay in hyperglycemia will decrease the lifetime threat of microvascular disease. We must intervene more vigorously at the earliest stage of hyperglycemia. In the patient with established hyperglycemia we must vigorously pursue experimentally established targets of glycosylated hemoglobin. Initially, the scientific validity of these primary and secondary prevention strategies will be based on end points primarily related to microvascular disease. Hopefully, this approach, coupled with a greater emphasis on cardiovascular risk factor reduction, will have a beneficial effect on macrovascular disease.

The most realistic initial goal of the general approach we have outlined is to delay rather than truly prevent the tertiary complications of the disease. Again, the greatest progress here is in the treatment of retinopathy, but there is at least some preliminary evidence that we may be able to delay ESRD.

Finally, we must pursue ways to better translate scientifically validated research strategies into good clinical practice. The initial result of this will be to decrease the enormous disease burden of diabetes. The next step will be to diminish the excess mortality imposed by NIDDM.

REFERENCES

1. National Diabetes Data Group: Classification and diagnosis of diabetes mellitus and other categories of glucose intolerance. Diabetes 28:1039 (1979).

2. P.H. Bennett, C. Bogardus, W.C. Knowler, S. Lillioja, Antecedent events for non-insulin-dependent diabetes mellitus. Adv Exp Med Biol 246:185 (1988).

3. R.A. DeFronzo, R.C. Bonadonna, E. Ferrannini, Pathogenesis of NIDDM, a balanced overview. Diabetes Care 15:318 (1992).

4. M.I. Harris, W.C. Hadden, W.C. Knowler, P.H. Bennett, Prevalence of diabetes and impaired glucose tolerance and plasma glucose levels in the U.S. population. Diabetes 36:523 (1987).

5. J. Pirart, Diabetes mellitus and its degenerative complications: a prospective study of 4,400 patients observed between 1947 and 1973. Diabetes Care 1:168,252 (1978).

6. R. Klein, B.E.K. Klein, S.E. Moss, M.D. Davis, D.L. DeMets, The Wisconsin Epidemiologic Study of Diabetic Retinopathy. X. Four-year incidence and progression of diabetic retinopathy when age at diagnosis is 30 years or more. Arch Ophthalmol 107:244 (1989).

7. A. Sasaki, N. Horiuchi, K. Hasewgawa, M. Uehara, Development of diabetic retinopathy and its associated risk factors in type 2 diabetic patients in Osaka district, Japan: a long-term prospective study. Diabetes Research and Clinical Practice 10:257 (1990).

8. E.T. Lee, V.S. Lee, M. Lu, D. Russell, Development of proliferative retinopathy in NIDDM. A follow-up study of American Indians in Oklahoma. Diabetes 41:359 (1992).

9. R.G. Nelson, J.A. Wolfe, M.B. Horton, D. J. Pettitt, et al., Proliferative retinopathy in NIDDM: incidence and risk factors in Pima Indians. Diabetes 38:435 (1989).

10. D.J. Ballard, L.L. Humphrey, L.J. Melton, P.P. Frohnert, C-p Chu, W.M. O'Fallon, P.J. Palumbo, Epidemiology of persistent proteinuria in type II diabetes mellitus: population-based study in Rochester, Minnesota. Diabetes 37:405 (1988).

11. R. Klein, E.L. Barrett-Connor, B.A. Blunt, D.L. Wingard, Visual impairment and retinopathy in people with normal glucose tolerance, impaired glucose tolerance, and newly diagnosed NIDDM. Diabetes Care 10:914 (1991).

12. S.M. Haffner, D. Fong, M.P. Stern, J.A. Pugh, H.P. Hazuda, J.K. Patterson, W.A.J. Van Heuven, R. Klein, Diabetic retinopathy in Mexican Americans and non-Hispanic whites. Diabetes 37:878 (1988).

13. K.F. Hanssen, The DCCT Research Group, and P. Brunetti, Is there a need for continuation of the DCCT in 1988. Diab & Nutri & Metab (clinical & experimental) 1:151 (1988).

14. R. Klein, B.E.K. Klein, S.E. Moss, M.D. Davis, D.L. DeMets, The Wisconsin Epidemiologic Study of Diabetic Retinopathy. II. Prevalence and risk of diabetic retinopathy when age at diagnosis is less than 30 years. Arch Ophthalmol 102:520 (1984).

15. W. Berger, G. Hovener, R. Dusterhus, R. Hartmann, B. Weber, Prevalence and development of retinopathy in children and adolescents with Type 1 (insulin-dependent) diabetes mellitus. A longitudinal study. Diabetologia 29:17 (1986).

16. R. Klein, B.E. Klein, S.E. Moss, M.D. Davis, D.L. DeMets, The Wisconsin epidemiologic study of diabetic retinopathy. III. Prevalence and risk of diabetic retinopathy when age at diagnosis is 30 or more years. Arch Ophthalmol 102:527 (1984).

17. M.I. Harris, R.E. Klein, T.A. Welborn, M.W. Knuiman, Onset of NIDDM occurs at least 4-7 yr before clinical diagnosis. Diabetes Care 15:815 (1992).

18. National Institutes of Health: Diet and exercise in noninsulin-dependent diabetes mellitus. Consensus Development Conference Statement 6(8):1 (1986).

19. S.D. Long, K. O'Brien, M. Swanson, K. MacDonald, N. Frazier, W.J. Pories, J.F. Caro, Weight loss prevents the progression of impaired glucose tolerance to non-insulin-dependent diabetes mellitus: a ten year longitudinal study. Diabetes 41(Suppl 1):72A (1992).

20. J. A. Marshall, F.R. Hamman, J. Baxter, High-fat, low-carbohydrate diet and the etiology of non-insulin-dependent diabetes mellitus: the San Luis Valley Diabetes Study. Am J Epidemiol 134:590 (1991).

21. G.K. Dowse, P.Z. Zimmet, H. Gareeboo, K.G.M.M. Alberti, J. Tuomilehto, C. Finch, P. Chitson, H. Tulsidas, Abdominal obesity and physical inactivity as risk factors for NIDDM and impaired glucose tolerance in Indian, Creole, and Chinese Mauritians. Diabetes Care 14:271 (1991).

22. A. Schranz, J. Tuomilehto, B. Marti, R.J. Jarrett, V. Gaubauskas, A. Vassallo, Low physical activity and worsening of glucose tolerance: results from a 2-year follow-up of a population sample in Malta. Diabetes Research and Clinical Practice 11:127 (1991).

23. S.P. Helmrich, D.R. Ragland, R.W. Leung, R.S. Paffenbarger, Physical activity and reduced occurrence of non-insulin-dependent diabetes mellitus. New Engl J Med 325:147 (1991).

24. J.E. Manson, E.B. Rimm, M.J. Stampfer, G.A. Colditz, W.C. Willett, A.S. Krolewski, B. Rosner, C.H. Hennekens, F.E. Speizer, Physical activity and incidence of non-insulin-dependent diabetes mellitus in women. Lancet 38:774 (1991).

25. J.E. Manson, D.M. Nathan, A.S. Krolewski, M.J. Stampfer, W.C. Willett, C.H. Hennekens, A prospective study of exercise and incidence of diabetes among US male physicians. JAMA 268:63 (1992).

26. R. Bressler and D. Johnson, New Pharmacological Approaches to Therapy of NIDDM. Diabetes Care 15:792 (1992).

27. M.I. Harris, Impaired glucose tolerance in the U.S. population. Diabetes Care 12:464 (1989).

28. E. Eschwege, J.L. Richard, N. Thibult, P. Ducimetiere, J.M. Warnet, J.R. Claude, G.E. Rosselin, Coronary heart disease mortality in relation with diabetes, blood glucose, and plasma insulin levels, the Paris prospective study ten years later. Hormone and Metab Res 15(suppl):41 (1985).

29. R.J. Jarrett, M.J. Shipley, Type 2 (non-insulin-dependent) diabetes mellitus and cardiovascular disease - putative association via common antecedents; further evidence from the Whitehall Study. Diabetologia 31:737 (1988).

30. S. Yusuf, J. Wittes, L. Friedman, Overview of results of randomized clinical trials in heart disease. II. Unstable angina, heart failure, primary prevention with aspirin, and risk factor modification. JAMA 260:2259 (1988).

31. The Lipid Research Clinics Investigators, The Lipid Research Clinics Coronary Primary Prevention Trial. Arch Intern Med 152:1399 (1992).

32. U.S. Renal Data System: 1991 Annual Data Report. U.S. Dept. of Health and Human Services, National Institutes of Health. 1991, p. 17

33. C.C. Cowie, F.K. Port, R.A. Wolfe, P.J. Savage, P.P. Moll, V.M. Hawthorne, Disparities in incidence of diabetic end-stage renal disease by race and type of diabetes. New Engl J Med 321:1074 (1989).

INDEX